Sound Reinforcement Eng

Sound Reinforcement Engineering
Fundamentals and Practice

Wolfgang Ahnert and Frank Steffen

CRC Press
Taylor & Francis Group
Boca Raton London New York

CRC Press is an imprint of the
Taylor & Francis Group, an **informa** business

CRC Press
Taylor & Francis Group
6000 Broken Sound Parkway NW, Suite 300
Boca Raton, FL 33487-2742

First issued in paperback 2017

CRC Press is an imprint of Taylor & Francis Group, an Informa business

No claim to original U.S. Government works

ISBN-13: 978-0-415-23870-0 (hbk)
ISBN-13: 978-1-138-56974-4 (pbk)

Typeset in Sabon by
Mathematical Composition Setters Ltd, Salisbury, Wiltshire

British Library Cataloguing in Publication Data
A catalogue record for this book is available from the British Library

Library of Congress Cataloging in Publication Data
Ahnert, Wolfgang.
  [Beschallungstechnik. English]
  Sound reinforcement engineering / Wolfgang Ahnert & Frank Steffen.
    p.   cm.
  Includes bibliographical references and index.
  1. Loudspeakers. 2. Acoustical engineering. 3. Public address
systems. I. Steffen, Frank. II. Title.
TK5983.A3613 1999
621.382'84--dc21   98-29583
                                                                   CIP

Visit the Taylor & Francis Web site at
http://www.taylorandfrancis.com

and the CRC Press Web site at
http://www.crcpress.com

Contents

Preface to the English Edition

The topic of sound reinforcement engineering has been addressed in Anglo-American technical literature in a way that sometimes differs utterly from its treatment in the German-speaking area. It remains the case that the theories and methods used for investigating correlations are exposed in very different ways. These differing points of view originate not only from the use of different technical terms, but also from different perspectives and approaches to sound reinforcement problems. The increasing globalization of equipment offered by international microphone and loudspeaker manufacturers has made it essential to counter contradictions that have arisen in this field and to introduce universally valid terms to allow for the efficient exchange of ideas in the solution of sound reinforcement problems.

The book *Beschallungstechnik – Grundlagen und Praxis* (Sound Reinforcement Engineering – Foundations and Practice), published in Germany in 1993, summarized the knowledge of that time regarding sound reinforcement problems and solutions in the German-speaking area. The present English edition endeavours to adapt this material to Anglo-American linguistic and practical usage. At the same time the topics have been updated so as to reflect modern trends in this special field. The chapters of the book that deal with fundamental concepts (Chapters 3, 5 and 6) were revised and the problems of sound reinforcement measurement (Chapter 7) have been given special consideration by the addition of material on recently developed methods. Moreover, the examples given in Chapter 8 of the German edition have been enhanced by examples of international solutions.

The present edition owes its publication to the commitment of Hirzel-Verlag Stuttgart/Leipzig, and primarily to Frau Sabine Körner, who encouraged the translation of the German edition. This translation is supported by a grant from Inter Nationes, Bonn. An English translation was produced by Herr Hans-Joachim Kaminski, whose 15 years of collaboration with the authors enabled him to match acoustical terms in both languages. Our English expert colleague Peter Mapp then edited the whole manuscript painstakingly and transformed it into modern technical English. The definition and adequate adaptation of linguistic usage required several full-day meetings between him and the authors. Thanks to his efforts, and those of Lionel Browne, who copy-edited the manuscript for publication, it is now possible that a technical book originating from the German-speaking area should be available for the English-speaking world.

Last but not least we extend our thanks to E & FN Spon and in particular to Tim Robinson for their committed handling of the project.

Berlin, 28 May 1998

Technical Translation Editor's Foreword

The aim of the translated edition has been to retain the style and format of this unique German reference book. This at times can lead to some unusual phrasing but enables the subtlety of the text to be more fully appreciated and understood. However, on occasion it has not been possible to provide a direct translation and I have taken the liberty of rewriting certain sections. In doing so, I hope that the distinctive nature of the original German book has been retained and possibly even enhanced.

It is interesting to note on occasion, the different approaches to sound system design and calculation practised in Germany (and in the former Eastern Bloc countries) as compared to those more commonly found in the USA and UK. In particular a new technical term 'Directivity Factor' has had to be invented for this translation in order to express the different calculation procedure as there was no direct equivalent English term for the German expression.

The book is readable on a number of levels. Apart from the rigorous mathematical treatment, many sections offer a detailed practical approach and are equally suitable for designers and architects as well as engineers and acousticians. The numerous case histories and examples offer a unique insight into the world of sound reinforcement engineering and practice. The authors have provided a fresh and different view on this complex subject and I would anticipate the book becoming a standard reference text on the subject.

Peter Mapp
Colchester, England

Preface to the German edition of 1993

The introduction of the thermionic valve (1915) and the loudspeaker (1912) made it possible to increase the power of sound signals, and to reproduce them as acoustic signals. Thus sound reinforcement engineering was created as a new technical discipline. Its first application was speech amplification, offering the possibility of addressing larger gatherings from a central location. In this respect there existed not only a technical and economic demand, but also a wide variety of social and artistic needs.

Soon the discipline developed into a wide range of applications, including:

* sound effect generation in theatres;
* music transmission in larger rooms and in open spaces;
* paging on railway stations;
* generation of background music in restaurants.

Advances in technical acoustics, especially in electroacoustics and room acoustics, had a stimulating effect on further development. During recent years further advance came about in the form of musical electronics: at first this merely made use of the means offered by sound reinforcement, but soon it began to formulate its own unique requirements. Today it is the increasing use of computer technology that is creating new possibilities for sound reinforcement engineering. Modern sound reinforcement systems can no longer be realized without close cooperation between the acoustician and the systems engineer.

The authors of this book have collaborated in the design of a number of large sound reinforcement systems, and thus have acquired valuable experience. Complemented by numerous ideas taken from international literature, this experience has been compiled in this book. We have prepared the material so that practitioners – be they project architects, users of systems working on particular events, sound engineers in charge of systems, or customers – can work with it immediately.

In contrast to the earlier book *Fundamentals of Sound Reinforcement Engineering* by Ahnert and Reichardt, published in 1981, the present volume considers the practical aspects of the discipline much more closely. However, it is still necessary to explain some fundamentals, especially in Chapters 3 and 5. This is, on the one hand, because some details have come to be presented from a new point of view since the earlier publication, and on the other hand in order to consider certain aspects that have not been explained so far, particularly the damaging effects of sound reinforcement systems on hearing.

Space has also been given to acoustical measuring equipment, which is playing an ever-increasing role in the calibration and commissioning of sound reinforcement systems. Chapter 7 deals with several recently introduced measurement techniques. In this way we hope to have filled a gap that has existed for some time, both in German-speaking countries and elsewhere; nevertheless we are well aware that this area is undergoing rapid development, and hence needs continual updating.

In Chapter 7 – as well as in Chapter 4, which covers the systems and equipment that are the basic components of sound reinforcement engineering – we had to take account of the limited size of the book, so that many topics have been treated only briefly. This is why we have made reference to special literature. This concerns mainly electroacoustical transducers, technical equipment for sound studios, and technical acoustics. In nearly all instances explanations of sound engineering problems have been given preference over descriptions of technical equipment, since this is more in tune with the specific objectives of this book.

We should like to thank Rolf Fischer for his careful revision of the manuscript, and G. Schattkowsky for the multiple typing of the manuscript.

Berlin/Blankenfelde, June 1992   Wolfgang Ahnert and Frank Steffen

1 Introduction

The integration of a sound reinforcement system into the overall concept of a project depends on how and to which extent the audience or the technical and artistic staff receive information via loudspeakers.

To achieve this it was necessary for electroacoustics to reach a sufficient level of development. Before this many of the modern tasks of sound reinforcement engineering could be fulfilled either not at all or only by means of room acoustics or other technical means. In churches, for example, the pulpits were provided with canopies that produced additional reflection to amplify and project the human voice. Acoustic funnels or speaking tubes were already widely used in antiquity [2.1], and in ship-building long tube systems were commonly used for transmitting speech. Many signals were transmitted by means of large and small bells or horns. Stage effects were produced by means of special devices such as wind machines, or metal plates for simulating thunder.

Many of these traditional devices are still in use; they fulfil their function adequately, and their often maintenance-free operation has been instrumental in enabling their technical survival. Their low adaptability and reduced transmission quality, however, do not favour their use in modern designs.

Apart from enabling the transmission of simple messages and information, or the generation of non-localizable stage effects, the development of electroacoustics paved the way for entirely new applications. These are due both to the wide variety of combinations and interconnection of different sources, making it possible for example to extend an information programme with background music and announcement signals, and to the possibility of directing the programme with optimal loudness to specific target areas.

The greatest advance, however, was achieved in the field of sound reinforcement and reproduction for cultural events. Open air, show and musical performances have nowadays become unimaginable without the use of sound reinforcement systems. The acoustic signals are either transmitted and re-radiated directly after pick-up (*real-time operation*), or first stored and then transmitted as well as radiated only when required (*playback operation*). With real-time operation special care is needed when the sound pick-up and the radiation of the amplified signal take place in the same area, as mutual interference (feedback) may occur. This sound reinforcement task requires observation of special measures aimed at safeguarding the stability of the system. The quality of reproduction expected in such performances now equals that usually attained by hi-fi systems in domestic living rooms. For entertainment performances, an entirely new branch of sound reinforcement engineering or electroacoustics has developed in the course of recent decades: *electronic music*. This enables totally electronic instruments to sound, or traditional instruments to be reproduced with an entirely new timbre. This technique, which also has its equivalent in classical music, is already exerting its influence on sound reinforcement engineering, and will do so even more in the future. Owing to its adaptation to specific instruments it differs, however, from conventional sound reinforcement equipment, which is assigned to fixed positions and has to be as universally applicable as possible within its area.

2 Functions of a sound reinforcement system

2.1 General requirements

A sound reinforcement system is usually expected to meet some or all of the following requirements [2.2]. It should be able to:

1 improve intelligibility and clarity;
2 extend the dynamic range;
3 improve the acoustic balance between the different parts of a performance (speech, vocal and instrumental music);
4 ensure an appropriate relationship between the visual and acoustical localization of original and 'simulation' sources (sound images), particularly if the action and reception areas are large and of complex geometry;
5 help to overcome difficult or complex acoustic environments;
6 include the audience in the performance activity;
7 modify the acoustical parameters of the reproduction room;
8 enhance the realization of spatial sound effects, such as moving sound sources through the auditorium or space;
9 modify human voices and instrumental sounds electronically, as well as use electronically produced noises and sounds as a deliberate means of presentation;
10 preproduce and preprogramme parts of the programme to simplify the technical processes.

Requirements 1–5 are generally applicable, but requirements 6–10 are of interest only for large sound reinforcement systems in cultural facilities. All these requirements constitute the basis for the sound reinforcement project concept, which precedes the specification phase, and serves as an intermediary between the client's wishes and their technical realization. To achieve an optimum solution, detailed and specific knowledge is essential, not only of the technological requirements that the system must meet, but also of the room acoustics in which it is to operate.

2.2 Integration into the architectural design

In formulating the concept for a sound reinforcement system, one of the first things to consider is how the system will be integrated into the architectural design of the room. For a new building or a refurbishment of a hall or an open-air theatre, the electro-acoustics consultant lays down the following parameters, in cooperation with a specialist room-acoustics consultant if necessary: the desirable reverberation time, the spaciousness that can be expected, and any measures required for avoiding possible echoes. This problem will be dealt with in Chapter 3. He also has to consider sightlines and visibility for the audience. This is important, because the loudspeakers must be directed towards the audience, and therefore tend to be located in areas of architectural or interior design sensitivity. Visual screening or 'hiding' of loudspeakers is possible only to a limited degree, and entails several acoustical problems, which will be dealt with later on [2.1]. The requirements may be quite varied. In sports facilities, for example, it is very important to avoid impairing visibility of the playing area. There are also requirements laid down by international sport associations regarding physical obstruction of the field of play.

In modern multi-purpose halls a visible arrangement of the loudspeakers is usually permissible (Figure 2.1), but the overall architectural design must not be disturbed, nor must the lighting and projection facilities be impaired. Particularly complex conditions arise in highly prestigious or listed rooms of historical or architectural importance. Figures 2.2a and 2.2b show as an example how loudspeakers were concealed in the reconstructed listed auditorium of the Semper Opera in Dresden. Similar solutions are often adopted in the foyers and restaurants of large hotels or in museum

Figure 2.1 Loudspeaker arrangement in the Haus der Kultur in Gera.

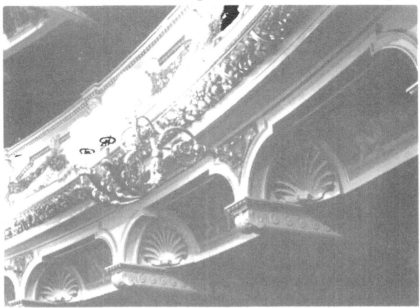

Figure 2.2a Candelabra concealing loudspeakers in the auditorium of the Semper
 Opera in Dresden, position 1.

buildings. Acceptable solutions can as a rule be achieved only through close
cooperation between the sound reinforcement consultant and the architect.
Sometimes the client makes the initial decisions on priorities. To avoid
subsequent installation deficiencies, the acoustics expert should also be
consulted in connection with possible architectural modifications. The
following acoustical parameters are relevant:

- the position and radiating direction of the loudspeakers;
- the type of loudspeakers to be used (dimensions, mass, size of the
  radiating surface);
- the type of enclosure necessary for the loudspeakers, the damping of
  their internal resonances, and the arrangement of sound-absorbing
  materials to avoid near-field interference;
- the possible need for acoustically transparent covering of the
  loudspeakers. It is usually necessary to take measurements with
  samples of the selected material, since – especially with textile
  materials – mere knowledge of thickness and mesh density is not
  sufficient to determine the acoustical transparency.

In historic or listed buildings it is not always possible to conceal the
loudspeakers. In the Semper Opera, for example, the proscenium
loudspeakers had to be arranged on extending frames that can be lowered

Figure 2.2b Candelabra concealing loudspeakers in the auditorium of the Semper Opera in Dresden, position 2.

from the ceiling, thus becoming visible when in use (Figure 2.3; see also [2.3]).

The influence of microphones on the architectural design is usually of minor importance. Nevertheless they should be included at an early stage in the architectural conception of a room, especially if microphone hoists are used.

Most problematic is the arrangement of a sound control facility in the auditorium. Such a *hall mixing console* must be located at an acoustically representative point within the audience area, to allow a reliable assessment of the overall acoustic impression to be made for as much of the audience as possible. In order to assess the balance between the two sides of the hall during multi-channel or stereo transmissions, the console should be placed on

Figure 2.3 Proscenium loudspeakers of the Semper Opera in working position.

the axis of acoustic symmetry of the receiving area. It is extremely difficult to integrate such a console inconspicuously, as is commonly demanded for all auxiliary equipment. This is why objections are often raised against the installation of such a mixing console. However, final mixing of the transmitted signal within the receiving area is necessary in all cases in which:

- the use of a large number of microphones is likely;
- there is only a short rehearsal time available for optimizing the acoustic pattern;
- the acoustic impression of the room is to be optimized during the performance;
- non-stationary sound or acoustic effects are to follow a visible source;
- a distributed discussion system is to be operated, with microphones for contributors in the auditorium.

These conditions are nearly all relevant for medium-sized and large multi-purpose halls. In this respect it has turned out to be advantageous for sparsely staffed installations to provide the hall mixing console location with additional reproduction equipment (generally a studio tape recorder), so that playback of recordings can also be transmitted from this position.

The exposed situation of such a console, the loss of good audience seats, the need to provide easy access without disturbing the audience nor being disturbed by them, as well as the avoidance of visibility or sightline impairment, impose a number of design problems, which have to be solved in close cooperation between the architect and the electroacoustics consultant. Figure 2.4 shows an example of the location of a relatively large hall control console.

With smaller halls and cultural centres in which less important performances are given, the situation may be different. Here the sound system is usually controlled from a separate control booth. However, connection points are also required for mixing consoles to be installed within the auditorium when necessary.

The conditions in theatres are similar. Nowadays connection points are provided for mobile mixing consoles to be installed within the auditorium. These are required for adjustment of the sound system during rehearsals, and for the performance of musicals and other productions that require a high standard of sound engineering.

Figure 2.4 Sound control console in the great hall of the former Palast der Republik in Berlin.

In concert halls it is normally sufficient to have connection points for mobile mixing consoles, which are required mainly for the reproduction of electronic music. Optimization of reproduction in the room via various loudspeaker channels requires reaction to a live performance, and so the best location has to be chosen for the connection points.

The architectural implications of mobile mixing consoles are not as serious as those of fixed consoles; nevertheless, their location should be determined through close cooperation between the architect and the electroacoustics consultant.

2.3 Communication with other areas

The control room of any sound reinforcement or public address system has to be in good communication with the information input sources or the recipients of the information broadcast by the system. Such information sources may, for example, be the control room of a railway station or industrial plant, or the management control suite of a sports centre or stadium. Other information input sources also often require direct communication with the control room of the sound system: for example, safety or works protection departments, or the janitor's lodge, for which direct microphone access may sometimes be required.

In cultural centres the cross-connections between the control room of the sound reinforcement system and the technical and administrative staff have to meet very special requirements, since here the electroacoustic equipment comprising the sound reinforcement system is used as a means of artistic creation. To ensure good cooperation between the artistic cast and the technical staff, intercom links need to be established to the following positions:

- the stage manager;
- the lighting staff;
- the projection room for cinema or video;
- the artistic director (the production director or the stage director during the performance);
- the conductor during rehearsal;
- the control rooms of co-users of an artistic production, such as broadcasters (radio and television), or the recording industry (these links are particularly important for the sound equipment, and will be dealt with later more in detail).

In most cases twin-wire ring or radial intercom networks are used. If communication is very intensive, four-wire links are installed, permitting simultaneous speaking of both users (*duplex intercom systems*).

Paging stations also exist that permit larger areas to be addressed from a microphone. The facility functions via the sound reinforcement system, and is frequently used during rehearsals. Calls by the stage manager or the safety control department to smaller or larger areas may also be required during a performance. During rehearsals, special communication facilities are placed at the disposal of the director and, for show and musical theatres, the conductor as well. Thus it is possible to establish a paging/announcement facility from the director's seat in the auditorium to the stage and to the hall, through which instructions may be given to the cast or the stage technicians. Fixed microphone stations are usually provided for this purpose, but some directors also demand portable microphones with a press-to-talk switch, so that they can be more mobile in their work. To reproduce these instructions loudspeaker clusters are provided in the stage proscenium directed towards the stage as well as towards the auditorium. For very large or high prosceniums, additional loudspeakers are used radiating across the stage opening.

For communication between the lighting technicians, which often takes place in parallel to the work of the director, separate communication facilities are generally used that are independent of the sound reinforcement system. Wireless intercom systems with earphones to provide freedom of motion for the lighting technicians work well.

Light signals are also used, particularly if an acoustic signal would be disturbing at the receiving end. This applies for instance to communication with a conductor, who has to be informed that a tape playback is about to start. Light signals are also convenient for communication with the sound-mixing staff in the auditorium during the performance.

2.4 Outside users

For cultural performances in particular, parts of the sound reinforcement system are increasingly being operated by users who do not belong to the house. This applies particularly to special sound reinforcement of electronic music, and also effects systems that are owned by performing artists and which are essential for the realization of their artistic performances. But also for recordings, or transmissions by radio, television or the record industry, parts of the sound reinforcement system are often jointly used, so that communications have to be established between the systems.

2.4.1 Connection of systems owned by performing artists

Since the introduction of electronic sound generators, and of direct pick-up from conventional instruments (such as guitars and percussion instruments), the timbre of these instruments is no longer determined by their construction and the way they are played, but by the electroacoustical processing of the picked-up signals and their radiation. As these factors have to be controlled by the artistes themselves, the electronic music equipment of the musicians has continually grown. This includes mixing consoles for controlling the balance between the individual voices of an ensemble, including vocal soloists and the accompanying backing singers.

Such systems, which are often characterized by considerable acoustic power, are capable of working with complete self-sufficiency in smaller halls and, for open-air performances, also with very large audiences. (Leased systems operated on their own account are often used for this purpose.) With large concert halls or multi-purpose halls things are quite different. In such rooms a portable sound reinforcement system arranged centrally in the stage area is often incapable of producing a satisfactory result throughout the hall. In the balcony and side areas particularly, there is often an unsatisfactory timbre and an insufficient sound level. In these cases the systems owned by the ensembles or groups should normally be used only for the area near the stage, or as a 'localizing source', for example in installations equipped with delay systems. To supply the more remote areas, the stage signal is patched into the permanently installed house system.

This means that it is necessary to install the necessary infrastructure and facilities, including power supply equipment for the stage system and the communication devices. Some parts of the system are therefore equipped with a wide range of decoupling elements for electrical isolation, impedance conversion and balancing of the lines, so that, for example, the replay units of the electronic musical equipment can be connected safely and free from contact or interference problems to the permanently installed supply and transmission systems. The mains supply of the stage outlets needs to be directly connected to the dedicated sound reinforcement system supply.

Music electronics continues to advance. Rock and pop, and entertainment music in general, are already unthinkable without it; it is also commonly used in 'serious' live electronic music. The interaction between sound reinforcement and music electronics will in the future give rise to many new solutions and ideas.

2.4.2 *Interaction between sound reinforcement systems and electronic media*

As already mentioned, sound reinforcement systems are often required to interface with systems of the electronic media (broadcasting, television, and commercial recording companies). In the simplest case, such as an on-the-spot report transmission, the signal is taken directly from the main output of a mixing console, in order to transmit extracts from a commentary, or the background noise of the event being reported. For large international events several co-users will need to be served simultaneously, which means that adequate distribution outlets have to be provided. This form of interface is commonly used in conference and sports centres. Another form of shared use can be found in multi-purpose and concert halls, and frequently also in theatres. Here the microphones provided for sound reinforcement purposes, which are often very numerous, are used in parallel. The signals obtained are routed via microphone distribution amplifiers, bypassing the mixing consoles of the sound reinforcement system and fed directly to the mixing equipment of the co-user, where an optimum sound balance is mixed for the medium concerned. This balance may well differ from that in the hall. Often up to four co-users may take the same low-level microphone signal.

Even closer interaction is needed in events organized jointly with the media. Here the sound reinforcement for the audience in the hall provides essential creative support in creating the right atmosphere, so that the audience are suitably enthusiastic about the performance. To this end certain playback signals required for the overall recording are also supplied to the audience in the hall by means of high-level playback lines, so that the co-producer's control equipment – usually a mobile unit – is linked to the fixed house system.

A large number of communication links are required to allow intercom communication between the transmission facilities and the sound reinforcement control centre, as well as with the other technical centres of the house, such as the lighting control. The connection points for the sound engineering and communication links are concentrated near the parking area for the mobile units. For small installations such as sports stadiums a lockable, weather-resistant connection board is usually sufficient. For larger facilities where frequent large-scale transmissions are expected, such as conference or cultural centres, special rooms are used that are accessible from outside. In addition to the connections for the programme and communication lines, these rooms also accommodate the power outlets for the mobile units and the video connections of the internal television system (Figure 2.5). Not all halls used for such productions are equipped with an

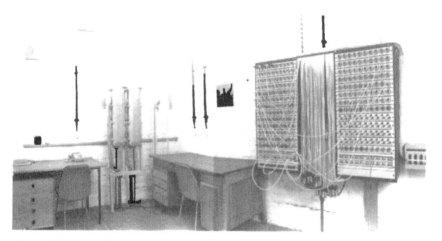

Figure 2.5 View of a connection room for mobile units.

adequate sound reinforcement system, and so additional mobile systems are often hired for large productions. These systems are usually installed and operated by the owner's personnel in cooperation with the users. So that the permanent in-house systems can operate in parallel, appropriate connection facilities have to be provided – before the power amplifiers for example. In large conference centres such as the ICC in Berlin, suitable communication units are dispersed around the building.

2.5 Interconnection with other systems

Sound reinforcement systems for cultural and entertainment purposes usually supply the acoustic information as an accompaniment for a visual performance. However, there are also applications in which the principal information is acoustic, but visual or lighting effects are desirable for enhancing the mood. A typical example is the discotheque, in which lighting signals are derived from the acoustic signal. Mixing consoles are provided with a light control output, which permits the connection of modulating equipment for various light sources. The modulation may be controlled by level, pitch or rhythm, in order to control spotlights, lighting moods and laser effects. This type of light-modulation equipment is also being increasingly used in live performances of pop and rock music, as well as with other kinds of entertainment.

Simple coordination of image and sound is also effected with sound-accompanied slide projections and multimedia shows. The equipment normally used for this purpose consists of a multi-track sound recorder that, as well as the sound track (or tracks), has one or more pilot tracks containing signals that are used to switch the slide projectors. Similar interconnections are used in modern planetariums, in which a central master computer coordinates the projection equipment, slide projectors and sound system according to a pre-set programme.

2.6 General hints

So we can see that the sound reinforcement system is often closely interconnected with the other technical, architectural and physical aspects of a facility. Failures or unsatisfactory operation of the sound reinforcement system are always seen as detracting from quality. If it functions correctly, this is usually taken for granted; if the system is very good, this is usually attributed to good room acoustics. The necessary expenditure and installation constraints are often undervalued, and this is why sound engineering does not enjoy the same conditions as, for instance, lighting, the effects of which are instantly visible. However, this underestimation may have negative consequences for the overall result. The electroacoustics consultant must therefore familiarize both the client and the person responsible for planning the project as early as possible with the requirements of an optimum sound reinforcement system. Economics, and both artistic and space considerations, mean that a system that uses all the facilities offered by sound engineering is not the optimum solution for all problems. It is essential to check, in each individual case, which means and procedures have to be used. This will be set out in subsequent chapters.

References

2.1 Ahnert, W. and Schmidt, W., *Akustik in Kulturbauten* (Acoustics in Cultural Buildings) (Berlin: Institut für Kulturbauten, 1980).
2.2 Steinke, G., Steffen, F. and Hoeg, W., 'Technologische Anforderungen an Beschallungssysteme großer Mehrzweckräume' (Requirements to be met by sound reinforcement systems of large multi-purpose rooms), Lecture, 6th Acoustics Convention, Budapest (1976).
2.3 Steffen, F., Schwarzinger, W. and Fels, P., 'Die tontechnischen Einrichtungen der Semperoper in Dresden' (The sound-technical facilities of the Semperoper in Dresden), *Technische Mitteilungen RFZ*, 30 (1986) 3, 49–56.

3 Room acoustics and auditory psychophysiology

In both planning and operating electroacoustical systems, certain requirements stemming from room acoustic considerations have to be met. This also applies to the design of the rooms in which the systems are to function. In order to obtain a sound pattern that is both aesthetically satisfactory and functionally adequate, it is also necessary to take account of the findings of auditory psychophysiology, such as the way the adjusted sound level affects the timbre balance between the lower and medium frequency ranges.

This is why this chapter does not only deal with the fundamentals of room acoustics, but also considers briefly such things as physical and emotional sound perception, and the problems of the effects of sound on the listener. We concentrate on those aspects of room acoustics and auditory psychophysiology that are relevant for loudspeaker reproduction. Measurement of the basic quantities described is covered in Chapter 7.

3.1 Fundamentals of room acoustics

3.1.1 Assessment of the quality of sound events

The listener to a concert or the visitor to a conference often pronounces judgement on the acoustic reproduction quality of a signal emitted by a natural source or via electroacoustical devices. This judgement is often very imprecise, in terms such as 'very good acoustics' or 'poor intelligibility'. Such assessments combine objective causes with subjective experiences acquired through listening to broadcast and television transmissions as well as CD, DAT and other hi-fi reproduction.

For speech one normally desires good intelligibility in an atmosphere unaffected by the room or by the electroacoustical means of amplification. The criteria for an assessment of music reproduction are far more sophisticated. 'Good acoustics' in this context means sufficient loudness, good sound clarity, and a spatial impression that is commensurate with the piece of music performed. Moreover, as far as the reproduction of traditional music is concerned, only the 'natural' timbre should be heard. ('Natural' timbre means, for example, that for reproduction in halls the level of the high-frequency components should decrease as the distance from the source increases.)

Definitions of the terms used in subjective assessment of speech and music reproduction have been laid down in the literature [3.1, 3.2], and in national and international standards [3.3, 3.4]. Originally these terms served mainly for the assessment of room acoustics, and are therefore important not only for enhancing understanding between the electroacoustician and the room acoustics expert, but also for assessing the electroacoustical reproduction itself. Some key terms are as follows:

- *Overall acoustic impression*: the suitability of a room for the (scheduled) acoustic performance.
- *Reverberation*: sound decay after sound excitation ceases.
- *Reverberation duration*: the duration of perceptibility of the reverberation.
  The reverberation duration depends on the objective reverberation time (which is a property of the room or the equipment), the output level (sound signal), the noise level or the threshold of hearing, and the ratio between the direct signal and the room signal. It is frequency dependent.
- *Clarity*: the temporal and tonal differentiability of the individual sound sources within a complex sound event.
- *Spatial impression*: the perception of the interaction of sound sources (ensembles) with their environs, including the listener.
  The spatial impression comprises several components, including: the room-size impression, the spaciousness, the reverberance, and the balanced distribution of reverberant sound.
- *Room-size impression*: the individually perceived and sound-event-dependent size of the acoustically perceived room.
- *Spaciousness*: the perception of the acoustic amplification of a source compared with its visual perception, especially in the lateral direction of the listener. The spaciousness depends both on the sound level of the original sound at the location of the listener, and on the intensity of reflections of the original sound, which arrive from lateral directions up to 80 ms after the direct sound.
- *Reverberance*: the perception that apart from the direct sound there is a reflected sound that is not perceived as a repetition of the signal.
- *Distribution of the reverberant sound*: perception of the distribution of the reverberant sound as a function of its direction of incidence (with the exception of discrete reflections).
- *Echo*: reflected sound arriving with an intensity and delay such that it can be discerned as a repetition of the direct sound.
- *Flutter echo*: periodical sequence of echoes.
- *(Local) diffusion*: evenness of sound field distribution with regard to intensity and direction of incidence.

- *Temporal diffusion*: a statistical measure of the temporal distribution of a sound field. (A high temporal diffusion means a low prominence of harmonic eigenfrequencies.)

3.1.2 Objective factors, criteria and quality parameters in rooms

3.1.2.1 Reverberation time and reverberation radius

When enquiring about the quality of the acoustics of a room, the expert as well as the acoustical layman will often still resort to the *hand clapping test*, to excite the volume of a hall with acoustic energy and listen to the process of reverberation decay. This is often followed by an estimate of the reverberation time [3.5], which is the oldest measurable variable in room acoustics and for many people the only one, even though it has been known for a long time that reverberation time on its own provides only a limited statement about the acoustic properties of a hall. Nevertheless it continues to be one of the main criteria, and so it is the first quantity that we shall consider.

The *reverberation time*, T, is defined as the time during which the mean steady-state energy density $w(t)$ of a sound field in an enclosure will decrease by 60 dB after the cessation of energy supply:

$$w_r(t) = w_{r0} \exp(-t/T) = 10^{-6} w_{r0} \mid t = T$$

As shown by corresponding transforms (see e.g. Kuttruff [3.6, p. 90]), the reverberation time depends on the volume and the damping treatment of the room:

$$T = \frac{0.163V}{4mV - S \ln(1-\alpha)} \approx 0.163 \frac{V}{A} \tag{3.1}$$

where V is the volume in m^3; A is the equivalent absorption area in m^2; α is the mean sound absorption coefficient (frequency dependent); S is the total surface of the room in m^2; and m is the absorption coefficient as a function of air absorption and frequency (Figure 3.1) in m^{-1}.

The *equivalent sound absorption area*, A, is calculated as

$$A = \alpha S = \sum_i \alpha_i S_i + \sum_n A_n + 4mV \tag{3.2}$$

where α_i is the sound absorption coefficient of the sub-area S_i, and A_n is the equivalent absorption area of objects and bodies.

The reverberation time cannot be reduced by electroacoustic measures, but there is a risk that it may be prolonged because of varying travel times

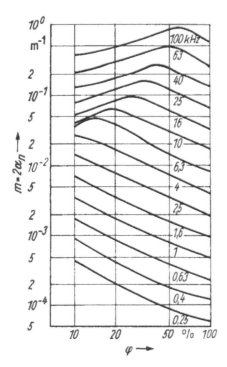

Figure 3.1 Absorption coefficient, *m*, of energy loss in air in the direction of sound propagation, *n*, as a function of relative humidity, at 20 °C.

between loudspeaker and listener (with a decentralized loudspeaker arrangement), or because positive feedback occurs. For this reason, and to reduce the risk of positive feedback (see Chapter 5), a somewhat shorter reverberation time is preferred for rooms where performances with sound reinforcement are frequently staged.

The *sound power absorbed* in a room, P_{ab}, can be inferred from the relation

$$\text{energy density, } w = \frac{\text{sound energy, } W}{\text{volume, } V}$$

under consideration of the differential quotient $P_{ab} = dW/dt$, as a measure of the energy loss in the room, from equations (3.1) and (3.2) (cf. [3.1], p. 239):

$$P_{ab} = \tfrac{1}{4} w_r c A$$

In steady-state conditions the absorbed power is equal to the sound power, P, fed into the room. Thus one obtains the *average sound energy density* in the diffuse sound field of the room as

$$w_r = \frac{4P}{cA} \tag{3.3}$$

where c is the velocity of sound.

While the sound energy density, w_r, is approximately constant in the diffuse sound field, the direct sound energy and thus also its density decrease in the near field of the source with the square of the distance from the source according to

$$w_d = \frac{P}{c} \frac{1}{4\pi r^2} \tag{3.4}$$

(Strictly speaking, this holds only for spherical sources [3.1, p. 120], but may at a sufficient distance also be accepted for most practical loudspeakers in which case the different energy is taken into account by the directivity characteristic; cf. section 5.2.)

In this range of dominating direct sound, the sound pressure loss is $p \sim 1/r$. (To be exact, this loss begins only outside the interference zone, the *near field*. The range of the near field is of the order of the dimensional magnitude of the source.)

If the energy densities of the direct sound and the diffuse sound are equal (i.e. $w_d = w_r$), equations (3.3) and (3.4) can be set equal to each other, and so one can infer a special distance from the source: the *reverberation radius*, r_H. For a spherical source it is (Figure 3.2)

$$r_H = \sqrt{\left(\frac{A}{16\pi}\right)} \approx \sqrt{\left(\frac{A}{50}\right)} \approx 0.141\sqrt{A} \approx 0.057\sqrt{\left(\frac{V}{T}\right)} \tag{3.5}$$

where r_H is in m, A is in m^2, V is in m^3, and T is in s.

In Figure 3.3 the variation of the overall energy density level, $10 \log w$ dB, is plotted as a function of the distance r from the source ($w = w_d + w_r$). In the direct field of the source one sees a decrement of 6 dB with each doubling of the distance. For a directional sound source with the *front-to-random factor* γ this can be expressed (see section 5.2) as

$$10 \log w_d \text{ dB} \approx 10 \log \gamma \text{ dB} - 20 \log r \text{ dB}$$

Thereafter follows, beyond the *reverberation radius* (for a spherical source)

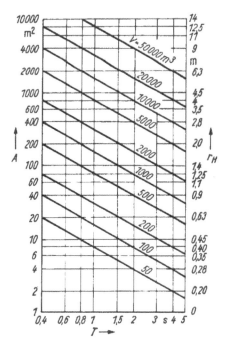

Figure 3.2 Reverberation radius, r_H, as a function of the equivalent absorption area, A, the volume, V, and the reverberation time, T, of a room.

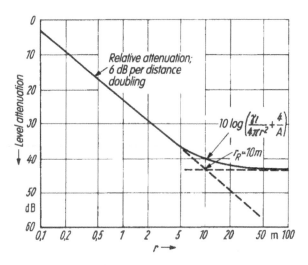

Figure 3.3 Sound level in enclosures as a function of the distance from the source.

or the *critical distance* (for a directional source; see section 5.4), a range in which a constant diffuse-field level $10 \log w_r$ dB $\sim -10 \log A$ dB prevails. In an absolute free field ($A \to \infty$) the free-field behaviour (6 dB decrement per distance doubling) would continue (dashed line in Figure 3.3).

Figure 3.3 shows that the reverberation radius can also be ascertained graphically. In this case, because of the directional effect of the source, one must use the *critical distance*, D_c. Hereafter the critical distance will be referred to as r_R. The critical distance thus obtained is $D_c = r_R = 10$ m.

3.1.2.2 Analysis of the energy–time curve

Compared with the reverberation time, the impulse response of a room provides considerably more information about the energy–time behaviour. Figure 3.4 illustrates this. On leaving a sound source (in this case an actor on stage), the direct sound is the first to reach the listener. Thereafter follow the initial reflections; as their density increases and level decreases, multiple reflections eventually occur, producing a diffuse reverberation.

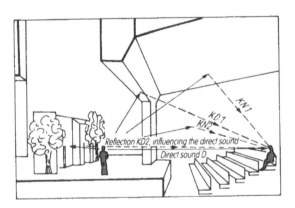

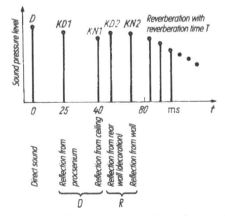

Figure 3.4 Possible sound paths in a theatre. D, direct sound effective; R, reverberant sound effective.

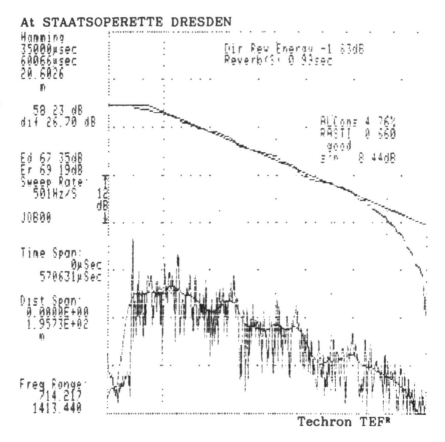

Figure 3.5 Reflectogram of an impulse sound test (TEF Analyser) in the Staatsoperette, Dresden.

This energy–time behaviour is assessed by sending an impulse into the room (impulse sound test), or by modern procedures such as time delay spectrometry or maximum length sequence measurement (sections 7.3.4 and 7.3.5). Figure 3.5 depicts a measured energy–time curve plotted as a function of time (time range 0–570 ms). For better understanding an integration curve approximately simulating the ear's inertia time is also shown. The resulting curve follows the envelope. Based on this weighting, which corresponds to actual aural perception, we can see that small gaps in the energy–time curve are not significant, for example for echo audibility. Critical conditions arise, however, if these peaks and dips show a periodical structure (see section 3.2.2.3). Figure 3.5 shows the reverberation decay curve, and some other energy values serving as quality criteria.

It is also possible to sum the energy portion in question over the whole time range. This energy curve can be obtained by integration of the square

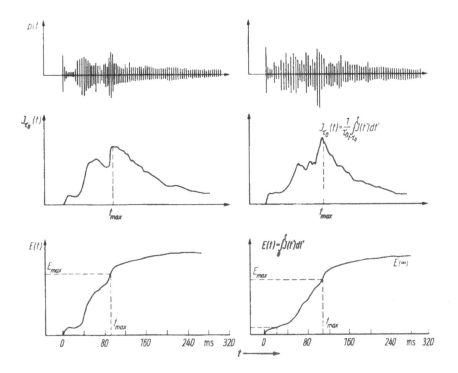

Figure 3.6 Time behaviour of the sound pressure, $p(t)$, of the sound intensity, $J_{\tau_0}(t)$ integrated with the inertia of the hearing system, and of the composite energy, $E(t)$.

of the sound pressure:

$$E = \int_0^t \tilde{p}^2(t') \, dt'$$

Figure 3.6 shows an energy curve of this kind. We can see that after a sufficiently long time (depending on the reverberation time; in practice one usually assumes around 500 ms) the reflected energy arriving at the listener's location becomes so weak that for $t \to \infty$ the final value E_∞ does not significantly increase further.

The energy–time curve provides a simple measure of the partial energy values that arrive at the listener's location at, say, 50 ms and 80 ms after the direct sound. Of the overall energy, E_∞, that arrives at the listener's location, only 5–15% can be attributed to the direct sound E_d, and 10% to the reverberation. About 80% of the energy is contained in the *initial reflections*. (We differentiate between the first part of the initial reflections – 5 to 40/50 ms – which effectively form part of the direct sound, and the later reflections – 40/50 to 80 ms – which effectively integrate with the reverberant sound. Compare Figure 3.4.) These are decisive for a large part of the subjective sound impressions, which change from one area of the audience to another along with the variation of these initial reflections. Sound reinforcement systems provide the opportunity to improve unfavourable reflectograms by compensating electroacoustically for the missing energy in those time intervals in which room-acoustical reflections do not occur. By means of signal repetition, for example via additional loudspeakers, it is thus possible to fill the gaps that occur in the sequence of room reflections.

All electroacoustically supplied energy will normally increase the sound level in the region of initial reflections. If this energy is slight, it does not contribute to an increase in loudness, but may nevertheless be useful for obtaining qualitative improvements: for example, with an appropriate time delay for eliminating echoes, enhancing spatial impressions, and avoiding mislocalizations.

The boundary between the early and late energy portions depends on the type of performance and on the reflection density build-up time; it lies at about 50 ms for speech and about 80 ms for symphonic music, after the direct sound. Early energy enhances clarity; late energy enhances the spatial impression. Laterally incident energy within a time range of 25–80 ms may even enhance clarity *and* spatial impression [3.7]. This is of crucial importance for the planning of sound reinforcement systems.

3.1.2.3 Criteria for intelligibility of speech and clarity of music

One of the main tasks for the sound reinforcement engineer is to improve the intelligibility of speech and enhance the clarity of music – especially if it is rhythmically accentuated. Optimization of the systems used depends on the following criteria.

ENERGY–TIME MEASURES USED FOR ASSESSING DEFINITION AND CLARITY

The definition measure C_{50} was derived from speech clarity D [3.8] as defined by Thiele:

$$C_{50} = 10\log\frac{\displaystyle\int_0^{50\text{ ms}} \tilde{p}^2(t)\,\mathrm{d}t}{\displaystyle\int_{50\text{ ms}}^{\infty} \tilde{p}^2(t)\,\mathrm{d}t}\ \mathrm{dB} \tag{3.6}$$

Extensive investigations were carried out to establish a measure for the clarity of traditional music. It was found that with symphonic and choral music it is not necessary to distinguish between temporal clarity and tonal clarity (the latter determines the distinction between different timbres)

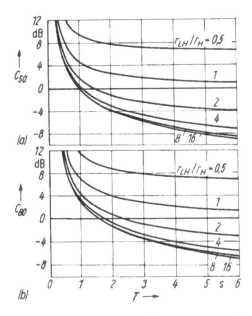

Figure 3.7 Expected values for the measures of (a) definition, C_{50}, and (b) clarity, C_{80}, as a function of reverberation time. Parameters: ratio between transmitter–receiver distance, r_{LH}, and reverberation radius, r_H.

[3.9]. Both are equally well described by the *clarity measure* C_{80}:

$$C_{80} = 10 \log \frac{\displaystyle\int_0^{80\text{ ms}} \tilde{p}^2(t)\, dt}{\displaystyle\int_{80\text{ ms}}^{\infty} \tilde{p}^2(t)\, dt}\ \text{dB} \tag{3.7}$$

It is possible to approximate the statistical definition and clarity measures by calculation [3.1]:

$$C_{50\text{stat}},\ C_{80\text{stat}} = 10 \log \frac{(r_H/r_{LH})^2 + 1 - \exp(-Ys/T)}{\exp(-Ys/T)} \tag{3.8}$$

where $C_{50\text{stat}}$ is a statistical measure of definition in dB; $C_{80\text{stat}}$ is a statistical measure of clarity in dB; r_H is the reverberation radius (with directional loudspeakers the critical distance, r_R, is to be used); r_{LH} is the distance from source to listener; T is the reverberation time in s; Y is a standardizing factor ($Y = 0.69$ for $C_{50\text{stat}}$ and 1.1 for $C_{80\text{stat}}$).

Figure 3.7 shows these statistical and thus predictable values as a function of reverberation time. The values decrease with increasing reverberation time. It can also be seen that for $T \leqslant 1$ s the statistical measure of definition, C_{50stat}, is > 0 dB, independently of the distance r_{LH}. For the measure of clarity, C_{80stat}, this holds true for reverberation times of $T \leqslant 1.6$ s. In rooms requiring high definition, or which are planned for the reproduction of preproduced sound signals, as is the case in a cinema or a discotheque, one should aspire to a statistical measure of definition of $C_{50} = 3-4$ dB, but for a multi-purpose hall, however, of $C_{50} = 0-4$ dB. This applies both to the acoustical design of these rooms and to the layout of the sound reinforcement system.

For reproduction of traditional music, clarity measures of $C_{80} = -2$ dB to $+6$ dB are optimal. The lower values apply to fast-moving music, whereas the higher values are more favourable for solemn and choir music.

ARTICULATION LOSS OF SPEECH

Peutz [3.10] and Klein [3.11] have ascertained that the *articulation loss of spoken consonants*, Al_{cons}, is decisive for assessing speech intelligibility in rooms. Proceeding along this line they developed a criterion for determining intelligibility. According to Peutz this is:

$Al_{cons} \leqslant 3\%$	ideal intelligibility
$Al_{cons} = 3-7\%$	very good intelligibility
$Al_{cons} = 7-12\%$	good intelligibility
$Al_{cons} > 12\%$	satisfactory intelligibility (limit value 15%)
$Al_{cons} > 25\%$	bad intelligibility

Al_{cons} increases with prolongation of the reverberation time and extension of the distance between loudspeaker and listener. By using directional loudspeakers it is possible to improve the intelligibility. The following equation applies approximately:

$$Al_{cons} \approx 0.652 \left(\frac{r_{LH}}{r_R} \right)^2 T\% \tag{3.9}$$

where $r_R \approx \sqrt{\gamma} \cdot r_H$ is the critical distance; r_H is the reverberation radius; γ is the front-to-random factor; r_{LH} is the distance from source to listener; and T is the reverberation time in s.

The correlations are shown in Figure 3.8. We can see that for $T \leqslant 1$ s the articulation loss Al_{cons} does not exceed a value of 15%, independently of the relation $r_{LH}/r_R \leqslant 5$. Peutz indicates that equation (3.9) applies only to distances of up to $r_{LH} \leqslant 5r_R$. It is possible to reduce the articulation loss to an acceptable value by decreasing the distance between the loudspeaker and the listener, r_{LH} (in a decentralized system), or by directional irradiation.

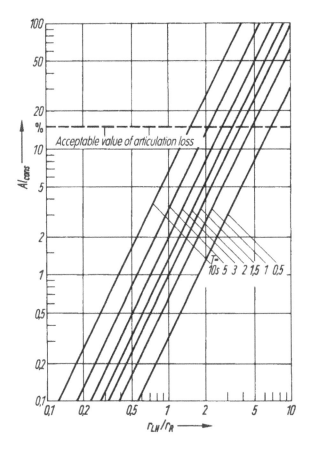

Figure 3.8 Articulation loss, Al_{cons}, as a function of reverberation time, T, and distance ratio, r_{LH}/r_R.

Resolving equation (3.9) for the front-to-random factor of the source, this front-to-random factor required for an articulation loss of no more than 7% (or less) is found to be

$$\gamma_L = 0.1T \left(\frac{r_{LH}}{r_H} \right)^2$$

Putting $r_{LH}/r_H = 5$ and $T = 2$ s results in $\gamma_L \approx 5$. Long reverberation times produce an increase of the articulation loss. Reverberation that is sufficiently long acts as *disturbing noise* on the subsequent useful signals, and thus reduces intelligibility.

Figure 3.9 shows the articulation loss Al_{cons} as a function of the signal and the noise level, and of the reverberation time, T [3.10]. The left diagram allows us to ascertain the influence of the *difference* $L_R - L_N$ (and of the reverberation time T, in s) on the Al_{cons} value, which gives $Al_{consR/N}$. Depending on how large the signal-to-noise ratio $(L_D - L_{RN})$ is, this value is then corrected in the right diagram in order to obtain $Al_{consD/R/N}$. The noise level and the signal level have to be ascertained as dB(A) values.

The smaller the signal-to-noise ratios are, the higher the Al_{cons} values rise, until they reach the limit of useful intelligibility.

As an example, an $Al_{consR/N}$ of approximately 18.2% for $L_R - L_N > 20$ dB (a good signal-to-noise ratio, S/N, is higher than 20 dB(A)) and $T = 2$ s is to be ascertained by means of the left diagram. Because of the strong direct sound of $L_D = L_R$ ($L_D - L_R = 0$ dB) the value of $Al_{consD/R/N}$ decreases when following the corresponding curve in the right diagram up to the abscissa value 0. One obtains $Al_{consD/R/N} \approx 8\%$.

Thus the figure also shows that no further practical improvement of intelligibility can be attained by increasing the signal-to-noise ratio to $\geqslant 20$ dB. (In practice this value is often considerably lower, because with high loudness values, say above 90 dB, and the resulting strong impedance variations in the middle ear, as well as the considerable bass boost caused by the frequency-dependent sensitivity of the ear, an intelligibility degradation is observed.)

DETERMINATION OF THE TRANSMISSION QUALITY BY RASTI

In a more general approach than that of Peutz and Klein, Houtgast and Steeneken [3.12] considered that speech intelligibility is reduced not only by reverberation and noise, but generally by all extraneous signals or signal alterations occurring between source and listener. They used the *modulation transmission function* (MTF) for determining this influence for acoustic purposes. The existing useful signal, S, is related to the prevailing disturbing signal, N (noise). The *modulation reduction factor* $m(F)$ thus ascertained is a quantity characterizing the influence on speech intelligibility:

$$m(F) = \frac{1}{\sqrt{[1 + (2\pi F \cdot T/13.8)^2]}} \cdot \frac{1}{1 + 10^{-((S/N)/10 \text{ dB})}} \tag{3.10}$$

where F is the modulation frequency in Hz; T is the reverberation time in s; and S/N is the signal-to-noise ratio in dB.

To this end, modulation frequencies of 0.63–12.5 Hz in third-octave bands are used. Moreover the modulation transmission function is

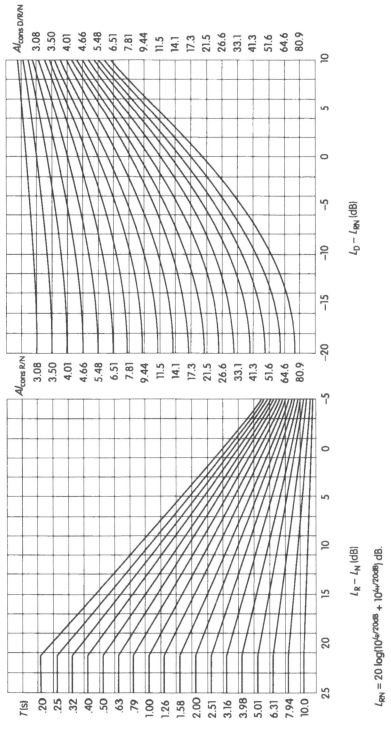

Figure 3.9 Articulation loss, Al_{cons}, as a function of the level ratio between diffuse sound, L_R, and direct-sound level, L_D, reverberation time, T, and noise level, L_N.

$$L_{RN} = 20 \log(10^{L_R/20dB} + 10^{L_N/20dB}) \, dB.$$

frequency weighted (WMTF – weighted modulation transmission function), in order to achieve a complete correlation with speech intelligibility. The modulation transmission function is divided into nine frequency bands, each of which is modulated with the modulation frequency [3.13].

The (apparent) effective signal-to-noise ratio, X, can be calculated from modulation reduction factors m_i:

$$X_i = 10\log\left(\frac{m_i}{1-m_i}\right) \text{ dB}; \qquad X = \frac{1}{9}\sum_{i=1}^{9} X_i$$

In order to make this computationally intensive and hence expensive procedure sufficiently practical for 'real-time' operation, the *RASTI procedure* was developed from it in cooperation with the firm Brüel & Kjaer [3.14] (see section 7.4). In this procedure the modulation transmission function is calculated only for two octave bands that are especially important for speech intelligibility (500 Hz and 2 kHz), and for nine modulation reduction factors. The resulting RASTI value (*rapid speech transmission index*) is by definition

$$RASTI = \frac{X + 15}{30} \tag{3.11}$$

Subjective investigations have established a positive correlation between the results of the RASTI procedure and intelligibility (Figure 3.10).

The procedures described, unlike the ones mentioned earlier, are well suited for checking the quality of a transmission both with and without an electroacoustic system (this is especially true for RASTI), but cannot be used for prediction or planning purposes. For this reason their main field of application is in the set-up and calibration of sound reinforcement systems (see Chapter 7).

SUBJECTIVE TESTING OF THE TRANSMISSION QUALITY OF SPEECH

The objective procedures require for reference purposes a subjective procedure for testing the transmission quality. This consists in the intelligibility of clearly pronounced words, often spoken in conjunction with an exciter sentence. In German intelligibility tests *logatoms* (monosyllabic non-words, e.g. 'grirk', 'spres') are employed for exciting the room. In English-speaking countries, however, test words as shown in Table 3.1 are used. Per test there are 200 to 1000 words to be used. The ratio between the number of correctly understood words and the total number read yields the *intelligibility of words*, expressed as a percentage. Values of 90–96% are considered to be excellent; values below 25–30% represent unintelligibility. When testing in reverberant rooms in particular it is often

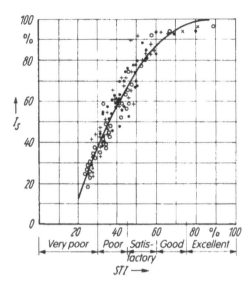

Figure 3.10 Relationship between STI (or RASTI) values and intelligibility of syllables, I_s.

Table 3.1 Examples of English words used in intelligibility tests

aisle	feet	loop	should	would
barb	file	mess	shrill	yaw
barge	five	met	sing	yawn
bark	foil	neat	sip	yes
baste	fume	need	skill	yet
bead	fuse	oil	soil	zing
beige	get	ouch	soon	zip
boil	good	paw	soot	
choke	guess	pawn	soup	
chore	hews	pews	spill	
cod	hive	poke	still	
coil	hod	pour	tale	
coon	hood	pure	tame	
coop	hop	rack	toil	
cop	how	ram	ton	
couch	huge	ring	trill	
could	jack	rip	tub	
cow	jam	rub	vouch	
dale	law	run	vow	
dame	lawn	sale	whack	
done	lisle	same	wham	
dub	live	shod	woe	
feed	loon	shop	woke	

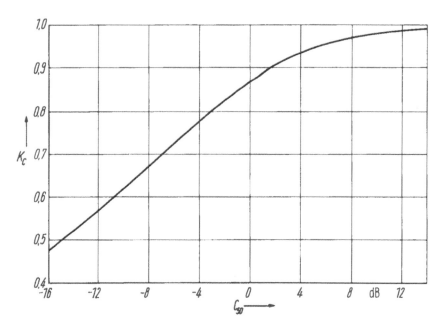

Figure 3.11 Intelligibility factor, K_c, as a function of the measure of definition, C_{50}.

advisable not to speak the syllables one by one, since in this way the masking sound decay of the previously spoken word does not occur. In such cases it has proved useful to speak an introductory sentence not belonging to the test (exciter sentence), such as 'Please mark the word "aisle"', 'The next is "barb"', before the word itself.

A number of researchers have investigated the dependence of syllable intelligibility on room acoustics [3.15–3.17]. The results show that syllable intelligibility, I_s, is determined by a factor K_x, which depends on frequency response limitations and the noise level [3.1, p. 61], and by masking through late-arriving sound energy, given as K_c:

$$I_s = 96 K_x K_c \ \% \tag{3.12}$$

Under certain circumstances it is possible to put $K_x = 1$. The factor K_c is related to the measure of definition, C_{50} (Figure 3.11), by the following equation:

$$K_c = \sqrt[s]{\left(\frac{1}{1 + 10^{-C_{50}/10 \ dB}}\right)} \tag{3.13}$$

For $C_{50} > 0$ dB the intelligibility factor becomes $K_c > 0.9$, so that according to equation (3.12) the syllable intelligibility, I_s, assumes values greater than 90%.

INTERRELATIONS BETWEEN THE PROCEDURES FOR CHECKING SPEECH INTELLIGIBILITY

The correlation between syllable intelligibility and articulation loss shown in Figure 3.12 results from equation (3.13) along with equation (3.9). For short reverberation times the syllable intelligibility is almost independent of the articulation loss. An inverse relationship sets in only with increasing reverberation time. With the usual Al_{cons} values between 1% and 50% syllable intelligibility may assume values between 68% and 93% (that is, a 25% fluctuation). According to Figure 3.11, measures of definition of between −8 dB minimum and +4 dB maximum are achievable under these conditions. Figure 3.12 also shows that for articulation losses < 15% (the limiting value of usable intelligibility), and independent of the reverberation time, the syllable intelligibility, I_s, always assumes values above 75%, which corresponds approximately to a measure of definition of $C_{50} > -4$ dB.

This correlation can also be seen in Figure 3.13, which shows the measured RASTI values according to equation (3.11) as a function of the articulation loss, Al_{cons}. Usable articulation losses of $Al_{cons} < 15\%$ require RASTI values of 0.4–1 (that is, between acceptable and excellent intelligibility). Through the equation

$$RASTI = 0.9482 - 0.1845 \ln Al_{cons}$$

it is also possible to establish an analytical correlation between the two values, and hence convert the values of one intelligibility measure into those of the other.

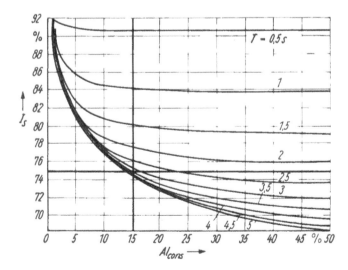

Figure 3.12 Syllable intelligibility, I_s, as a function of the articulation loss, Al_{cons}. Parameter: reverberation time, T. Preconditions: approximately statistical reverberation behaviour; signal-to-noise ratio $S/N \geqslant 25$ dB.

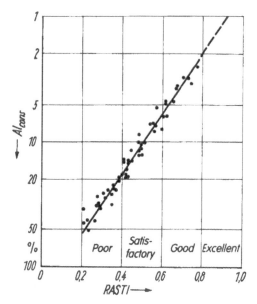

Figure 3.13 Relationship between Al_{cons} values and RASTI values.

3.1.2.4 Spatial impression with music

While the achievement of good intelligibility of speech requires the signal to be transmitted as free from room acoustic influence as possible, listening habits make the additional experience of a spatial impression desirable for choral and classical music. This spatial impression can be broken down into its two main components: spaciousness and reverberance.

Reverberance describes the subjective impression of the reverberation process. A distinction has to be made between the duration of reverberation and the reverberation level. The reverberation level characterizes the loudness of the reverberation process and thus its perceptibility: that is, the reverberation energy is related to the definition-enhancing early energy. One of the quantities that characterizes this relation is the *reverberance measure*:

$$H = 10 \log \frac{\int_{50 \text{ ms}}^{\infty} p^2(t) \, dt}{\int_{0}^{50 \text{ ms}} p^2(t) \, dt} \text{ dB} \tag{3.14}$$

For symphonic music, reverberance measures of 3–8 dB are optimal, according to Beranek [3.18]. With reverberance measures $H > 8$ dB, music is

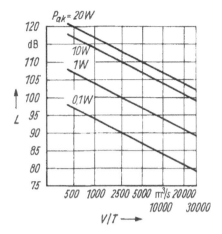

Figure 3.14 Sound pressure level, *L*, obtainable in the diffuse sound field as a function of the ratio between volume, *V*, and reverberation time, *T*. Parameter: sound power, P_{ak}.

perceived as being too reverberant, and with $H < 3$ as being too dry. Also other measures, such as the differently defined spatial impression measures, and initial reverberation time algorithms, take the reverberation level into account.

Figure 3.14 depicts the sound pressure level attainable in the diffuse sound field as a function of the ratio V/T, the parameter being the exciting sound power, P_{ak}. If, for example, a symphony orchestra ($P_{akmax} \approx 10$ W) plays in the symphony hall of the Schauspielhaus Berlin ($V/T \approx 8000$ m^3/s), levels of approximately 105 dB can be realized in the diffuse sound field. Higher diffuse sound levels and thus higher reverberation levels are unlikely to occur in practice.

The duration of reverberation results from the objective reverberation time recommended for particular applications (Figures 3.15 and 3.16). In a speech (drama) theatre of usual dimensions the reverberation time *T* should be between 1.0 and 1.3 s. In a concert hall of equal dimensions the nominal values are higher: between 1.6 and 2.1 s. Figure 3.17 shows the recommended reverberation time as a function of frequency. In rooms for music performances an increase of reverberation time is recommended for frequencies below 250 Hz. This increase produces a warm, round sound pattern.

Unlike reverberation time, a quantitatively accurate appraisal of the spaciousness-enhancing effect of initial reflections is still difficult today. Attempts in this respect are the *lateral sound level*, L_s, according to Kuhl [3.19], the *lateral sound degree*, *LE*, according to Jordan [3.20], and the various measures of spatial impression, *R*, according to Lehmann, Schmidt and Trautmann [3.7, 3.21, 3.22]. More detailed information can be found in the *Taschenbuch Akustik*, section 9.1 [3.23]. This also gives the

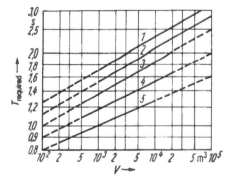

Figure 3.15 Recommended values of reverberation time at 500 Hz for different room types as a function of volume: 1, rooms for oratorios and organ music; 2, rooms for symphonic music; 3, rooms for solo and chamber music; 4, opera theatres, multi-purpose halls for music and speech; 5, drama theatres, assembly rooms, sports halls.

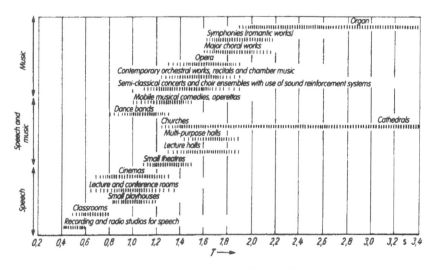

Figure 3.16 Optimum reverberation time at 500–1000 Hz (according to Russell Johnson).

calculation method for the frequently used *measure of spatial impression*, *R*, according to Lehmann [3.1, p. 66]. Lehmann found that when the measure of spatial impression, R, is between −20 dB and +20 dB, 14 grades of spaciousness can be subjectively distinguished (Figure 3.18). The measure of spatial impression, *R*, should be between 0 and 8 dB.

With reverberation times of $T \geqslant 2$ s, the measure of spatial impression, *R*, and the measure of clarity, C_{80}, are inversely proportional: that is, clarity decreases with increasing spaciousness and vice versa.

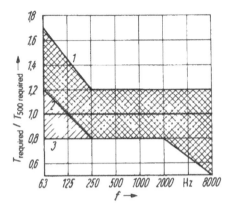

Figure 3.17 Tolerance ranges for the recommended reverberation time values according to Figure 3.15: range 1–2, music; range 1–3 speech.

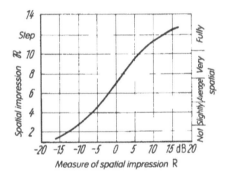

Figure 3.18 Subjective spatial impression, ℜ (in steps), and perception of spaciousness as a function of the measure of spatial impression, *R*.

Obtaining optimum results requires compromises to be laid down by the user or the building owner after consultation with the acoustician. From this basis, the more adaptable sound reinforcement engineering will be assigned certain tasks.

3.1.3 Objective variables influencing sound propagation in open spaces

3.1.3.1 Sound propagation

According to equation (3.4) the sound level decreases by 6 dB for each doubling of the distance. The level of a spherical source is expressed by

$$L_d = L_w - 20 \log r \, \mathrm{dB} - 11 \, \mathrm{dB};$$ (3.4a)

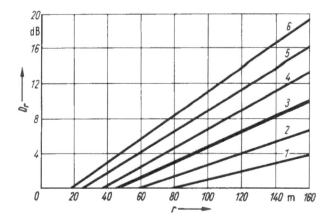

Figure 3.19 Additional propagation attenuation, D_r, with varying atmospheric situation as a function of distance r: range 1–2, very good (dusk); range 2–3, good (overcast); range 3–4, mediocre (mean solar radiation); range 4–5, poor (heavy solar radiation); range 5–6, very poor (desert heat).

where L_w is the sound power level; L_d is the direct sound pressure level; and r is the distance from source to listener.

At a distance of $r = 0.28$ m from a point source, the sound pressure level and sound power level are equal. At just 1 m distance (taken as the reference distance) the levels already differ by 11 dB: so with a sound power of 1 W (which gives $L_w = 120$ dB sound power level) the sound pressure level amounts to only $L_d = 109$ dB.

For open-air installations, where the distance between loudspeaker and listener may be very great, an *additional propagation attenuation, D_r,* which depends on temperature and relative humidity, must be considered. In this case the sound pressure level at the distance r is calculated for the assumed spherical source as

$$L_d = L_w - 20 \log r \, \mathrm{dB} - 11 \, \mathrm{dB} - D_r \, \mathrm{dB} \qquad (3.4\mathrm{b})$$

Curve 3 in Figure 3.19 is the average value of the empirically ascertained family of curves that should be used in practice. It shows that up to a distance of 40 m there is normally no need to consider additional attenuation. This applies, for example, to nearly all indoor rooms. The additional attenuation, D_r, increases with frequency (Figure 3.20). An electroacoustical system is therefore required especially to compensate for high-frequency losses.

Figure 3.21 shows the correlation of these frequency-dependent attenuation losses, D_r, with frequency and air humidity. It can be seen that, in bad weather, low frequencies are also affected by propagation losses.

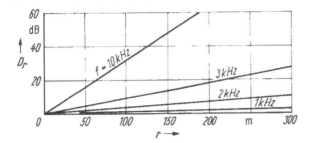

Figure 3.20 Atmospheric damping, D_r, as a function of distance, r, at 20 °C, 20% relative moisture and very good weather situation. Parameter: frequency, f.

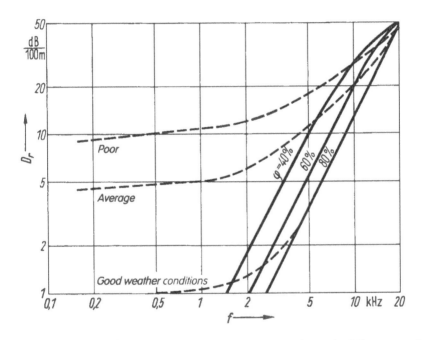

Figure 3.21 Additional propagation attenuation, D_r, with different weather conditions as a function of frequency: relative moisture.

Because of thermal expansion of air, the speed of sound increases by about 0.6 m/s per degree Kelvin. This implies that in a stratified atmosphere, in which the individual air strata have different temperatures, sound propagation is influenced accordingly (Figure 3.22) [3.24]. If the air is warmer near the ground and colder in the upper layer, upward diffraction of the sound

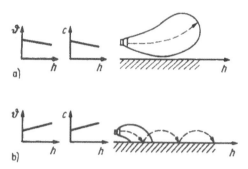

Figure 3.22 Sound propagation influenced by temperature: (a) negative temperature gradient, sound speed decreases with increasing height; (b) positive temperature gradient, sound speed increases with increasing height.

takes place: sound energy is lost from the near-ground transmission path, so that the propagation conditions deteriorate (Figure 3.22a). This situation can occur, for example, with strong solar irradiation on flat terrain, as well as in the evening over water surfaces that have been warmed up during the day. The opposite conditions prevail with cool air at ground level and warmer air in the upper layer, as is the case over snow or in the morning over water. Sound energy is diffracted from the upper layers down to the lower layers, where it boosts the energy and creates especially favourable conditions for propagation (Figure 3.22b).

Given that wind speed is relatively low compared with the speed of sound (in a storm it is approximately 25 m/s as against the average sound speed of $c = 340$ m/s), sound propagation is not normally significantly influenced by wind. However, because of the roughness of the earth's surface, wind speed is lower at ground level than in higher layers, so that sound propagation may be influenced by wind gradient in a similar manner as by temperature gradient (Figure 3.23). Thus squally or gusty wind conditions can severely restrict intelligibility. [3.24].

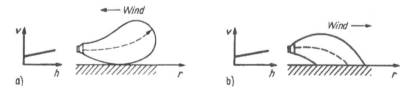

Figure 3.23 Sound propagation (a) against the wind or (b) with the wind; wind speed increasing with height in both cases.

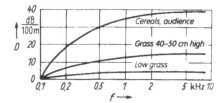

Figure 3.24 Sound attenuation with different surface conditions.

If sound is propagating over grassland, fields, or of course an audience, it may be heavily attenuated by diffraction into the absorbing surface (Figure 3.24). The loss at higher frequencies is particularly significant.

3.1.3.2 Criteria for intelligibility of speech and clarity of music

Given that sound reflections are largely absent in the open air (apart from courtyard and town-square situations), many of the quality criteria

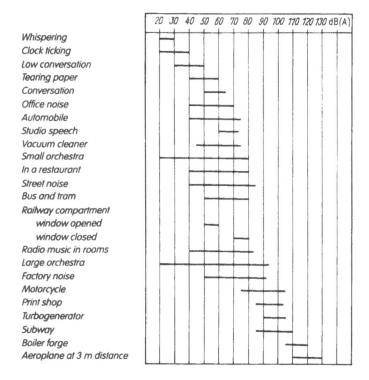

Figure 3.25 Loudness of various sound sources.

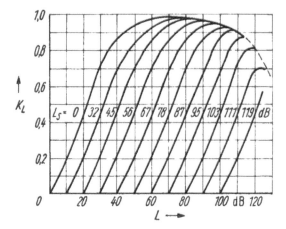

Figure 3.26 Intelligibility factor, K_L, as a function of signal level, L, and noise level, L_s.

developed for enclosures can be applied only if these 'reflections' are supplied by loudspeakers whose sound is correspondingly delayed with respect to the direct sound. The quality of reproduction is generally determined by the direct sound and the first initial reflections, so that one may reckon with a high definition or clarity. However, there are likely to be disturbing environmental noises, and losses of the higher-frequency signal components at larger distances. Intelligibility at the listener's location is thus determined by the ratio between useful and noise signals as well as by the timbre. Figure 3.25 shows a list of sound level values in dB(A) (that is, A-weighted) for various sound events. On airfields (for engine noise at close distance), for example, levels of 130 dB(A) are likely. Conversation, however, reaches values of 60–70 dB(A).

Based on these conditions, syllable intelligibility, I_s, can be calculated according to equation (3.12) by entering $K_c = 1$ and taking K_x (here $= K_L$) from Figure 3.26. According to this graph, with a noise level of say 67 dB and a useful level of 75 dB the factor $K_L = K_x = 0.8$. The resulting syllable intelligibility is 77%, which can be classified as good to very good.

3.2 Fundamentals of psychoacoustics and physiological acoustics

Given that a copious literature is available on the themes of psycho-acoustics and physiological acoustics [3.1, 3.23, 3.25, 3.26], we shall discuss in this context only those factors that are of particular importance for the planning and operation of sound reinforcement systems. As well as the relevant psychoacoustical and physioacoustical effects, we shall also deal briefly with hearing damage as a corollary of excessive loudness.

3.2.1 Sound perception and sensation

3.2.1.1 Pitch perception

The ability of pitch perception is limited to the frequency range between approximately 16 Hz and 16 kHz; the upper limit in particular may show considerable individual variations, mainly influenced by age. Young persons without hearing impairment (see also section 3.2.3) may still perceive sounds of up to 20 kHz, but with increasing age the upper perception limit can decrease to 10 kHz or lower.

For speech transmissions one should use the frequency range between 200 Hz and 4 kHz. Music, however, requires the use of a frequency range of at least 100 Hz to 10 kHz, or preferably 40 Hz to 15 kHz [3.27]. Like most other sensory perceptions, the perception of pitch differences does not depend on the absolute but on the relative frequency variation. This means that human pitch perception corresponds not to a linear but to a logarithmic frequency scale [3.28].

This is why one uses logarithmic frequency scales in acoustics as well as in music. For the different spheres of application, three frequency scales can be distinguished:

- The *physical-acoustical* frequency scale is based on a frequency ratio (interval F) between two frequencies f_1 and f_2 that forms an octave of $F = 1:2$ subdivided into half octaves $(1:\sqrt{2})$ and third octaves $(1:\sqrt[3]{2})$. In order to standardize measurement conditions, the limit and centre frequencies have been established according to IEC 225, edition 1966 [3.29].

$$F = \log_{10}(f_2/f_1)\text{dec} \equiv \log_2(f_2/f_1) \text{ octave}$$
$$\equiv 3\log_2(f_2/f_1)\text{third octave} \equiv 1200\log_2(f_2/f_1)\text{cent.} \qquad (3.15)$$

- The *musical* frequency scale is also based on the octave, but this time subdivided into 12 semitones. The audible frequency range comprises 10 octaves. Several attempts have already been made to assign fixed frequencies to the musical semitones, for example by setting the standard tone A at the frequency of 440 Hz. However, with large symphony orchestras in particular there is a tendency to set a slightly higher tuning, in order to achieve an especially brilliant sound.

With frequency ratios $f2/f1$ for which F assumes the numerical value 1, there result the units dec, oct, third, cent and the semitone, being:

$$F = \log 10 \text{ dec} = 1 \text{ dec} \qquad \text{for } f_2/f_1 = 10$$
$$F = \text{lb } 2 \text{ oct} = 1 \text{ oct} \qquad \text{for } f_2/f_1 = 2$$
$$F = 3 \text{ lb } \sqrt[3]{2} \text{ third} = 1 \text{ third} \qquad \text{for } f_2/f_1 = \sqrt[3]{2} \approx 1.26$$
$$F = 1200 \text{ lb } \sqrt[1200]{2} \text{ cent} = 1 \text{ cent} \qquad \text{for } f_2/f_1 = \sqrt[1200]{2} \approx 1.0006$$
$$F = 12 \text{ lb } \sqrt[12]{2} \text{ semitone} = 1 \text{ semitone} \quad \text{for } f_2/f_1 = \sqrt[12]{2} \approx 1.06$$

$$(3.16)$$

Thus there are

$$1 \text{ octave} = 3 \text{ third octaves} = 12 \text{ semitones} = 1200 \text{ cents} \qquad (3.17)$$

- The *physiological* scale is determined by the structure of the pitch perception organ – the basilar membrane. The scale subdivides into *critical bands* corresponding to about one octave in the lower frequency range (up to 200 Hz), but which thereafter get assigned to ever narrower relative bandwidths (Figures 3.27 and 3.28). The overall audible frequency range is divided into 25 frequency bands of 1 bark each. Approximately there are

$$1 \text{ critical band} = 1 \text{ bark} \approx 100 \text{ mel} \approx 35 \text{ pitch steps}$$
$$\equiv 1 \text{ coupling width} \approx 1.3 \text{ mm} \equiv 1000 \text{ hair cells}$$

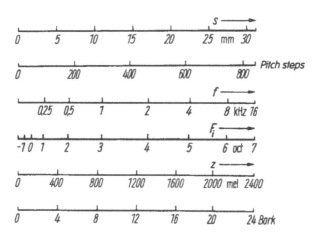

Figure 3.27 Distance, *s*, from the oval window on the basilar membrane in mm, and pitch steps (discernible pitch variations). Excitation maximum characterized by frequency, *f*, frequency index, F_i, and subjective pitch, *z*, in mel or bark.

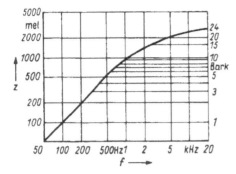

Figure 3.28 Tonality and frequency band scale as a function of frequency.

The minimum perceptible pitch variation – that is, the perception of a slight shift of a frequency interval F – is frequency dependent. In its most critical range between 1000 and 4000 Hz it is possible to distinguish pitch variations of as little as 0.7% by direct comparison. A long-term memory for pitches, however, occurs only in people with a sense of 'perfect pitch'. According to Feldtkeller and Zwicker [3.30], the hearing system is capable of distinguishing about 850 pitch steps.

In the lower frequency range in particular there exists a slight pitch shift depending on loudness. With a loudness variation of 50 dB it lies between 5% at 20 Hz and 1% at 1000 Hz. This effect should not be confused with the so-called 'equal loudness contour effect' (see section 3.2.1.2). It is based on the fact that greater loudness levels tend to excite larger areas on the basilar membrane into resonance, so that stimulus messages are also emitted by auditory nerves that are normally assigned to another frequency.

3.2.1.2 Loudness perception and masking effect

Loudness perception is limited downwards by the threshold of hearing and upwards by the threshold of pain.

The phenomenon of the threshold of hearing originates from the fact that a certain minimum sound pressure is required to produce a hearing impression. At 1000 Hz this minimum sound pressure, averaged over a large number of people, amounts to

$$\tilde{p}_0 = 2 \times 10^{-5} \text{ Pa} = 20 \ \mu\text{Pa}$$

or a sound intensity of

$$J_0 = 10^{-12} \text{ W/m}^2$$

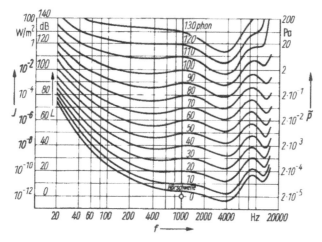

Figure 3.29 Curves of equal loudness level for pure sounds.

Following international standardization, these threshold values correspond to a sound pressure level of 0 dB. The hearing threshold depends strongly on frequency, and rises significantly for lower frequencies, as can be seen in Figure 3.29. (Because of the ear's decreased sensitivity to low-level, low-frequency sound we are not disturbed by natural sounds such as air turbulence or long-range, low-frequency transmission.)

The upper limit of sound perception is set by the pain caused by the reaction of the 'clipping protection' of the auditory system (disengagement of the ossicles). This limit lies at 10^6 times the sound pressure or 10^{12} times the sound intensity of the threshold value for 1000 Hz. But before this level is reached non-linear distortions occur, which commence at a level of about 90 dB. Loudness perception broadly obeys a logarithmic law (the Weber–Fechner Law).

One scale for loudness perception is the subjective *phon* scale according to Barkhausen [3.31]. It is established by comparing a given tone at one frequency with a 1000 Hz reference signal and adjusting the test tone to give a perceived loudness equal to that of the 1000 Hz reference. In this way one obtains curves of equal loudness level, which are similar to the curves shown in Figure 3.29 with some individual variations.

Note that the term 'phon' may be used only for such subjectively ascertained loudness values. Because of the arbitrary standardization, at 1000 Hz

$$L_N = L = 20\log(\tilde{p}_N/\tilde{p}_0) \text{ phon} = 20\log(\tilde{p}/\tilde{p}_0) \text{ dB}$$
$$= 10\log(J/J_0) \text{ dB} \tag{3.18}$$

The diminished sensitivity of the auditory system to low and high frequencies at low sound levels is approximately simulated for determined

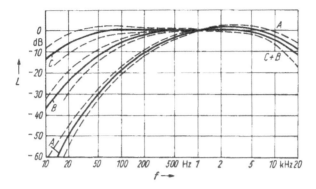

Figure 3.30 Frequency weighting curves recommended by the IEC for sound level meters.

loudness values by means of *weighting curves* in the sound level meter (Figure 3.30). According to IEC publication 651 (1979), the A-weighted curve corresponds approximately to the sensitivity of the ear at 30 phons, whereas the B-weighted and C-weighted curves correspond more or less to the sensitivity of the ear at 60 phons and 90 phons respectively [3.32]. Since current sound level meters do not have any weighting curve assigned to a determined sensitivity adjustment, the sound level values ascertained show at the most by chance an absolute relation to the corresponding loudness values. The A curve, which is switchable in any sound level meter, however, is also of importance for sound reinforcement engineering for measuring the sound level distribution of speech and information sound systems, particularly in noisy environments, but also in the vicinity of outlets of ventilating and air-conditioning systems (for example, air outlets in the backs of chairs, which are quite common nowadays). The A-weighting curve is recommended in such cases in order to avoid mismeasurement caused by air turbulence or low-frequency sound (see section 7.2).

If several tones or noises of different neighbouring frequencies and different loudness levels occur simultaneously, under certain conditions the acoustical stimulus of the lower-level sound may be inaudible, even though its sound level lies above the threshold of hearing. In this case the weaker noise is *masked* by the louder one [3.28, 3.33–3.35]. The audibility threshold of the weaker acoustical stimulus is determined by the *masked threshold of audibility*. Figures 3.31 and 3.32 show the masking effect under different excitation conditions. This effect is based on the fact that in the basilar membrane there is excited not only the narrow range corresponding to the stimulating frequency but, as loudness levels increase, the neighbouring ranges as well. As can be seen from the figures, this affects higher frequencies more than lower ones. This effect is of great importance

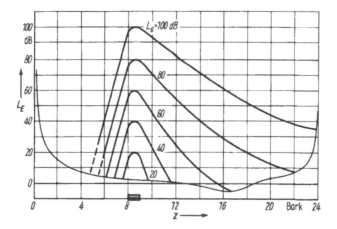

Figure 3.31 Excitation level, L_E, on the basilar membrane caused by narrow-band noise with a centre frequency of 1 kHz and a noise level of L_G (indicated on the horizontal axis between 8 and 9 bark): z, subjective pitch.

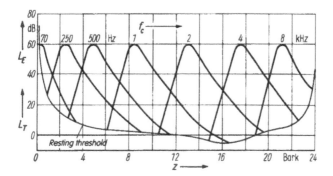

Figure 3.32 Excitation level, L_E, versus the subjective pitch, z (position on the basilar membrane), when excitation is by narrow-band noise of $L_G = 60$ dB and a centre frequency of f_c. L_T, resting threshold.

for sound reinforcement engineering. Its influence explains the fact that narrow-band frequency response notches, which often occur because of loudspeaker acoustic radiation interference, are normally inaudible, whereas higher-frequency peaks in the response may give rise to considerable timbre changes due to masking of the neighbouring ranges. Also, the inaudibility of weak disturbing noises in the presence of a strong, useful modulation may be attributed to the masking effect.

In this respect it is also interesting to know how to 'add' sound levels of the same spectrum. Since the addition is that of the common logarithm of energy,

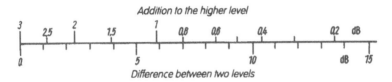

Figure 3.33 Calculation of the overall level from two individual levels.

it is necessary to add the $\tilde{p}^2$-proportional energy contents. This is easy if n sound sources have the same level and the same spectrum, resulting in

$$L_{\text{tot}} = L + 10 \log n \text{ dB} \tag{3.19}$$

With sound stimuli arriving at the listener's position additionally to the direct sound from sound reinforcement systems (either simultaneously or briefly delayed or staggered), the sound components to be added nearly always have different levels. For ascertaining the overall level one may use the nomogram given in Figure 3.33.

3.2.1.3 Timbre and perturbing noises

The *timbre* is normally attributed to the influence of the sound spectrum. This spectrum produces on the basilar membrane, as can be seen in Figures 3.31 and 3.32, an excitation pattern including the corresponding masking. This pattern is passed on to the central nervous system, where the sounds of speech and musical instruments are 'recognized' from these excitation patterns. The predominance of certain harmonics is essential in this respect (Figure 3.34). The excitation patterns are subject to constant change. Often the transition from one pattern to another is considered as the typical indicator of a speech sound, as is the case for example with diphthongs (ae, au, eu, iu, ai), and with initial and final consonants that are correctly recognized only in conjunction with the associated vowels, and not just by their own spectrum alone.

Timbre refers essentially to the linear distortions of the frequency response that become apparent, for example, in harmonics formation or in the emphasis and de-emphasis of treble and bass. But they may also occur in the case of brief repetitions (i.e. by residual tone formation) or on account of moderate non-linear distortions of individual instruments or voices (for instance of the trumpets or choir voices).

Timbres are easily recognized when it is possible to compare them directly with the original signal, which may quite easily be the case with sound reinforcement. Absolute recognition of timbres, however, is very difficult, especially if some habituation has taken place, and possible only for listeners having the *sense of absolute pitch*.

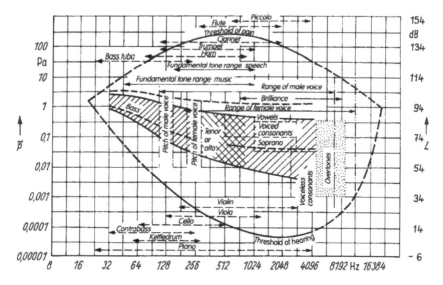

Figure 3.34 Position of the fundamental tones and harmonics in the auditory area (according to [3.15]): *L*, sound level; $\bar{p}$, sound pressure.

Noises that are completely unrelated to the original sound are not classed with timbres. They include switching clicks, 'frying' sounds, heavy non-linear distortions, and pitch variations caused by wow and flutter in analogue mechanical recording and replay devices. The listener assesses such additions as 'new signals', recognizing them instantly. In section 5.6 we shall discuss timbre variations in electroacoustical systems.

3.2.1.4 Time behaviour of the auditory system

For our auditory system there have been ascertained *characteristic time periods* that indicate how much time goes by before a tone is recognized, how quickly sound events become audible, how blending of sounds takes place, and how quickly sound stimuli decay in the auditory system.

The *transient periods* of sound events are especially important. For any musical instrument, any original signal and also the human voice, the initial transient is an essential characteristic that enables recognition. If we take a single note on the piano and edit out the initial transient, making only the lingering and decaying sound audible, the instrument will no longer be recognizable as a piano. Analogous tests can be made with nearly all other instruments: they cannot be recognized by the spectrum of the 'sustained' tone alone. The initial and final tones of speech show particularly short transient times, of 2–40 ms. Vowels last 50–300 ms, but are subject to heavy spectral changes during this period.

The initial transients of musical instruments have also been investigated [3.36]. Brass instruments (such as the trumpet) have very short rise-time periods of about 20 ms; woodwind instruments (including the flute) achieve steady state within 50–100 ms. String instruments' initial transients also build up very quickly when they are plucked (within 20 ms); when they are bowed, however, the transient period ends only after 100–150 ms.

During sound transmission, these transient periods must not be influenced or distorted in their development, and in sound reinforcement systems they must not be distorted by the loudspeakers. This can normally be achieved without any problem for the reproduction of conventional instruments, but when transmitting percussive instruments with direct acoustic pick-ups, or signals produced by electronic sound generators, fulfilment of requirement is difficult and demands a high technical standard from both the loudspeakers and their arrangement.

Pitch recognition depends on a somewhat longer 'characteristic period', since one has first to perceive several periods before being able to determine a pitch. For medium and high frequencies the characteristic period amounts to 4–8 ms, but for low frequencies periods of 25 ms are required (for 100 Hz). If the available period is shorter, the sound is perceived only as a click. Longer characteristic periods are required to recognize pitch variations of a sound event that is already perceived. In this case the auditory system tries to recognize the old sound for as long as possible.

Shorter than the transient time is the recognizability of differences in the envelope of the transient process. Its magnitude more or less equals the time constant (often referred to as the relaxation time) that a human being requires for perception by all the sense organs. It differs somewhat from individual to individual, but is about 20–25 ms.

According to Niese [3.37] the ear can be considered as an inertia-based energy store, so that the inertia evaluation may be simulated by a simple RC element with time constant $\tau_0 - RC$. A squaring network converts the sound-pressure-proportional input signal into a sound-intensity-proportional voltage (Niese model). For impulse evaluation with sound level meters a value of $\tau_0 = 35$ ms has been internationally agreed. Since the charging and discharging processes of the inner ear are equal, the model can also be used for the decay time. After disconnection of an excitation, the integration time amounts to about 25 ms, with individual variations. As a compromise, the IEC Commission [3.38] derived from this the integration time constant of the impulse sound level meter at 35 ms. This time is a good compromise between blending of the impulses (long boundary condition) and individual evaluation (short impulse with intervals). Thus the time constants for build-up and decay are not only frequency independent but also level independent (see also [3.39] and [3.40]).

This also makes it possible to determine the perceptibility of an echo: if, after an impulse or a switching-off, the process in question is repeated after more than 35 ms, the echo is perceptible, and no longer integrated in the charging process. Important in this respect are comparable heights of the reflection or impulse levels, and the duration of the impulses together with the time between them, as well as the occurrence of reflections in the interval between them. See also Dietsch [3.41].

Only a consciously perceptible signal repetition is referred to as an echo. An empirical value in this regard is a delay of 100 ms. This time is required for recognizing a sound (syllable) as a repetition. Repetitions with shorter delays produce a disturbing effect, because they reduce intelligibility and cause a timbre change. Here a limit duration of 50 ms is assumed as a guide value for the *blurring threshold* in speech (100 ms corresponds to a path length difference of 34 m; 50 ms to a path length difference of 17 m).

Haas has investigated more closely the blurring threshold with speech [3.42, 3.43]. With a reverberation time of 0.8 s (in a seminar room), a speaking rate of 5.3 syllables per second, and a level difference of 0 dB between the direct and delayed speech signals, he found the blurring threshold to be 68 ms. This is the value at which 50% of the experimental subjects to whom a repetition with a 68 ms delay had been presented, rated the reproduction as 'disturbing' (for more details see [3.1], p. 39).

Furthermore the time constant of the ear enables the deduction that there exists a minimum or 'internal' reverberation time of the human ear, fluctuating individually around 350 ms or 0.35 s. This time results in the previously mentioned Niese model with a decay from 0 to −60 dB, if the decay period from 0 to 6 dB amounts to 35 ms. Thus reverberation times of $T < 0.35$ s are no longer perceptible.

3.2.1.5 Acoustic localization

Acoustic localization is achieved mainly as a result of the binaural structure of the auditory system. In the median plane (front, above, behind) the travel time differences between the two ears do not contribute to localization, because both ears are equidistant from the median plane. *Monaural localization* of a sound event is nevertheless possible because of so-called directional frequency bands. These result from frequency-dependent shadowing by the pinna, and allow the assignment of originating directions 'behind', 'above' or 'in front' to narrow-band noises (see [3.26], p. 108 ff.).

Binaural localization is easier to explain. Lateral sound incident from outside the median plane produces in the medium and upper frequency ranges a pressure increase in the directly affected ear, whereas the shadowed ear is accordingly less affected. The auditory system is thus more sensitive to lateral sound incidence than to frontal incidence. The direction of incidence is inferred from time and level differences at the ears. Figure 3.35 shows the travel time differences and Figure 3.36 the level differences at the ears.

Investigations [3.44] have revealed that with frequencies below 300 Hz direction is inferred mainly from travel time differences, and with frequencies above 1000 Hz from level differences.

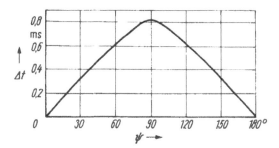

Figure 3.35 Inter-aural time difference, Δ*t*, of the sound reaching the ears, as a function of the angle of incidence.

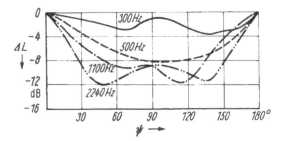

Figure 3.36 Variation of the sound level difference with discrete frequencies and horizontal motion of the sound source around the head.

In investigations on the 'influence of a simple echo on the intelligibility of speech', Haas established that the ear always localizes first the signal of the source from which the sound waves first arrive [3.42]. This holds true, however, only for finite time and level differences. The *precedence effect* derived from this and other investigations thus establishes that the perception of the direction of incidence is determined by the first arriving signal. This is still true if the secondary signal (the repetition of the first one) has a level up to 10 dB higher, and arrives within 30 ms (Figure 3.37). Only with travel time differences of *t* > 40 ms does the ear start to slowly notice separate reflections, but it still continues to localize by the primary event. The blurring threshold (see above) here also lies at about 50 ms. With longer delay times (from 30 ms upwards) one first notices timbre changes and then with times ≥ 50 ms, echoes ('chattering effect').

The precedence effect is frequently used for localizing sources in sound reinforcement systems. This will be covered in Chapter 6.

If identical signals from two loudspeakers show a mutual delay of only 1 ms (3 ms maximum) they merge into one signal as far as localization is concerned. There results a *phantom source* between the two directions of

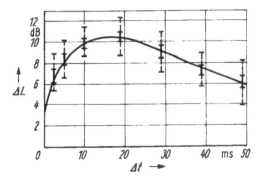

Figure 3.37 Critical level difference, ΔL, between reflection and undelayed
sound producing an apparently equal loudness impression of both
(speech) signals, as a function of the delay time, Δt.

incidence, and we speak of a *summing localization effect*. If there is neither
a delay nor a level difference between the two signals, the phantom source is
located on the bisecting line between the directions towards the two
loudspeakers. Delay or attenuation of one of the two loudspeaker signals
produces a displacement of the phantom source towards the loudspeaker
radiating the earlier or the louder signal (Figure 3.38).

In the localization of the phantom source, the *elevation effect* is also of
some importance: the phantom source is located not on the connecting line

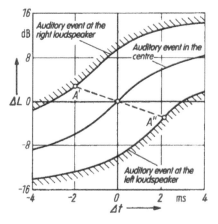

Figure 3.38 Apparent situation of the phantom source at the listening position
between two loudspeakers located on an operating line A'–A'' in
the usual stereophonic arrangement (about ± 30° out of the centre
line), as a function of inter-aural signal arrive time difference, Δt
(positive when left side is earlier), and of the level difference, ΔL
(positive when right side is louder).

between the two loudspeakers, but slightly higher. The angle of elevation depends on how far to the side of the listener the loudspeakers are located: the further round to the side, the greater the apparent elevation of the source. If the listener is exactly between the two loudspeakers, the phantom source even shifts to the top.

If there is a phase shift of 180° between the two signals, the phantom source ceases to be localizable, in spite of simultaneous or nearly simultaneous sound incidence. This can easily be achieved in electro-acoustical systems by reversing the polarity. Therefore we also speak of the *polarity reversal effect* or the *phase changing effect* (see also [3.27]). This produces a 'broader' source, a diffusion-related impression. The polarity reversal effect and phantom source occur with very small delays (maximum 5 ms), and must therefore be related to the summing localization effect. The precedence effect, in contrast, is related to delays of between 1 and 30 ms. Further details of other threshold values, such as for perceptible and disturbing echoes, can be found in [3.1] pp. 45 ff.

3.2.2 Perceptibility of distortions

Any sound event, on its way from the point of generation to the auditory system, is subject to signal-distorting influences. This 'transmission path' may consist of a sound reinforcement system inclusive of the acoustical paths, or an amplifier system with earphones and/or intermediate storage on disk, tape or the like. Also, the ear itself causes linear and non-linear distortions. Since the latter are always present at the corresponding sound levels for the individual listener, they are hardly registered, or only by an increase of the loudness impression.

3.2.2.1 Linear distortions

Here linear distortions are understood to be distortions (variations) of the *amplitude–frequency response*. In an ideal transmission the amplitude of the response level is frequency independent, but in reality deviations occur of greater or lesser magnitude. Such deviations of frequency responses down to 1 dB can be perceived by direct *A/B* comparison. Since this comparison is missing in normal hearing, however, only presences (frequency-response boosts) of 3–4 dB and absences (level dips) of ⩾ 5 dB become audible (the audibility fluctuates according to the width of the boost or dip). *Frequency response clipping* also becomes noticeable as linear distortion. Investigations into syllable intelligibility (cf. equation (3.12)) have revealed that in cases of frequency-response clipping it is necessary to introduce intelligibility coefficients K_{ft} and K_{fh} (Figure 3.39) indicating how intelligibility is reduced by the absence of low and (especially) high frequencies.

In most cases it has been established that in electroacoustical systems the *phase–frequency response* does not really need to be considered. This

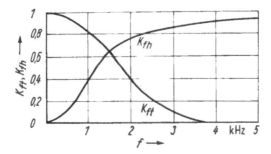

Figure 3.39 Syllable intelligibility factors K_{ft} and K_{fh} as functions of the frequency transmitted.

means that even heavy variations of the phase response are normally inaudible. They become audible only if they make themselves noticeable by amplitude distortions during the build-up of a pulse or the initial transient, or if they lead to interferences.

3.2.2.2 Non-linear distortions

If there is no linear correlation between the excitation and the reproduction rendered by a system, one speaks of non-linearity. Non-linearities in the transmission channel bring about new frequencies *not* existing in the original sound, and thus produce a distortion of the original. Their audibility depends largely on the frequency range in which the original signal is located, and on the loudness of reproduction. This is due to the heavily frequency-dependent behaviour of the threshold curve of equal loudness as well as to the loudness-dependent masking effects. The audibility of non-linear distortions also depends on the position of the resulting harmonics relative to the originating fundamental tone. The auditory system is least sensitive to integer multiples of the fundamental-tone harmonics. In this case timbre variations are noticed only when the newly formed harmonics amount to between 10 and 30% of the originating fundamental tone (see also section 5.5.).

The ear reacts in a far more critical way, however, if double or complex tones give rise to sum or difference tones that are not harmonically related to the individual partial tones. Especially disturbing in this respect are harmonics situated at a frequency ratio of 2 : 3 from the fundamental tone (in musical terms, an interval of a fifth).

In the reproduction of musical instruments, distortions are more disturbing with instruments that produce few harmonics (such as the flute). Here the audibility limit for second-harmonic distortions corresponds to a harmonic distortion of 1%, and for third-harmonic distortions of only

0.3%. The *distortion factor, K,* is defined as

$$K = \sqrt{\left(\frac{\displaystyle\sum_{2}^{n} U_n^2}{\displaystyle\sum_{1}^{n} U_n^2}\right)} \times 100\% \qquad (3.20)$$

where U_n is the amplitude of the nth partial harmonic, and U_1 is the amplitude of the fundamental frequency.

In addition to harmonic distortion, two further measures have been introduced for assessing non-linear distortion: *intermodulation distortion* for the medium-frequency range, and *difference tone distortion* for high frequencies (see [3.23], pp. 1293 ff.).

Inexperienced listeners often interpret the presence of non-linear distortions as an excessive loudness level.

3.2.2.3 *Comb-filter distortions*

In addition to linear distortions characterized by impairment of the frequency response, and non-linear distortions characterized by the appearance of new frequency components, a third form of distortion plays an important part in sound reinforcement engineering. These distortions may cover both characteristics, depending on the time behaviour they are subjected to. They are classified under the catch-all title of 'comb-filter distortions', and correspond widely to what in room acoustics is referred to as lack of temporal diffusivity.

Comb filters are produced either by the interaction of a direct sound component and a briefly delayed reflected sound, or by the interaction (addition) of two similar (audio) signals when one of them has been electronically (or mechanically) delayed. The following example explains the effect in more detail. If a very short but powerful pulse (ideally a Dirac pulse) is emitted by an ideal loudspeaker, the resulting spectrum is frequency independent over the whole frequency range in the direct field of the loudspeaker (provided the loudspeaker has an ideal transient behaviour and ideal frequency response). The impulse response of the room in the diffuse field may be frequency dependent, since after interaction of the stimulus with the room, the acoustic properties of the room are mapped here. The above-mentioned relationship for the direct field can be described in mathematical terms as follows. The signal $g_1(t)$ is assumed to be a Dirac pulse

$$g_1(t) = \delta(t) \qquad (3.21)$$

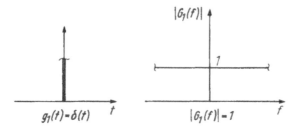

Figure 3.40 Fourier transform of a Dirac impulse as a frequency-independent value.

The amplitude spectrum (square of Fourier transforms) is then

$$| G_1(f) |^2 = 1 \tag{3.22}$$

Figure 3.40 illustrates the correlation: the Dirac pulse is shown undistorted.

If the acoustic signal $g_1(t)$ (here the Dirac pulse $\delta(t)$) is electrically delayed by Δt (this can also be achieved by means of two loudspeakers offset by $c\Delta t$, Figure 3.41) and superimposed on the original signal, interference occurs with coherence of both signals. Compared with the original signal, the spectrum of the resulting summation signal shows boosts and dips offset from one another by equal frequency intervals $f = 1/\Delta t$. The depth of the dips (cancellations) depends on the ratio of amplitudes, q, between the superimposed components. Because of this spectrum, which can also be realized by 'comb-like' filtering with adjacent pass and stop bands of equal frequency intervals, these distortions are called *comb-filter distortions* [3.6, 3.45].

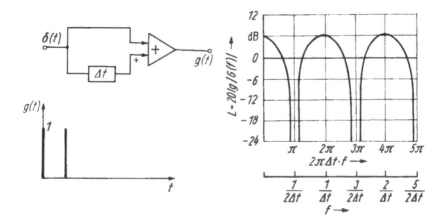

Figure 3.41 Frequency response in the shape of comb filters after superposition of two signals of equal level.

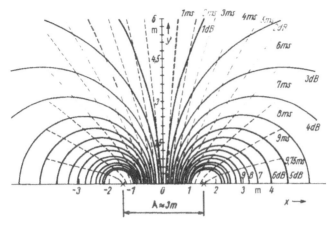

Figure 3.42 Curves of equal sound level difference and equal delay time difference for two loudspeakers located at a distance of $A \approx 3$ m, in the x–y plane.

Mathematically we have now the following context. The summing signal, $g_1(t)$, is now

$$g_1(t) = \delta(t) + q\delta(t - \Delta t) \tag{3.23}$$

where $\delta(t)$ is the Dirac impulse, and $20 \log q$ is the level attenuation in dB between the two signals.

The amplitude spectrum is obtained by the Fourier transform

$$|G_1(f)|^2 = 1 + q^2 + 2q \cos \omega\Delta t \tag{3.24}$$

If the levels of the pulses are equal ($q = 1$) we have

$$|G_1(f)|^2 = 2(1 + \cos \omega\Delta t) \tag{3.24a}$$

After one transformation the modulus of the amplitude spectrum, $|G_1(f)|$, becomes

$$|G_1(f)| = 2|\cos(\omega\Delta t/2)| \tag{3.24b}$$

Figure 3.41 shows this context.

Figure 3.42 shows that the comb-filter effect is increasingly audible outside the range of effectiveness of the summing localization effect (section 3.2.1.5) and with approximately comparable level characteristics of the two loudspeakers, $L_1 \lessgtr L_2 \pm 5$ dB. Figure 3.43 shows, as a function of the separation A of the loudspeakers, the displacement of the area in which timbre changes caused by comb-filter effects may be audible.

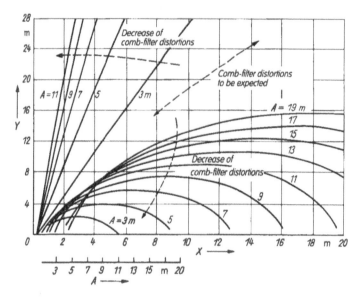

Figure 3.43 Curves of equal sound level difference, $\Delta L = 5$ dB, and equal
delay time difference of $\Delta t = 3$ ms for two loudspeakers, as a
function of the distance A between the loudspeakers (analogous to
Figure 3.42, but only right-hand side shown).

This effect is, however, not perceptible with normal stereo listening at
home (cf. section 5.5.3.5 [5.15]).

We can see in Figure 3.41 that the first zero in the comb-filter-like
distorted frequency response results at $f_1 = 1/(2\Delta t)$. With travel times of
just 1 ms ($\cong 34$ cm distance between the acoustical centres of two
loudspeakers) it is already at 500 Hz, and with longer times it gets shifted
towards lower frequencies, so that the whole frequency range is included.

If we add not just *one* time-delayed reflection but a whole level sequence,
as is usually the case with artificial reverberation systems with unfavourable
mode distribution, there follows by analogy with the time function in
equation (3.23) a sequence of

$$g_1(t) = \sum_{n=0}^{\infty} q^n \delta(t - n\Delta t) \tag{3.25}$$

the resulting amplitude spectrum of which is

$$|G_1(f)|^2 = \frac{1}{1 + q^2 - 2q \cos \omega \Delta t} \tag{3.26}$$

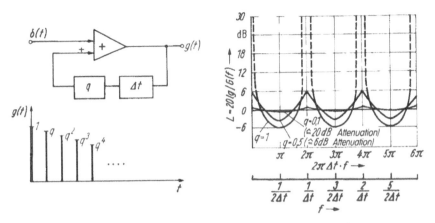

Figure 3.44 Frequency response of level sequences for various level attenuations.

The relation is shown in Figure 3.44. We can see that the peaks of the comb-filter curves occur with the same spacing of $1/\Delta t$ as in Figure 3.41, but are much more pointed. Whether such a comb-filter effect (that is, a comb-filter-like distorted impulse response caused by room or electroacoustical effects) produces an audible timbre change or not thus depends on the delay time Δt and on the level attenuation 20 log q dB.

Kuttruff [3.6] writes in this respect that the absolute threshold for an audible timbre change rises with increasing delay: that is, when the spacing between the comb-filter peaks becomes narrower. This is shown in Figure 3.45. It can also be seen that a very high but very dense irregularity of frequency curves in rooms ($1/\Delta t \rightarrow 0$) does not entail timbre changes, because the different delays produce a high temporal diffusivity. Above a

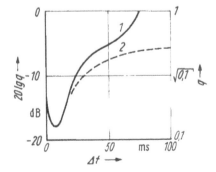

Figure 3.45 Critical values of q or 20 log q dB provoking just-audible timbre changes, as a function of the delay, Δt [3.6]: 1, white noise applied to comb filter according to equation (3.23); 2, white noise applied to comb filter according to equation (3.25).

certain threshold value of $\Delta t \approx 25$ ms (corresponding to frequency spacings of $1/\Delta t < 40$ Hz), the comb-filter impression changes into a sensation of roughness. This also happens if delay times of artificial reflections are a multiple of a fundamental delay of this order of magnitude. The origin lies in the short-time spectral analysis carried out by the ear.

In conclusion we can say that phase-locked signal repetition of any kind may give rise to travel time distortions with ensuing timbre changes – that is, comb-filter distortions. Repetitions in the millisecond range may in this way lead to the described comb-filter distortions (sound echoes) or produce roughness (flutter echoes). If the spacing of the signal repetitions is bigger, perceptible linear distortions may occur in the lower frequency range.

Therefore, if differently delayed signals are emitted by one loudspeaker, it is necessary to make a test for timbre change. Detected changes can be reduced, for instance, by phase rotation at the output of the delay equipment. Other references (e.g. [3.46]) show that the disturbing occurrence of the comb-filter effect can be significantly reduced by an adequate combination of delay times and all-pass filters, as well as by a phase rotation at the outputs of the delay equipment.

Other unfavourable technical conditions may also give rise to audible comb-filter phenomena, such as the following:

- An unfavourable microphone arrangement, in which shortly after the arrival of the direct sound the microphone receives strong reflection: for example, if boundary microphones are incorrectly arranged, or with microphones arranged over a speaker's desk. Here the comb-filter effect can be prevented by covering the reflecting surfaces with absorbing materials.
- Very phase-true signal repetition caused by digital delay circuits. To render the comb-filter distortion inaudible in this case, a delay of $\Delta t = 30$ ms requires, according to Figure 3.45, an attenuation of 8–10 dB between the original and the delayed signal. Measurements have shown that 9 dB is the required signal sequence attenuation.
- With deficiencies in the reverberation equipment, permitting a periodical repetition of the decay process, 'metallic' timbres may become audible.

The comb-filter effect, which is described here as a repetition effect, is known in room acoustics as a *lack of temporal diffusivity*. In physical terms we can explain this effect by saying that the system under consideration (room, reverberation equipment, etc.) features only harmonically interrelated eigenfrequencies. The subjective effect of this phenomenon presents itself with very short repetition times (narrow spacing of the eigenfrequencies) in the form of an additional tone (residual tone). In this the repetition sequences (Δt) form the periods of this tone, $f = 1/\Delta t$, or serve as the fundamental tone of the perceived harmonics. Typical examples are the 'roaring' sound of a unidimensional space (such as a pipe) or the metallic sound of poorly tuned reverberation equipment.

When the repetition times (or the spacing of the eigenfrequencies) become large enough for the resulting residual tone to be situated outside the hearing range, the effect will be perceived as a 'sound echo', and with still greater separations, as a flutter echo (see also section 5.6.3.3).

3.2.2.4 Modulation distortion

Listeners can readily distinguish between amplitude modulation and frequency modulation distortion. Both types presuppose modulation of a carrier oscillation by a modulating frequency. It has been found that the value of the modulating frequency is particularly decisive for the perceptibility of modulation distortions. These distortions include also the *wow and flutter* caused by fluctuations in sound recording equipment (disk, tape). To avoid perception of these fluctuations, the relative frequency variations should as a rule not exceed 0.7% [3.47].

3.2.3 Effect of high loudness levels on the hearing system

So far in this section we have dealt with the mechanism of acoustic perception. We are now going to briefly discuss the detrimental effect of excessive loudness levels of transmitted music and – less frequently – speech. This aspect is becoming increasingly important, because in rock and pop music in particular, but in other forms of entertainment music as well, high sound levels are presented over long periods and with only short interruptions. Since in this case the 'aesthetic' (i.e. wanted) stimulus rather than the 'disturbing noise' constitutes the source of impairment, this influence has not yet been sufficiently realized. Recent investigations [3.48, 3.49] show, however, that music of any kind may lead to hearing impairment whenever detrimental sound levels are produced over long periods at close range to the ear (see particularly [3.50]).

Violinists in symphony orchestras have been found to suffer after a long period of activity a reduction of hearing in the left ear, attributed to holding the instrument [3.49]; but other investigations do not suggest any hearing impairment in orchestral musicians [3.51].

3.2.3.1 Mechanism of hearing impairment

To help in understanding the mechanism that causes hearing impairment, we shall first briefly discuss the implications of the impairment, and its examination.

A negative influence on the auditory system is detected by a shifting of the hearing threshold – that is, the deviation from a standardized threshold (see section 3.2.1.2) – after exposing it to a certain acoustic excitation.

Normally there first occurs a *temporary threshold shift*, TTS, which disappears after a period of recovery. If the ear is subjected to very long or very high and frequent acoustic excitations, however, it may undergo a *permanent threshold shift*, PTS, indicating actual impairment. The degree of impairment of a person's hearing is characterized by the *integrated temporal threshold shift*, ITTS, according to [3.52] and [3.53]:

$$\text{ITTS} = \int_{t_e + t_R} (\text{TTS}) \, dt \tag{3.27}$$

where t_e is the period of exposure to sound, and t_R is the recovery time.

This impairment mechanism, which works according to a dosage principle, was established on the basis of investigations by Kraak, Fuder, Kracht and others [3.53]. It reveals that permanent impairment is particularly to be expected after exposure to frequent and prolonged acoustic excitation. Thus there is an especially high risk of impairment if the auditory system is exposed to a high sound level during working hours as well as during leisure time. By means of this dosage principle it was possible to establish a correlation between ITTS and noise dosage for the 'noise' that corresponds in spectral terms to loud rock and pop music:

$$\text{ITTS} = 17.6 \text{ dB/Pa} \cdot B_L; \tag{3.28}$$

where B_L is the noise dosage $\int_{t_e} |p_a(t)| \, dt$ and $p_A(t)$ is the A-weighted sound pressure.

Further investigations revealed that a PTS can also be derived from the dosage principle. An interrelation applicable to music and relating to 4 kHz is as follows:

$$\text{PTS}_{4 \text{ kHz}} = 50 \log(B/B_0) \text{ dB} \tag{3.29}$$

where $B = B_L + B_a$ is the overall dose including age-equivalent dose (B_a), and $B_0 = 2 \times 10^7$ Pa's.

For our purposes it is also possible to determine the noise dosage by

$$B_L = 0.8 \sum_i 10^{L_{Ai}/20 \text{ dB}} p_0 \Delta t_i \tag{3.30}$$

where L_{Ai} is the sound level of the A-weighted sound pressure during exposure time Δt_i; $p_0 = 2 \times 10^{-5}$ Pa is a reference sound pressure; and $\sum_i \Delta t_i$ represents the above-mentioned sound exposure period, t_e.

Figure 3.46 shows the PTS for 4 kHz as a function of the noise dosage caused by music with very high loudness levels. The sound levels involved when plotting the diagram varied between 80 and 105 dB(A).

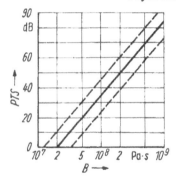

Figure 3.46 Permanent threshold shift, PTS, at 4 kHz as a function of the noise dosage, *B*, for stationary noise from 79 to 105 dB(A): ———, regression line; ------, limits of the ± 2 s range (s, residual standard deviation).

Some examples will illustrate the interrelations.

1. Youths visit a discotheque on average once or twice a week over a period of 5 years. There their ears are exposed to an average sound level of 115 dB(A). This adds up to an overall exposure of about 400 days of 5 h each: that is, 2000 h. According to equation (3.30) this represents an accumulated noise dosage of $B_L \approx 6.5 \times 10^7$ Pa.s. From equation (3.29) or from Figure 3.46 this results in a permanent threshold shift of PTS ≈ 25 dB, implying that the threshold of hearing has been shifted irreversibly by this value. (From 40 dB on one speaks of hardness of hearing!)
2. In symphonic concerts levels of $L_A = 115$ dB are seldom reached. Here the average levels are $L = 85$ dB(A). For the same permanent threshold shift of PTS ≈ 25 dB as in the previous example to be reached, there would need to be a noise exposure of this intensity for more than 40 years, and this at 300 days per year for 5 h each. This explains why few professional hearing impairments have been proven in symphonic musicians. For pop and rock musicians it is quite a different story, as can be demonstrated by the final example.
3. Let us assume that a rock musician is exposed for 200 days per year for 6 h each to 120 dB(A). This amounts to an annual noise dosage of $B_L \approx 6.9 \times 10^7$ Pa.s, and results over a period of 10 years in a permanent threshold shift of PTS ≈ 70 dB. This means professional disability as a result of deafness.

3.2.3.2 Reasons for high sound levels in sound reinforcement systems

It is undeniable that in performances of entertainment music, and particularly in juvenile rock music functions as well as discotheques, very high sound levels prevail, which are even increased in the course of the performance for enhanced emotional effect. At close range to the performance long-term sound levels of over 90 dB(A) are produced, which can often exceed 110–120 dB(A) in close proximity to the loudspeaker stacks. Spectral analysis of such performances reveals in most cases a very bass-accentuated signal, which is intentionally produced in order to obtain the desired 'vibration of the abdominal wall'. With regard to hearing impairment, this bass emphasis may even be considered as a positive factor, since most of the irreversible impairments occur in the range between 5000 and 6000 Hz [3.54, 3.55]. Reddel and Lebo [3.54] report that, according to visitors to such events, the music has to be so loud in order to completely engross the listeners. One often finds that with bands of poor performance or low-quality reproduction, the music tends to be louder so that the resultant distortions in the ear cover up the faults, whereas with better music performances the levels produced are normally lower. Moreover, the musicians grow increasingly hard of hearing in the course of their professional lives, so that they are forced to play louder, or must have a louder accompaniment, to hear each other. This results in a cumulative process that leads to ever higher loudness and thus inevitably to hearing impairment.

A problem to be considered, especially in rooms with good irradiation conditions from the platform, consists in establishing a balance between a weak-level soloist and the stronger orchestra group. Here a 'build-up' of the sound level may occur through crosstalk between the microphones. This results in a spiralling process.

3.2.3.3 Possibilities for reducing excessive sound levels

In the operation of sound reinforcement systems it is very difficult to avoid harmful sound levels of over 90 dB by technical means. Restriction of the sound level by limiting the installed amplifier and loudspeaker power is not feasible, since in the interest of a high overdrive safety these components of a system should have a power handling reserve of at least 10 dB. (The increasing use of digitally recorded programmes, which require even more 'headroom', makes an even greater power reserve necessary.) Nevertheless it is essential to adjust the levels of a system so that the power handling reserve cannot be abused for excessive loudness enhancement. In several countries there have been installed in discotheques so-called 'loudness indicators', which show by means of a visual display that certain predetermined maximum sound levels have been exceeded, and thus indicate to the listener that he or she is being exposed to a hearing impairment. There are also automatic loudness limiters, which are inserted before the power amplifiers and impose a preset attenuation when a certain maximum level is exceeded. The functioning of such devices has constantly to be supervised by the responsible authorities.

An excessive sound level caused by too strong an irradiation by the orchestra from the stage can be counterbalanced by corresponding room-acoustical measures. For example, it is possible to adjust the stage reflectors so that the sound is directed towards the flies. Other possibilities include changing the arrangement of the orchestra on the platform, or even enclosing the orchestra in a nearly complete sound-absorbing screen, as described in [3.56].

Apart from its immediate effects, excessively loud music is also disturbing for those nearby not partaking in the 'sound event'. Youth clubs and discotheques must therefore take special building noise control measures to prevent noise pollution of the environment. In this respect legislation has established exact limit values, justifying intervention when they are exceeded. This, however, is better done at the source of the noise pollution: that is, in the electroacoustical system.

Internationally, regulations regarding the permitted maximum sound levels produced by sound systems in entertainment venues and buildings vary considerably. For example, in Germany the limitations of the sound levels produced by sound reinforcement systems in theatres and multi-purpose halls are given in DIN 15905 [3.57]. This also contains the levels that sound reinforcement systems are permitted to produce, as well as measuring and assessment instructions. A minimum distance of 3 m is prescribed between the listeners' seats and the (main) loudspeakers. Finally it would also be feasible to include the actual threshold-shift value in the certificate of qualification of musicians belonging to rock and pop groups.

3.3 Implications of sound transmission mechanisms for sound engineering

As can be seen from section 3.1, conferring good acoustic properties on a room has been and continues to be the task of *room acoustics*. Since this is essentially achieved by architectural design, room acoustics are in some countries also named 'architectural acoustics'. Architectural measures, however, are limited, since:

- the sound sources used have only a finite sound power;
- the room-acoustical means interfere heavily with the architectural design, and thus cannot always be optimally employed (for example, if the preservation of historical monuments is involved);
- room-acoustical measures require a considerable architectural expenditure, and can often be optimally designed for only one application;
- a provisional design renders only a limited success in spite of requiring a high expenditure.

For these reasons the *methods of sound reinforcement engineering* are increasingly being used to influence certain room-acoustical parameters and thus improve the acoustic properties of a room. This applies to improving intelligibility as well as to increasing spaciousness. In this respect it is often said that a system can be considered to be functioning optimally when it is impossible to tell whether a sound event has been produced by the original source only or with the support of the electroacoustical system.

Another task of sound reinforcement engineering consists in realizing certain sound events, for example in presenting unnaturally modified signals, influenced as little as possible by the properties of the room. To this effect it is necessary to repress the acoustical properties of the room as far as possible by using directional sound. By means of complementary room-acoustical measures it is also possible to create a 'dry' acoustic atmosphere.

Of course it is not possible to reduce reverberance – that is, the reverberation time of a room – by means of a sound reinforcement system, but it is possible to increase the direct-sound component by means of directional loudspeakers. The sound reinforcement system is also capable of furnishing 'short-time reflections', which improve, for instance, the definition and clarity of speech and music respectively. The following sound-field components can be manipulated:

- direct sound;
- initial reflections influencing the direct sound;
- initial reflections influencing the reverberation;
- reverberation.

Figure 3.47 shows an attempt to elucidate these relationships. As we can see, a visitor to a concert or a theatre is exposed, among other things, to visual and acoustical impressions. The acoustical impressions are transmitted to the visitor via the room acoustics: that is, the sound pressure variations at the listener's location are used in assessing the quality of a sound event. If the quality of the transmitted sound is bad, a manipulation that simulates sound pressure variations may bring about an improvement. Important examples of intentionally influencing the signal in this way are the achievement of a desired sound blending, and the spatial impression in the reproduction of symphonic and choral music.

As has already been shown, during a sound signal's travel from emission at the source until its arrival at the listener, it is subject to a number of influences that affect it. Sometimes it is affected to such an extent that its content is no longer directly recognizable; sometimes it is changed in such a way that the listener is involuntarily supplied with information that enables him or her by reference to previous experience (association patterns) to judge the quality of the transmission path. From certain linear distortions we can, for instance, draw conclusions about the angle of origin of the sound (see section 3.2.1.5), or the distance of the listener from the source. It is known that blind persons may reach a high grade of perfection in the evaluation of such signal components. If we perceive a sound event whose

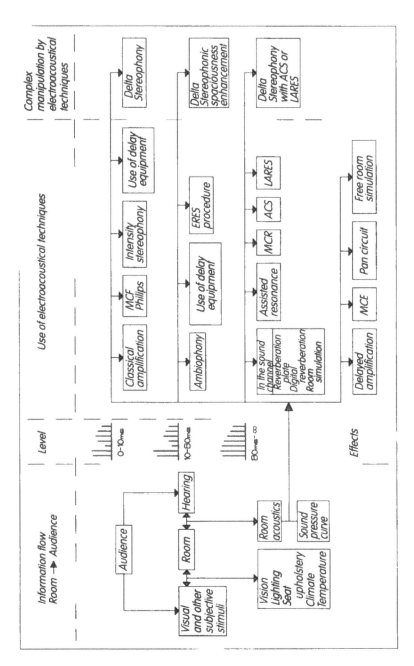

Figure 3.47 Possibilities of manipulating room-acoustical conditions.

localization and transmission path do not coincide with the visible source, however, we reject it as unnatural or distorted.

These conditions must be known to the acoustician, and possibly also to the client, if optimum reproduction of visible and known sources is to be achieved.

For better assessment of the transmission conditions we shall now briefly explain the mechanisms involved.

3.3.1 Sound propagation in the room

Any sound source irradiates into the room a particular amount of energy in a particular spectral distribution and with a particular directional characteristic. Part of this energy falls on the listener (or the microphone); the remaining energy propagates omnidirectionally (depending, of course, on the directional characteristics), thus falling on absorbing or reflecting surfaces. Subject to numerous reflections in the room, the total energy builds up the diffuse sound field. A certain number of these reflections arrive at the listener, who perceives them as *short-time reflections*. The spectral structure of the diffuse field, and thus the spectral impression for the listener – beyond three times the diffuse field distance of the source – is therefore determined by the frequency dependence of the irradiated and absorbed sound power, and not by the pressure of the direct sound. The directivity of any sound source, be it the human voice, an instrument, or a loudspeaker, is in most cases highly frequency dependent [3.58]. As directivity increases in proportion to the size (dimensions) of the sound-emitting surface relative to the wavelength of the radiated sound, the front-to-random factor increases with frequency, and thus there is a reduction of the overall irradiated sound power.

The sound level, L_r, in a diffuse sound field is given by

$$L_r = L_d - 10 \log[A(f)\,\gamma(f)] \text{ dB} + k \tag{3.31}$$

where L_d is the free-field sound level; $A(f)$ is the equivalent sound absorption area of the excited room; $\gamma(f)$ is the front-to-random factor of the source as a function of frequency; $k = 17$ dB $- 20 \log \Gamma(\phi, f)$ dB $+ 20 \log r$ dB; r is the distance between source and listener; and $\Gamma(\phi, f)$ is the angular directivity ratio as a function of angle and frequency.

This proportionality implies two conclusions:

- In rooms with a long reverberation time, the equivalent sound absorption area is determined essentially by air absorption, which increases heavily with frequency (see section 3.2.3.1). Timbre therefore becomes very bass-accentuated in a diffuse field.
- Since the different frequency dependences of the front-to-random factors of the various instruments and voices cannot be simulated by a

single loudspeaker, directly picked-up voices cannot be reproduced without timbre change in a diffuse field.

There is a particular problem in multi-purpose halls where it is necessary to alternate between voice reinforcement and music amplification. These two functions pose different requirements, which the electroacoustical equipment and the room acoustics have to correspond to:

- It is necessary to aim for a reverberation time of $T = 1.3-1.5$ s. If this is not possible for some reason or other, the equivalent sound absorption area, A, should not be less than $0.1-0.2$ times V, the volume of the room (A is in m^2 and V is in m^3). The smaller values are to be used for larger rooms ($V \geqslant 1000$ m^3), and the larger ones for smaller rooms ($V < 1000$ m^3).
- The sound absorption areas (including those formed by the audience) should be distributed over all room dimensions. In this one should try to make them act as scattering areas.
- If the room contains a stage house (fly tower), its reverberation time must be shorter than that of the adjacent hall. In this way one avoids a prolongation of the reverberation time in the auditorium, even with intensive stage monitoring (see Chapter 6).

Also, if the acoustic conditions are in essence fulfilled by room-acoustical means, one has to ensure when designing the sound reinforcement system that the sound emitted by the loudspeakers is not reflected as a closed or even focused wavefront to the source location (the place from where the original sound is emitted). This is especially true when working with a delay system: that is, when the sound energy is higher in the area of the reflecting surface than near the source.

In multi-purpose halls, the sound reinforcement system is thus expected to provide the soft sound attack for a good musical performance as well as the hard attack for speech transmission (at congresses, for example), and moreover to transform the large space impression into the best possible intelligibility (see section 6.4.3). Concert halls have their limitations in this regard. Here, apart from concert and jazz performances, it is possible only with a high technical expenditure to realize show concerts, revues and performances of light entertainment by making use of the sound reinforcement system alone. The inherent acoustic properties of the concert hall are very influential on the overall sound pattern in the hall, and restrict substantially the amount by which the sound pattern can be influenced by the electroacoustical system. For this reason one should, for a multi-purpose hall, refrain from trying to meet the criteria for 'concerts of the highest sound quality'. In this case it is better to try to design a hall 'suitable for performing symphonic concerts of good quality', in order to make it possible to present other performances of sufficient quality. A better solution, although considerably more expensive, is the construction of two halls of different acoustic layout.

To achieve adaptability to conflicting usage requirements it is also possible to work in a hall with 'variable' acoustics.

3.3.2 *Sound propagation in the open*

The open-air applications of sound reinforcement engineering range from the transmission of information for large areas to artistic applications in open-air theatres. In recent years sound amplification at important pop music events has come to play an important role (see Chapter 6).

The specifics of sound propagation in the open were described in section 3.1.3.1. Compared with the indoor situation, the essential difference is the *lack of reverberation*. This means that the definition-enhancing sound-field component can be increased as much as is desired, but without the masking effect of reverberant sound one gets to perceive disturbances.

If the upper and lateral boundary surfaces of a room are dropped or moved to a large distance, as is the case in the open, the number of reflections is greatly reduced. Often there remain only some initial reflections from the floor or from the rear and side walls, as is the case in ancient amphitheatres or in some open-air stages. Such early reflections are advantageous for the loudness and definition of the transmitted signal, but all reflections that reach the listener more than 50 ms after the original signal are perceived as disturbing echoes. This risk is especially great in the open, where intermediate reflections are lacking. But also short-time reflections having a high coherence with the direct sound may give rise to disturbing comb-filter effects.

To mitigate the increased risk of echo, the sound reinforcement measures indicated in Chapter 6 have to be taken. If this does not suffice, it is often necessary to resort in addition to some often rather expensive room-acoustical measures. These include echo elimination by means of sound-absorbing or scattering surfaces [3.59]. Absorbers are only effective against echoes, however, if they are fixed in front of detrimentally reflecting surfaces; if they suppress intermediate reflections, they may even have an echo-conducive effect. For open-air installations, absorption damping enters into consideration only in exceptional cases.

If buildings are grouped around the area covered by the sound reinforcement system, they are normally far enough from the listener that reflections from them are perceived as echoes. To reduce these reflections the loudspeakers should be directed accordingly or, if possible, the building should receive a sound-scattering surface [3.60].

Echoes are often generated by the very sound emitted by the system. The more it becomes necessary to distribute the loudspeakers locally, the greater becomes the risk that signals from remote loudspeakers will produce echoes of those emitted from near loudspeakers.

The characteristic drop of the high frequencies with increasing distance of the listener from the source also occurs in the open. This is partly because the listener is often not located in the main radiating direction of the sources, which are very directional in the upper frequency range, and partly because of the strong divergence attenuation caused by the distances between source and listener, which in the open are usually considerable. Here the frequency dependence described in section 3.1.3.1 shows a noticeable effect.

3.3.3 Further aspects of the use of sound reinforcement systems

First, the listener's expectations should be mentioned. In the open, for instance, the listener does not expect too much reverberation, but also indoors it sounds unnatural if too much reverberation prevails at close range to a stage. It may also be contradictory to expectation if the definition and clarity of the sound pattern at the rear of the hall are enhanced so much by excessive sound amplification that one gets the impression of sitting close to the stage. This effect may also occur if the rear hall area is served by decentralized loudspeakers that reproduce the high frequencies with unnaturally high energy because of the reduced distance between loudspeaker and listener. It may be necessary to provide these decentralized loudspeakers with filtering devices to adapt their reproduction characteristics to the natural room impression.

With large stages, *monitoring* between the actors becomes a significant problem. Soloist and orchestra may, in certain circumstances, hear each other only through monitoring loudspeakers. Monitoring is required especially if heavily delayed room signals in the action area largely impair the reciprocal hearing of the performers. Because of the 'Lee effect' [3.61] performers may even be hampered by the delayed room response of their own sound. (The Lee effect refers to the difficulty experienced when talking or singing while also hearing a delayed version of the speech or vocal – i.e. trying to talk against a strong echo.) The monitoring signals have to be transmitted undelayed, or with only an insignificant delay. Only in this way is it possible to achieve a 'rounded' acoustic ensemble performance, preventing the 'disintegration' of the sound pattern.

The use of stage monitors may cause problems if there are loudspeakers already arranged on the stage (especially in large open-air theatres) that are interconnected by delay networks. In this case it is possible that the signals from the monitoring loudspeakers, if audible in the auditorium, may give rise to mislocalizations.

References

3.1   Ahnert, W. and Reichardt, W., *Grundlagen der Beschallungstechnik* (Foundations of sound reinforcement engineering) (Berlin: Verlag Technik, 1981).

3.2   Kuhl, W., 'In der Raumakustik benutzte hörakustische Termini' (Hearing-acoustical termini used in room acoustics), *Acustica*, 39 (1977) 1, 57–58.

3.3   DIN 1320 Entw. 04.90 *Akustik*; DIN 1304 Entw. 09.86 *Begriffe-Zusätzliche Formelzeichen für Akustik* (Berlin: Deutsches Institut für Normung EV, 1992).

3.4   *Begriffe und Definitionen zur subjektiven Bewertung der akustischen Eigenschaften von Rundfunk- und Fernsehstudios mit Publikum, Konzertsälen für Musikdarbietung* (Concepts and definitions for the subjective evaluation of the acoustical properties of radio and television studios with audience, concert halls for music performances). OIRT-Empfehlung 68/1 (Prague, 1980).

3.5   Sabine, W.C., *Collected Papers on Acoustics* (Cambridge: Harvard University Press, 1923).

3.6   Kuttruff, H., *Room Acoustics* (London: Applied Science Publishers, 1973).

3.7   Lehmann, U., 'Untersuchungen zur Bestimmung des Raumeindrucks bei Musikdarbietungen und Grundlagen der Optimierung' (Investigations on the determination of spatial impression with music performances and bases for optimization), Dissertation, Technical University of Dresden (1974).

3.8   Thiele, R., 'Richtungsverteilung und Zeitfolge der Schallrückwürfe in Räumen' (Directional distribution and time sequence of sound reflections in rooms), *Acustica*, 1 (1953) Suppl. 2, 291–302.

3.9   Reichardt, W., Abdel Alim, O. and Schmidt, W., 'Definitionen und Meßgrundlage eines objektiven Maßes zur Ermittlung der Grenze zwischen brauchbarer und unbrauchbarer Durchsichtigkeit bei Musikdarbietungen' (Definitions and measuring base of an objective measure for ascertaining the boundary between useful and useless transparency), *Acustica*, 32 (1975) 3, 126.

3.10 Peutz, V.M.A., 'Articulation loss of consonants as a criterion for speech transmission in a room', *Journal of the Audio Engineering Society*, 19 (1971) 11, 915–919.

3.11 Klein, W., 'Articulation loss of consonants as a basis for the design and judgement of sound reinforcement systems', *Journal of the Audio Engineering Society*, 19 (1971) 11, 920–925.

3.12 Houtgast, T. and Steeneken, H.J.M., 'Envelope spectrum and intelligibility of speech in enclosures', paper presented at the IEEE-AFCRL 1972 Speech Conference.

3.13 Houtgast, T. and Steeneken, H.J.M., 'A review of the MTF concept in room acoustics and its use for estimating speech intelligibility in auditoria', *Journal of the Acoustical Society of America*, 77 (1985), 1060–1077.

3.14 RASTI-Sprachübertragungsmesser Typ 3361 (The RASTI speech transmission meter type 3361), Datasheet, Brüel & Kjaer, Naerum, Denmark.

3.15 Knudsen, V.O., *Architectural Acoustics* (New York: Wiley, 1932).

3.16 Kürer, R., 'Studies on parameters – determining the intelligibility of speech in auditoria', *Proceedings of the Symposium on Speech Intelligibility*, Liège (1973).

3.17 Aschoff, V., *Hörsaalplanung. Grundlagen und Ergebnisse der Audiologie* (Lecture hall planning. Foundations and results of audiology) (Essen: Vulkan-Verlag, 1971).

3.18 Beranek, L.L., *Music, Acoustics and Architecture* (New York: Wiley, 1962).

3.19 Kuhl, W., 'Räumlichkeit als Komponente des Raumeindrucks' (Spaciousness as a component of spatial impression), *Acustica*, 40 (1978), 167.

3.20 Jordan, V.L., *Acoustical Design of Concert Halls and Theatres* (London: Applied Science Publishers, 1980).

3.21 Schmidt, W., 'Raumakustische Gütekriterien und ihre objektive Bestimmung durch analoge oder digitale Auswertung von Impulsschalltestmessungen' (Room-acoustical quality criteria and objective determination of the same by analogue and digital evaluation of impulsive sound test measurements), Dissertation, Technical University of Dresden (1979).

3.22 Trautmann, U., 'Die meßtechnische Erfassung der Räumlichkeit in Sälen für musikalische Darbietungen' (The measuring-technical registration of spaciousness in halls for music performances), Dissertation, Technical University of Dresden (1986).

3.23 Fasold, W., Kraak, W. and Schirmer, W., *Taschenbuch Akustik* (Vade mecum of acoustics), 2nd edn (Berlin: Verlag Technik, 1991).

3.24 Herrmann, U.F., *Handbuch der Elektroakustik* (Handbook of electroacoustics) (Heidelberg: Hüthig, 1983).

3.25 Cremer, L. and Müller, H.A., *Die wissenschaftlichen Grundlagen der Raumakustik* (The scientific foundations of room acoustics), Vol. 1 (Stuttgart: Hirzel, 1978).

3.26 Blauert, J., *Räumliches Hören* (Stereophonic hearing) (Stuttgart: Hirzel, 1974).

3.27 CCIR Rec. 644, *Recommendations and Reports of the CCIR*, Vol. X, Part 1 (Genf, 1986).

3.28 Zwicker, E. and Feldtkeller, R., *Das Ohr als Nachrichtenempfänger* (The ear as communication receiver), 2nd edn (Stuttgart: Hirzel, 1967).

3.29 *Octave, half-octave and third-octave band-filters intended for the analysis of sound and vibrations*, IEC Publication No. 225 (1966).

3.30 Feldtkeller, R. and Zwicker, E., 'Die Größe der Elementarstufen der Tonhöhenempfindung und der Lautstärkeempfindung' (The magnitude of the elementary stages of pitch sensation and loudness sensation), *Acustica* (1953), suppl. 1, 96–100.

3.31 Barkhausen, H. 'Ein neuer Schallmesser für die Praxis' (A new sound meter for practice), *Zeitschrift für Technische Physik* (1926), *VDI Zeitschrift* (1927), 1471 ff.

3.32 Reichardt, W., 'Subjective and objective measurement of the loudness level of single and repeated impulses', *Journal of the Acoustical Society of America*, 47 (1970) 6, 1557–1562.

3.33 Zwicker, E., 'Ein Verfahren zur Berechnung der Lautstärke' (A procedure for calculating loudness), *Acustica*, 10 (1960), 304.

3.34 ISO Rec. 532, 1975, *Method for calculation of loudness level*.

3.35 ISO Standard 523, 1975-07-15, *Method for calculating loudness level*.

3.36 Reichardt, W. and Kussev, A., 'Ein- und Ausschwingvorgang von Musikinstrumenten und integrierte Hüllkurven ganzer Musikinstrumentegruppen eines Orchesters' (Initial and decay transients of musical instruments and integrated envelopes of whole instrument groups of an orchestra), *Zeitschrift für Elektrische Informations- und Energietechnik*, 3 (1973) 2, 73–88.

3.37 Niese, H., 'Untersuchungen zur geschlossenen Darstellung der Lautstärkebildungsgesetze bei beliebig komplexer Geräuschanregung' (Investigations on the consistent representation of loudness increase with excitation by noise of any complexity), *Hochfrequenztechnik und Elektroakustik*, 70 (1961), 5.

3.38 *Precision sound level meters. Additional characteristics for the measurement of impulsive sounds*. IEC Publication 179A (1973).

3.39 Niese, H., 'Die Trägheit der Lautstärkebildung in Abhängigkeit vom Schallpegel' (The inertia of loudness formation as a function of sound level), *Hochfrequenztechnik und Elektroakustik*, 68 (1959) 5, 143.

3.40 Port, E., 'Die Lautstärke von Tonimpulsen verschiedener Dauer' (The loudness of sound impulses of varying duration), *Frequenz* (1959) 8, 242.

3.41 Dietsch, L., 'Objektive raumakustische Kriterien zur Erfassung von Echostörungen und Lautstärken bei Sprach- und Musikdarbietungen' (Objective roomacoustical criteria for registering echo disturbances and loudnesses in speech and music performances), Dissertation, Technical University of Dresden (1983).

3.42 Haas, H., 'Über den Einfluß eines Einfachechos auf die Hörsamkeit von Sprache' (On the influence of a single echo on the audibility of speech), *Acustica*, 1 (1951) 2, 49–58.

3.43 Meyer, E. and Schodder, R., 'Über den Einfluß von Schallrückwürfen auf Richtungslokalisation und Lautstärke bei Sprache' (On the influence of sound reflections on directional localization and loudness of speech), Lecture 11 July 1962, Second Physical Institute, University of Göttingen, offprint.

3.44 Jeffers, L.A. and McFadden, D., 'Differences of interaural phase and level detection and localization', *Journal of the Acoustical Society of America*, 49 (1971) 1169–1179.

3.45 Bürck, W., *Akustische Rückkopplung und Rückwirkung* (Acoustic feedback and repercussion) (Würzburg: Triltsch Verlag, 1938).

3.46 Wöhle, W. and Schroth, G., 'Zusatzeinrichtung für Schallverzögerungs-anlagen' (Auxiliary device for sound delay equipment), DDR-Patent No. 203321.

3.47 Zwicker, E. and Zollner, M., *Elektroakustik* (Electroacoustics), 2nd edn (Berlin: Springer, 1987).

3.48 Kraak, W., 'Investigations on criteria for the risk of hearing loss due to noise'. In: *Hearing Research and Theory* (New York: Academic Press, 1981), Vol. 1, pp. 187–303.

3.49 Irion, H., *Musik als berufliche Lärmbelastung?* (Music a professional noise nuisance?). Research report no. 174. Bundesanstalt für Arbeitsschutz und Unfallforschung, Dortmund (Bremerhaven: Wirtschaftsverlag, 1978).

3.50 Kraak, W., 'Individual risk of hearing damage caused by noise', AES Convention of the Audio Engineering Society, Hamburg (1989), Preprint no. 2794 (D-3)

3.51 Kuhl, W., 'Keine Gehörschädigung durch Tanzmusik, sinfonische Musik und Maschinengeräusche beim Rundfunk' (No hearing damage by dancing music, symphonic music and machine noise in broadcasting), *Rundfunktechnische Mitteilungen*, 20 (1976) 1, 1–3.

3.52 Kraak, W., 'Die Bestimmung der gehörschädigenden Wirkung des Lärm nach dem Prinzip schädlichkeitsäquivalenter Wirkungen' (Determination of the hearing demands caused by noise in the principle of equivalent harmfulness), *Wissenschaftliche Zeitschrift der Technischen Universitaet Dresden*, 26 (1977) 6, 1209–1213.

3.53 Kraak, W., Kracht, L. and Fuder, G., 'Die Ausbildung von Gehörschäden als Folge der Akkumulation von Lärmeinwirkungen' (The formation of hearing damage as a consequence of the accumulation of noise impact), *Acustica*, 38 (1977) 2, 102–117.

3.54 Reddel, R.C. and Lebo, C.P., 'Ototraumatic effects of hard rock', *California Medicine* (1972) 1–4.

3.55 Dishoek, V.H., 'Die Vorbeugung von Berufstaubheit' (Prevention of professional deafness), *Münchener Medizinische Wochenschrift*, 98 (1956), 1625–1628.

3.56 Völker, E.J., 'Beschallung des Publikums bei einer Fernsehveranstaltung' (Sound irradiation of the audience in a television performance), *Frequenz*, 11 (1970), 325–339.

3.57 DIN 15905 T, 5th edn, 10.89 *Tontechnik in Theatern und Mehrzweckhallen. Maßnahmen zum Vermeiden einer Gehörgefährdung des Publikums durch hohe Schalldruckpegel bei Lautsprecherwiedergabe* (Sound equipment in theatres and multi-purpose halls. Measurements for the avoidance of hearing impairment of the audience by high sound pressure levels with loudspeaker reproduction).

3.58 Meyer, J., *Akustik und musikalische Aufführungspraxis* (Acoustics and musical performance practice) (Frankfurt (Main): Verlag Das Musikinstrument, 1972).

3.59 Ahnert, W. and Richert, W., 'Phasengitterdiffusoren in der Raumakustik' (Phase grid diffusors in room acoustics), Lecture 7, Tagung Arbeitsgruppe Akustik der Physikalischen Gesellschaft der DDR, Berlin (1988).

3.60 Kress, S. and Rietdorf, W., *Wohnen in Städten* (Living in towns). (Berlin: Verlag für Bauwesen, 1973).

3.61 Lee, B.S., 'Effects of delayed speech feedback', *Journal of the Acoustical Society of America*, 22 (1950) 824.

4 Components for sound reinforcement engineering

The different devices, equipment and systems will be briefly described, and their technical characteristics will be dealt with as appropriate, with reference to their technological importance within the framework of sound reinforcement engineering. The technical parameters will be indicated for the all-important electroacoustic transducers, since they are required for calculations on sound reinforcement systems.

4.1 Loudspeakers

Loudspeakers are classified according to the type of driver unit used, into electrodynamic, electrostatic, and piezoelectric loudspeakers. In sound reinforcement engineering, electrodynamic loudspeakers are used almost exclusively. (An exception is the piezoelectric tweeter systems that are mainly used for loudspeakers in music electronics.) The transducer systems are mostly cone loudspeakers or compression-driven horn-type loudspeakers. In the following sections we shall consider the different types of dynamic loudspeaker that are most frequently used in sound reinforcement engineering.

4.1.1 Operating principles

4.1.1.1 The electrodynamic loudspeaker

Figure 4.1 shows the basic structure of a dynamic loudspeaker: the sound-radiating diaphragm, the moving coil on its bobbin core, and the permanent magnet. The front fastening of the diaphragm, which is effectively rigid at low frequency, is also called the *rim suspension* or just the *surround*. At the connecting point with the moving-coil body, the diaphragm is centred by the *spider*. To allow the diaphragm to move freely, these fastening elements are manufactured from an elastically deformable material that provides the necessary damping in addition to its rigidity.

The current-carrying coil, which moves in the magnetic field, is dimensioned according to mechanical, electrical and thermal conditions.

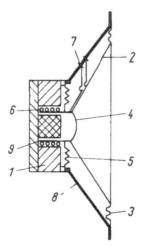

Figure 4.1 Schematic representation of the structure of a dynamic loudspeaker: 1, magnet system; 2, diaphragm with mass m_M; 3, suspension surround; 4, dome; 5, spider or centring diaphragm; 6, moving coil; 7, connecting wire; 8, frame; 9, moving-coil carrier.

The main points of practical importance in this respect are heat dissipation, and the use of a heat-resistant glue to ensure connection between coil and coil carrier under conditions of high stress. Figure 4.2 shows the equivalent circuit diagram of an electrodynamic loudspeaker, and the frequency response of such a loudspeaker is depicted in Figure 4.3. In the lower frequency range there is a rise up to the fundamental resonance frequency f_0. The fundamental

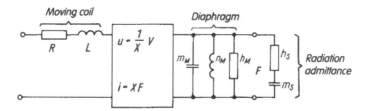

Figure 4.2 Equivalent circuit of an electrodynamic loudspeaker: m_M, mass of diaphragm and dome; n_M, compliance of suspension surround and spider; h_M, friction losses caused by vibration of diaphragm, surround and spider; h_s and m_s, real and imaginary components respectively of the radiation admittance (co-vibrating air mass); R, ohmic resistance of the moving coil; L, inductance of the moving coil; $X = 1/(Bl)$, transmission ratio of the electrodynamic coupling quadripole; B, air-gap flux density of the magnetic system; l, effective conductor length in the magnetic field; v, velocity of the diaphragm; F, driving force.

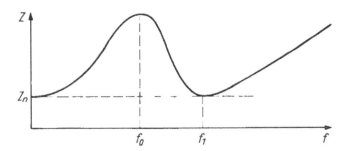

Figure 4.3 General frequency response of the input impedance Z of an electrodynamic loudspeaker: f_0, fundamental resonance frequency; Z_n, rated impedance.

resonance is determined essentially by the transformed mass of the diaphragm and the compliance of the diaphragm suspension. Above the fundamental resonance frequency the impedance decreases once more down to a frequency f_1, at which the tensions induced by the moving-coil inductance and the transformed mass of the membrane or diaphragm just compensate one another. It follows that in this frequency range the impedance is determined by the active component of the moving-coil impedance. In the higher frequency range the impedance rises again because of the inductance of the moving coil.

To determine the fundamental resonance we have to consider the mass of the air that oscillates in phase with the diaphragm, by adding it to the mass of the diaphragm. This reveals that the fundamental resonance is influenced by the design features of the loudspeaker such as ports and enclosures:

$$f_0 = \frac{1}{2\pi\sqrt{(nm)}} \tag{4.1}$$

The fundamental resonance constitutes the lower limit of the transmission range. By damping the fundamental resonance heavily it is possible to flatten the resonance curve to such an extent that a small margin below the fundamental resonance can also be used for transmission.

The loudspeaker's maximum power absorption lies in the region of the minimum above the fundamental resonance. It is therefore determined by the ohmic resistance of the moving coil. The input impedance at this point is called the *nominal impedance* of the loudspeaker [4.1] (see Figure 4.2). The nominal impedance of loudspeakers is now standardized at 4, 8 or 16 Ω. The real impedance must not be below this value, nor exceed it by more than 20%.

4.1.1.2 Transmission behaviour

The transmission behaviour of loudspeakers is determined by various factors, which are briefly explained below.

The *nominal load capacity*, P_n, of a loudspeaker is the permissible electrical power specified by the manufacturer according to the design characteristics. It is determined by an endurance test using a special noise signal [4.2, 4.3]. This test signal is either switched on for 1 min and off for 2 min during a period of 300 h, or constantly applied for 100 h. In this test a limiter must be used to ensure that sporadic peaks of the test signal do not produce overvoltages.

The spectral envelope of the test signal was originally ascertained from the amplitude statistics of conventional programme material. The increasing use of electronic instruments and the transmission of closely miked instruments however, mean an increase in the high-frequency content. For this reason, the test signal was modified somewhat at the beginning of the 1980s in favour of a stronger accentuation of the upper frequency range. This has been taken into consideration by the new standards since the publication of the IEC Publication 268-1C (2nd edn) of 1982 (Figure 4.4). It must still be borne in mind, however, that because of the specified test signals tweeters are declared to have a much higher nominal load capacity than is permissible according to their sinusoidal rating.

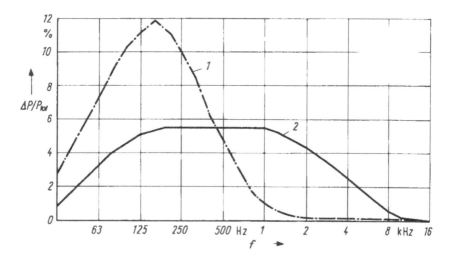

Figure 4.4 Programme simulation for testing the nominal power rating of loudspeakers according to IEC 268-1C (1982) in comparison with previous requirements. Power spectrum of a noise signal: P, power per third octave; P_{tot}, total power; 1, according to DIN 45573 T2 ed. 1976; 2, according to IEC Publ. 268-1C of 1982 and DIN 45573 T2 ed. 1983.

The ratio between the sound pressure, $\tilde{p}$, and the voltage, $\tilde{u}$, required to this effect at the radiator is called the *sensitivity*, T_S:

$$T_S = \frac{\tilde{p}}{\tilde{u}} \qquad (4.2)$$

We can distinguish between the *free-field sensitivity*, T_d, and the *diffuse-field sensitivity*, T_r. The free-field sensitivity is normally indicated for a point on the reference axis at a distance of 1 m from the loudspeaker. This can be expressed by

$$T_d = \frac{\tilde{p}_d}{\tilde{u}} \frac{r}{r_0} \qquad (4.3)$$

where $\tilde{p}_d$ is the sound pressure in the direct field at distance r from the reference point, r is the selected measuring distance, r_0 is the 1 m reference distance, and $\tilde{u}$ is the alternating voltage applied to the loudspeaker.

If one deviates from this measuring condition – for example, by measuring at an angle to the reference axis or at a distance other than 1 m – this has to be indicated [4.1]. The diffuse-field sensitivity has to be ascertained in a diffuse field, for instance in a reverberant chamber. In order to eliminate the room property characterized by the equivalent absorption area of the room, a correction factor has to be used:

$$T_r = \frac{\tilde{p}_r}{\tilde{u}} \frac{r_H}{r_0} \qquad (4.4)$$

where $\tilde{p}_r$ is the sound pressure in the reverberant field, and r_H is the reverberation radius of the measured room with the equivalent sound absorption area A (cf. equation (3.5)). T_s and T_d or T_r are given in Pa/V.

The *sensitivity level*, G_S, is defined as a logarithmic quantity of sensitivity:

$$G_S = 20 \log \frac{T_S}{T_0} \text{ dB} \qquad (4.5)$$

The *reference sensitivity*, T_0, is preferably 1 Pa/V. If another value is chosen, it has to be indicated.

The graphic representation of the sensitivity level as a function of frequency is called the *frequency response*.

One of the quantities most frequently used in sound reinforcement engineering is the *rated* or *characteristic sensitivity*. In combination with the nominal load capacity it enables, among other things, the maximally achievable sound pressure on the main reference axis of a loudspeaker or a loudspeaker system to be established. According to the definition given in the German standard DIN 45570 it is the ratio between the sound pressure $\tilde{p}_d$ averaged over a specified frequency range (usually 250–400 Hz) and measured on the reference axis at a distance of 1 m from the acoustic centre of the radiator (usually the centre of the radiating surface of the tweeter), and the square root of the power supplied. By standard, this power is referred to the nominal impedance Z_n of the radiator ($P_S = \tilde{u}^2/Z_n$). Thus the rated sensitivity is

$$E_K = \frac{\tilde{p}_d}{\tilde{u}} \sqrt{Z_n \frac{r}{r_0}} = \frac{\tilde{p}_d}{\sqrt{P_S}} \frac{r}{r_0} \qquad (4.6)$$

where r is the measuring distance on the reference axis in the free field, r_0 is the 1 m reference distance, and P_S is the apparent electrical power supplied.

Because of its reference to power, this expression is also known as the *rated power sensitivity*. According to DIN 45570 T1 the logarithmic version of this expression is called the *characteristic sound level*, L_K; it is also called the *sensitivity/dB*. It is defined as follows:

$$L_K = 20 \log \frac{E_K}{E_{K0}} \text{ dB} = 20 \log E_K + 94 \text{ dB} \qquad (4.7)$$

where $E_{K0} = 20 \text{ μPa}/\sqrt{\overline{W}}$.

Examples of the sensitivity of a range of loudspeaker systems are given in Table 4.1.

Table 4.1 Sensitivity of various types of loudspeaker

Loudspeaker type	Characteristic sound level, L_K (dB)	Rated sensitivity, E_K (Pa/$\sqrt{W}$)
Compact box loudspeaker	84	0.3
Bass reflex loudspeaker		
in compact design	86	0.4
as large loudspeaker	94–100	1–2
Sound column	88–106	0.5–4
Broadband sound column	94–97.5	0.5–1.5
Horn radiator	94–109.5	1–6
Broadband horn radiator		
combination	94–106	1–4
Narrow-band horn radiator	<118	< 16

An important parameter for appreciating the sound-field conditions in rooms is the *front-to-random factor, γ*. This characterizes the relationship between the acoustic power that would be radiated into the room by an omnidirectional radiator having both the same free-field sensitivity and the same acoustic power as the actual radiator to be assessed:

$$\gamma = \frac{\oint_S \tilde{p}_0^2 \, dS}{\oint_S \tilde{p}^2(\vartheta) \, dS} = \frac{S}{\oint_S \Gamma^2 \, dS}$$

where S is the surface of a sphere, Γ is the angular directivity ratio of the loudspeaker, and dS is a surface element of the reference sphere.

A measurement method for ascertaining the front-to-random factor was established in the IEC publication 268-5 (1972) [4.3] (see also [4.4], pp. 107 ff.):

$$\gamma = \left(\frac{\tilde{p}_d r}{\tilde{p}_r r_H} \right)^2 \tag{4.8}$$

where $\tilde{p}_d$ is the sound pressure measured in the free field at a distance r on the reference axis of the radiator, and $\tilde{p}_r$ is the sound pressure measured in the diffuse field of a reverberating room with reverberation radius r_H (according to equation (3.5)).

The *directivity factor, $Q(\vartheta)$*, is often used for this term, but it is a function of the angle ϑ (see also equation (4.14)).

The logarithmic equivalent of the front-to-random factor is the *front-to-random index*:

$$C = 10 \log \gamma \text{ dB} \tag{4.8a}$$

This corresponds to the difference between the free-field and the diffuse-field sensitivity levels:

$$C = G_d - G_r \tag{4.8b}$$

or, as expressed by the sound levels measured at 1 m distance in the direct field of the loudspeaker (L_d) and in the diffuse field (L_r) of a room having reverberation time T and volume V:

$$C = L_d - L_r + 10 \log(T/V) + 25 \text{ dB} \tag{4.8c}$$

where T is in seconds and V is in m^3. An equal input power P_{el} is assumed.

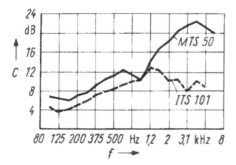

Figure 4.5 Frequency dependence of the front-to-random index of the sound
columns ITS 101 (Musikelectronic GmbH Geithain) and MTS 50
(Weder Elektrohaus GmbH, Meissen).

Because of the dimensions of the radiators, the wavelength of the radiated
sound in the lower frequency range is large compared with the radiating
surface. Thus the resulting directivity is either very low or insignificant.
With rising frequency the relationship changes and the directivity increases.
As long as all the areas of the diaphragm oscillate in phase, the directivity of
broadband loudspeakers experiences a steady increase. With composite
loudspeakers (such as two- or three-way units), however, the directivity
increases until the loudspeaker with a smaller radiating surface takes over.
In the cross-over range the directivity at first decreases and then resumes
increasing with the frequency.

Given that in the diffuse field the frequency dependence of the directivity
exerts a great influence on the timbre, the aim is to develop radiators whose
directivity is as frequency independent as possible. Also, in the design of
loudspeaker systems the aim is to obtain an arrangement in which the
resulting front-to-random factor shows only a slight frequency dependence.

Figure 4.5 shows the frequency-dependent front-to-random index of two
loudspeaker systems. The column loudspeaker MTS 50 shows the typical
increase of the front-to-random index with frequency. This could be greatly
reduced by suitable arrangement of the individual loudspeakers and the
inclusion of additional tweeters, as for the sound column ITS 101.

For sound reinforcement purposes it has been proven in practice that
slight increases of the front-to-random index of the loudspeaker system of
approx. 3 dB/octave are appropriate, because most natural sound sources
show a similar increase, giving rise to a corresponding timbre change.

For approximate calculation or measurement of the front-to-random
index one has to cover at least the range between 500 and 1000 Hz.
Manufacturers are increasingly quoting the frequency dependence of
directivity in the data sheets for their products.

By means of the front-to-random index, C, and the nominal power rating,
P_n, it is also possible to describe the characteristic sound level of a

loudspeaker system:

$$L_K = L_W + C - 10 \log P_n - 11 \text{ dB} \qquad (4.7a)$$

where L_W is the sound power level.

The *efficiency* of a loudspeaker system is determined by the ratio between the radiated acoustic power and the supplied electrical power:

$$\eta = \frac{P_{ak}}{P_{el}} = \frac{E_K^2}{\rho_0 c} \frac{4\pi r_0^2}{\gamma_L} \times 100\% \qquad (4.9)$$

where $\rho_0 c$ is the characteristic acoustic impedance of air $= 408$ Pa s/m^3 at 20 °C.

Combining all constants we obtain the following approximation:

$$\eta = 3 \frac{E_K^2}{\gamma_L} \% \qquad (4.9a)$$

This relationship can be seen in Figure 4.6. In practice the efficiency of loudspeaker systems lies between 0.1 and 10%. As for the rated sensitivity, the efficiency is often referred to the nominal impedance Z_n of the radiator, and designated as *nominal efficiency*, η_n:

$$\eta_n = \frac{\tilde{p}_d^2 Z_n}{\gamma_L \tilde{u}^2} \frac{4\pi r^2}{\rho_0 c} \qquad (4.10)$$

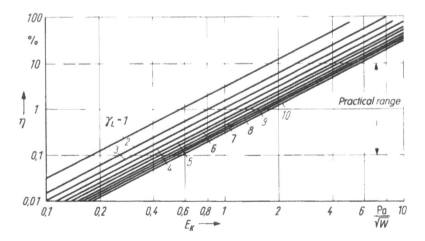

Figure 4.6 Efficiency of a loudspeaker as a function of rated sensitivity and front-to-random ratio.

where $\tilde{p}_d$ is the sound pressure measured on the reference axis at distance r, u is the voltage measured at a real resistance Z_n, and $\rho_0 c$ is the characteristic sound impedance of air $= 408$ Pa s/m^3.

Equations (4.9) and (4.10) suggest that because of the frequency dependence of the front-to-random index (see Figure 4.5) and the generally insignificant frequency dependence of the free-field sensitivity, the efficiency of a loudspeaker system may also depend heavily on the frequency. Even if the directivity of a loudspeaker is rendered largely frequency independent, the radiated output and thus the efficiency decrease above a certain limiting frequency. Since the wavelength of the bending wave within the membrane is getting shorter, subdivisions occur on the diaphragm so that only part of it continues to contribute to sound radiation by moving in phase with the moving coil. This holds true for a cone membrane as well as for a dome tweeter and the diaphragm of a compression driver loudspeaker. The limit frequency above which the radiated power decreases at a rate of $1/f^2$ lies according to [4.5] at

$$f_g = \frac{1.1c}{2\pi a} \tag{4.11}$$

where c is the speed of sound, and a is the radius of the diaphragm.

As the diameter of the diaphragm increases, the radiated sound power and thus the efficiency of a loudspeaker also increase, but the upper limit frequency decreases. To give a single-value rating of the efficiency of a loudspeaker one should therefore take the average value that results from the fundamental resonance according to equation (4.1) and the limit frequency for partial diaphragm oscillations according to equation (4.11).

4.1.1.3 Directional properties

All loudspeakers used in practice show a more or less pronounced directional dependence of radiation which is in almost all cases – like sound-beaming behaviour – frequency dependent. This angular dependence of sound radiation is characterized by three quantities, which we shall now consider in detail.

The *angular directivity ratio*, Γ, for a frequency or a frequency band is the ratio between the sound pressure p radiated at an angle ϑ from the reference axis, and the sound pressure p_0 generated on the reference axis at an equal distance from the acoustic centre (the acoustic centre is located at the position of the moving coil of the loudspeaker or, for a composite loudspeaker, of the tweeter):

$$\Gamma(\vartheta) = \frac{\tilde{p}(\vartheta)}{\tilde{p}_0} \tag{4.12}$$

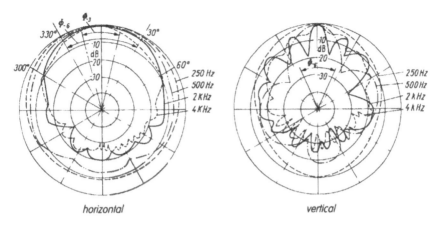

horizontal   vertical

Figure 4.7 Polar plot of the angular directivity gain of the ITS 101 sound column, with indication of the radiation angles.

In general $\Gamma(\vartheta) \leqslant 1$; it is only if the maximum directional characteristic does not occur at $\vartheta = 0°$ that $\Gamma(\vartheta)$ may be greater than 1.

The logarithmic equivalent of the angular directivity ratio is the *angular directivity gain*:

$$D(\vartheta) = 20 \log \Gamma(\vartheta) \text{ dB} \tag{4.12a}$$

Figure 4.7 shows the directional characteristic of the radiator ITS 101 in a polar plot of the directivity gain. We can see the main maximum at $0°$, and several secondary maxima at higher frequencies.

An important parameter for direct-sound coverage is the *angle of radiation*, Φ. This represents the solid-angle contour within which the directivity gain drops by a maximum of 3 dB or 6 dB (or another value to be specified) as against the reference value. Curves of equal directivity gain are denoted by Φ_{-3}, Φ_{-6} or generally Φ_{-n}: the higher the directivity, the smaller the angle of radiation (see Figure 4.7).

Because of the curves of equal directivity gain and the sound distribution loss, the incidence of direct sound from a loudspeaker on a surface may produce elliptical curves, each of which represents a given contour of the direct sound coverage. These curves are important in the planning of sound reinforcement systems in terms of *coverage areas* (Figure 4.8).

In building acoustics, to combine the influence of the directional effect and that of the distribution between directional and omnidirectional energy, one uses the *directivity deviation ratio* [4.4, 4.6]:

$$\Gamma^*(\vartheta) = \sqrt{\gamma} \cdot \Gamma(\vartheta) \tag{4.13}$$

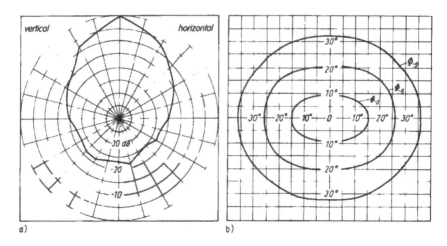

Figure 4.8 Plot of the directional effect of the Electro-Voice DML-1122 loudspeaker: (a) directivity characteristic; (b) coverage area. Frequency
2 kHz; front-to-random index 15 dB; maximum sound level at 1 m
distance 125 dB.

This quantity is also very important in sound reinforcement engineering: it
characterizes the reverberation component caused by the loudspeaker in the
excited room.

The square of this quantity is the well known angle-dependent *directivity
factor*, Q:

$$Q = g(\vartheta) = \gamma\Gamma^2(\vartheta) \tag{4.14}$$

The logarithmic equivalent of the directivity factor $Q(\vartheta)$ is the *directivity
index*, *DI* (also angle dependent):

$$DI = H(\vartheta) = 10\log g(\vartheta) \text{ dB} = 10\log Q(\vartheta) \text{ dB} \tag{4.14a}$$

In the German literature one uses for the directivity factor Q the
reverberation directional value, $g(\vartheta)$ [4.4]; the directivity index, *DI*, is called
the *reverberation directional index*, $H(\vartheta)$.

4.1.1.4 Transmission range (band width)

According to the DIN 45750 standard [4.1], the transmission range of a
loudspeaker is the frequency range that is usable or is preferably used for
sound transmission. It is generally characterized by that region of the
transmission curve in which the level measured on the reference axis in the

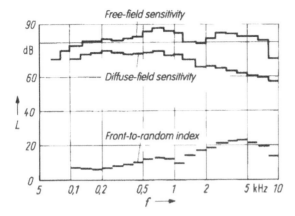

Figure 4.9 Frequency dependence of the front-to-random index, as compared with the free-field and diffuse-field sensitivities.

free field does not drop below a reference level. The reference value is averaged over a bandwidth of one octave in the region of highest sensitivity (or in a wider region as specified by the manufacturer). In the determination of the upper and lower limits of the transmission range, no peaks and troughs are considered whose interval is less than 1/8 octave.

This definition implies that the transmission range of loudspeakers must be checked before they are used in sound reinforcement systems. With loudspeakers intended for indoor use it is also necessary to consider the front-to-random factor: that is, the influence of the diffuse-field component on the formation of the resulting sound pressure.

For special-purpose loudspeaker systems, such as studio monitoring equipment, narrower tolerances of the free-field sound pressure are indicated for the transmission range. Thus OIRT recommendation 55/1 [4.7] allows for the range from 100 Hz to 8 kHz a maximum deviation of ±4 dB from the average value, whereas below (down to 50 Hz), and above (up to 16 kHz), the tolerance field widens to −8 dB and +4 dB.

Figure 4.9 shows the behaviour of the free-field sensitivity, diffuse-field sensitivity and front-to-random index of a loudspeaker (cf. equation (4.8)).

Moreover the transmission range is influenced, especially in the lower frequency range, by the installation conditions or the arrangement of the loudspeaker. Figure 4.10 shows that the arrangement of the loudspeaker system has a considerable influence on the transmission curve. This is because with arrangement of the radiator in front of, below or above a reflecting surface, interference of the direct sound with strong reflections gives rise to comb-filter-like cancellations, which can be demonstrated by a narrow-band analysis of the resulting signal (see section 7.3). These cancellations are particularly pronounced if a speaker with rear ports is

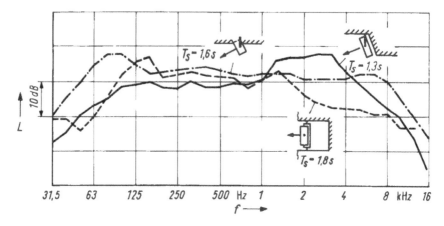

Figure 4.10 Dependence of the reproduction curve of a TZ 133 loudspeaker on the mounting conditions. T_s, reverberation time of the hall.

positioned in front of a wall, or if the reflections come from a distance of about 1.5 m out of a room corner, between the ceiling and the wall for example.

As a rule one can say that peaks and troughs that are *not* measurable in a third-octave band filter analysis are normally not perceived by the ear (unless they show pronounced periodic structures: see section 5.5.3.5).

Good bass emission is achieved if the radiating plane is embedded in a reflecting surface, such as a wall or a ceiling. In this case there may also be a certain angle between the radiating plane and the surrounding surface.

4.1.2 Types of radiator

The various tasks of sound reinforcement engineering require different types of radiator. These vary in the size and shape of their enclosures, the form of sound conduction, and the types of driving system used, as well as in the arrangement and combination of all these components. In this way it is possible to obtain a variety of directional characteristics of sound radiation, sound concentrations, sensitivities, frequency ranges and dimensions, which facilitate solutions for diverse applications.

4.1.2.1 Broad-band single radiators without preferential directivity

Among the simplest radiators are single loudspeakers of small dimensions and rating, as used for decentralized information systems, for instance for covering large flat rooms or for producing room effects in multi-purpose halls. Integration of these loudspeakers into a wall or an enclosure avoids an

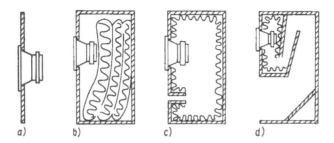

Figure 4.11 Different measures for suppressing the 'acoustic short-circuit':
(a) baffle panel; (b) closed cabinet; (c) bass reflex cabinet;
(d) transmission cabinet.

'acoustic short-circuit' by suppressing pressure compensation between the front and rear sides of the diaphragm. To this end a baffle panel or an open or closed cabinet may be used (Figure 4.11).

When using a *baffle panel* there should be a distance of at least one-eighth of a wavelength between the centre of the loudspeaker and the nearest edge of the baffle.

With a closed cabinet, the oscillating (moving) part of the loudspeaker operates in one direction against the relatively stiff air cushion of the cabinet. Compared with the unenclosed loudspeaker, this essentially reduces the compliance of the loudspeaker–enclosure system (see Figure 4.2). Loudspeakers for such *compact cabinets* are therefore provided with an especially soft diaphragm suspension, which means that they cannot be easily used for other purposes. By reducing the compliance the volume of the cabinet influences not only the fundamental resonance, but also the efficiency of the loudspeaker in the lower frequency range.

The acoustical conditions are more favourable with vented enclosures: the *bass reflex cabinets* or *phase reversal cabinets*. Such loudspeakers are nowadays used less as decentralized broadband radiators and more for high-power, large loudspeaker arrays: they will therefore be dealt with in detail in the following section.

For loudspeakers that are to be arranged in a decentralized distribution, cone diaphragms with a diameter of 100–200 mm are used. Smaller drivers show a very reduced rated sensitivity, whereas larger ones produce a directivity in the higher frequency range that is undesirable for these purposes. Loudspeakers with larger diaphragm diameters are therefore provided with additional scattering bodies (diffusers) arranged in front of the diaphragm centre. For the same reason, treble dome tweeter diaphragms are more suitable for radiating high frequencies than treble cones.

Essential criteria for loudspeakers that are intended mainly to be arranged in large numbers and distributed equally are an acceptable price and ease of installation (see also section 6.2.1.1).

4.1.2.2 Bass reflex loudspeakers

In order to improve sound radiation by enclosed loudspeakers at the lower limit of the transmission range, bass reflex cabinets were developed at the beginning of the 1950s [4.8]. These are Helmholtz resonators consisting of a hollow space enclosing an air volume with compliance n_v, and an opening (port) in which and in front of which an air mass m_v oscillates in phase. The resonant frequency of this mass–spring systems is

$$f_B = \frac{1}{2\pi}\sqrt{\left(\frac{1}{m_v n_v}\right)} = \frac{c}{2\pi}\sqrt{\left[\frac{S}{V_B\left(L+\dfrac{\pi}{2R}\right)}\right]} \tag{4.15}$$

where S is the area of the opening, L is the depth of the opening, R is the radius of the opening, c is the speed of sound, and V_B is the volume of air enclosed by the box.

By placing the resonant frequency in the region of the fundamental resonance of the loudspeaker (see equation (4.1)) it is possible to extend the transmission range by about half an octave downwards. The impedance frequency response acquires two neighbouring maxima (Figure 4.12). The cabinet is normally tuned by dimensioning the mass m_v; the effectiveness of this tuning can be checked by means of the impedance characteristic. The mass can be influenced by

- varying the diameter of the port;
- varying the depth of the port (for example, by inserting a tube that can be varied in length);
- inserting an additional mass in the port (such as an undriven loudspeaker diaphragm, a so-called passive diaphragm).

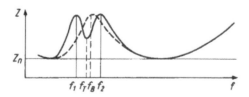

Figure 4.12 Frequency dependence of the input impedance of a bass reflex loudspeaker: f_B, basic resonance of loudspeaker and bass reflex cabinet; f_T, tunnel resonance of the bass reflex opening; f_1, f_2, coupling resonances.

Below their cabinet resonance f_B, bass reflex systems show an acoustic short-circuit between the port opening and the loudspeaker diaphragm. In the region of the cabinet resonance a phase rotation occurs, so that above this resonance the air in the opening oscillates in phase with the driving membrane.

One former frequent disadvantage of the bass reflex cabinet was a pronounced resonance rise entailing a strong lingering decay in the region of the cabinet resonance. Keibs [4.8] pointed out that sufficient damping of the resonance system was necessary to linearize the frequency response of the bass reflex system. Thiele [4.9] and Small [4.10] subsequently gave more specific dimensioning suggestions by considering the loudspeaker–cabinet system as a filter of the fourth order that can be linearized by means of appropriate attenuation and coupling of the different resonances. To this end it is necessary to know the quality of the individual resonances, and loudspeaker manufacturers nowadays indicate the quality (or the q factor) of the individual components.

To ensure a good transient response with small enclosure dimensions, bass reflex systems are nowadays often heavily overdamped. This entails a drop in the transmission curve at the lower limit of the transmission range. This drop has to be electrically equalized by bass emphasis of the assigned amplifiers or – for loudspeakers in 100 V installations – by means of passive equalization of the radiators. In this form the bass reflex loudspeaker is becoming increasingly popular in sound reinforcement engineering, especially for transmission of the lower frequency range. Combined with broadband treble horns, bass reflex loudspeakers are nowadays frequently used as standard radiators in high-quality sound reinforcement systems (Figure 4.13).

4.1.2.3 Loudspeaker arrays with in-line arrangement of radiators (loudspeaker line or sound column)

Many tasks in sound reinforcement engineering require radiators that are capable of producing a high sound level at large distances from their point of installation, but have minimal influence on microphones located close to them. To achieve this, they need to have well-defined directional characteristics and directivity. A radiator type suitable for this purpose is a loudspeaker array consisting of a stack of identical loudspeakers operated in phase. In the plane at right angles to this arrangement a pressure addition occurs, whereas in the areas above and below this plane a cancellation is produced by interference due to the different arrival times of sound components originating from the individual loudspeaker drivers (Figure 4.14). The directional characteristic that is achieved in this way has a width (or horizontal dispersion) that is essentially determined by the radiating characteristics of the individual drivers, whereas vertically it

Figure 4.13 Sound reinforcement loudspeaker combining a bass reflex unit with a pressure-chamber horn (Electro-Voice S-200).

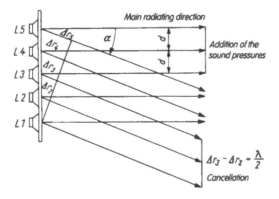

Figure 4.14 Operating principle of a sound column.

demonstrates a beam form. This is especially suitable for covering a wide auditorium in front of a stage area that effectively remains well screened from the emitted sound radiation.

The drawbacks of an in-line loudspeaker arrangement are as follows:

● The directional characteristic is frequency dependent, and becomes effective only with a line length of $\lambda/2$ of the radiated sound.

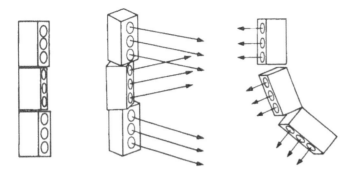

Figure 4.15 Swivelling the main radiating direction of component parts of a sound column to linearize the radiation behaviour.

- With longer in-line arrays, at a specific listening distance a high-frequency fall-off occurs, caused by interference of the outer loudspeakers. With an array 2.4 m long and a listening distance of 12 m, according to Kammerer [4.11] cancellations may occur at frequencies as low as 2.8 kHz.
- The increase of directivity in the desired direction is accompanied by undesirable side-lobes, which radiate sound energy in unwanted directions and thereby also lead to a degree of sound scattering.

All these frequency-dependent properties of the column or line-source loudspeaker make it possible for timbre changes to occur over the width and depth of the auditorium. In order to eliminate or limit this drawback, the line of speakers is subdivided in the upper frequency range. This is done either by physically bending the column 'banana-fashion' or by offsetting the elements of the column by up to 10° to the right and the left (Figure 4.15). In both cases the column operates over its whole length in the lower frequency range, but for the higher frequencies the elements are only partially active.

A recent suggestion for reducing the frequency dependence of the directional characteristic and directivity of sound lines and columns consists in supplying the sound signal with different phases and levels to the individual loudspeakers. According to Möser [4.12] it is thus possible to achieve an extensively frequency-independent directivity and a frequency-independent power spectrum (Figure 4.16). However, the efficiency of the column loudspeaker or sound line is essentially reduced in the lower frequency range.

This phase and level control was developed by the Philips company for large-area loudspeaker arrays. It put into practice various Bessel distributions for levels and phases, which produce defined directional characteristics of the array without causing excessive lobing of those characteristics.

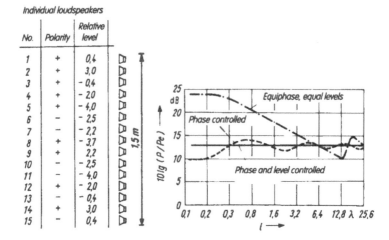

Figure 4.16 Effect of phase and level control on the linearization of the front-to-random ratio of a sound column: *P*, acoustic power of the overall in-line loudspeaker arrangement (sound column); P_e, acoustic power of an individual loudspeaker.

Such Bessel arrangements are nowadays used for powerful centralized loudspeaker arrays in pop and rock music in order to achieve optimum coverage of the varying auditorium areas at different venues.

To reduce unwanted high-frequency side-lobes and potential sound scattering, broadband line loudspeakers in particular are provided with additional treble loudspeakers, mostly in the centre of the sound column. This is a further way of subdividing the line.

In order to obtain a directivity effect also in the frequency range below the nominal effectiveness of the loudspeaker column, loudspeaker arrays are frequently designed as *pressure gradient radiators*. The enclosures are provided with one or more rearward openings of defined acoustic resistance, making them operate as phase-shifting elements. In this way the rearward signal is cancelled and a cardioid directional characteristic is achieved [4.13]. By enlarging the rearward opening it is also possible to realize a bidirectional (figure of 8) characteristic (Figure 4.17). However, to improve high-frequency radiation it is necessary to provide additional treble loudspeakers at the rear.

It is not always possible to arrange the loudspeaker lines optimally in front of the auditorium. For this reason a sound column was developed with electronic delay circuits between the individual drivers [4.14]; a virtual rotation of the loudspeaker line is achieved by appropriate selection of the delay time increments. It is necessary to tilt the individual loudspeakers in the longitudinal axis of the line, as this corresponds to the main radiating

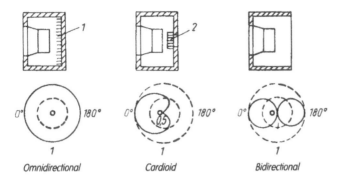

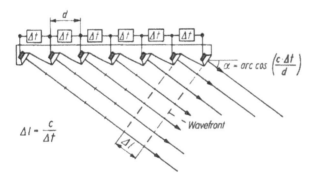

Figure 4.17 Effect of different loudspeaker enclosures on the directional characteristic: 1, damping material; 2, defined flow resistance.

Figure 4.18 Modification of the directional characteristics of a sound column by inserting electrical delay elements between the individual loudspeaker units.

direction of the line, so as to minimize the timbre changes that would otherwise occur (Figure 4.18). The required delay time Δt depends on the spacing of the individual systems and on the intended rotation of the virtual line.

4.1.2.4 Horn-type loudspeakers

Another way to achieve a desired directional characteristic is to arrange sound-conducting surfaces in front of the loudspeaker driver system. As such arrangements are mostly of horn-like design, they are called horn loudspeakers. In addition to influencing the directional characteristics, the horn arrangement also enables a better adaptation of the drive system to the wave impedance of the air, and thus improves the efficiency. At the same time, the characteristic sensitivity is also enhanced purely by the increase in the directivity (see Table 4.1). Thanks to this high characteristic sensitivity, and the potentially directional characteristic and directivity, this loudspeaker design is well suited for sound reinforcement in large auditoriums, where the emitted frequency range and different target areas (coverage areas) nearly always require the use of different types of loudspeaker.

The high distortion coefficient produced by the design, and frequently caused by feedback of the resonance of the air column in the horn on the drive system, may be considered as a drawback. This effect occurs mainly in the lower frequency range because the horn configuration is insufficiently optimized in this area, and also because of reflections from neighbouring acoustic surfaces.

The optimum horn loudspeaker requires a broadband driving system and a horn shape that is tailored to the whole frequency range of interest. Such horns may have diverse forms, and numerous indications as to their dimensioning are given in [4.5]. To minimize the reflections produced at the horn mouth, the opening must merge continuously and smoothly into the surrounding medium. This can be achieved, for instance, by an *exponential horn shape*. To realize such horns also for the low-frequency range, they must have very large dimensions. An appropriately shaped horn for a lower-limit frequency of 50 Hz, for example, requires a mouth diameter of 2 m. Radiators of such dimensions are suitable for only a few applications, as for instance in a cinema behind the projection screen.

The length of the horn also depends on the required directivity and the frequency to be radiated. It may be reduced by having a fairly large exciting surface (which, however, can be optimized for only a narrow frequency range), or by folding the horn. Moreover, the horn should be as rigid as possible, and should have a high acoustic attenuation. Since this cannot be realized especially for the low- and medium-frequency bands, the horn walls should at least be free from pronounced self-resonances, and should have a high internal damping so as to show a largely frequency-linear transmission behaviour.

For the reasons mentioned it is not sensible to construct a broadband horn for the overall transmission range. A better solution is to have several horn loudspeakers that complement one another.

BASS HORNS

Because of the large dimensions involved, the design of bass horns requires particularly extensive compromises. Practical models of bass horns usually have a horn shape in one dimension only; at right angles to this dimension sound conduction is by means of parallel surfaces. The horn shape can be arranged vertically or horizontally, as shown by the two examples in Figure 4.19. The illustrated designs also show two variants of loudspeaker arrangement in front of the horn. In Figure 4.19a the horn is excited by the rear side of the loudspeaker, while the front of the loudspeaker radiates

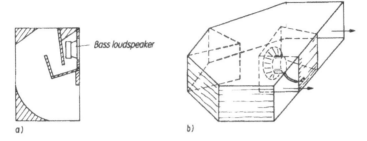

Figure 4.19 Two configurations of bass horns: (a) bass reversal cabinet (transmission cabinet); (b) bass horn in folded enclosure (W-box).

directly. The horn therefore has to be dimensioned so that it causes a phase rotation of 180°, since otherwise an acoustic short-circuit would result. This design type, which is called a *transmission horn*, can with sufficient approximation be realized only for a limited frequency range. Figure 4.19b shows a folded horn suited for being designed in an acceptably handy shape. The driving systems for bass horns are usually large-area cone diaphragm loudspeakers, which with some types are arranged in bass reflex enclosures. Because of the host of compromises that have to be made in the design of such bass horns, and also because of their driving systems, they are applicable only for a narrow frequency range below about 300 Hz. But even here their use is acceptable only because the human ear shows a reduced timbre recognition in this region [4.15].

The power-handling capacity of such bass horns, which are mainly used in electronically amplified music, is about 100–500 W.

MEDIUM-FREQUENCY HORNS

The greatest variety of driver and horn designs is available for horn loudspeakers for the medium frequency range of about 300 Hz to 3 kHz.

The drivers used are mostly dynamic pressure-chamber (compression driver) systems connected to the horn proper (i.e. the flare) by means of a throat, the so-called *throat-adapter*. To increase the driving power, several pressure-chamber systems can be connected via a Y *adapter* to one horn (Figure 4.20a). In this case one must take care that the phases and travel times are equal at the intersection of the two channels, so that no cancellations can occur. The Electro-Voice company is believed to have solved this problem by introducing their 'manifold' technology (Figure 4.20b) in which thanks to a novel adapter configuration there no longer occur any longitudinal cancellations, and only very slight transverse cancellations [4.16].

For covering close, medium-range and distant auditorium areas, horns of different directivity are available. These horns are nowadays often designed

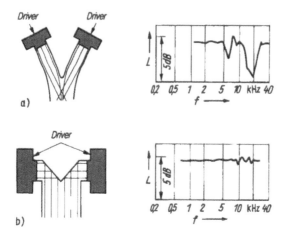

Figure 4.20 Manifold technique: (a) Y adapter with treble interference; (b) same without treble interference. Left, general arrangement; Right, composite frequency response.

so that their directivity and directional characteristic are largely frequency independent, as Figure 4.21 shows by means of an example. Such CD *horns* (constant-directivity horns) exhibit different exponential and hyperbolic curvatures over their whole length. The advantage of the frequency-independent directivity, which comes to bear particularly when ascertaining optimum coverage areas, is however impaired by a sensitivity drop in the upper frequency range. (This frequency dependence has to be compensated by electrical equalization.)

Further reductions of the frequency dependence of horns are achieved by the *diffraction horns* (diffraction at the gap), which are of very narrow design, and by means of a multiple subdivision in the shape of the so-called *multicellular horns* (Figures 4.22 and 4.23).

Another design variant aimed at reducing directivity, and used also for treble horns, consists in the arrangement of *acoustic lenses* in front of the horn. The different speed of sound propagation between the angled vanes of the lens and in the undisturbed field results in the desired dispersion (Figure 4.24). The angling of the vanes has no effect on the main direction of propagation.

TREBLE HORNS

For the upper frequency range, two main types of horn loudspeaker are produced: horn radiators, with similar design characteristics as those of the medium-frequency horns, and functioning in the frequency range from 1 to 10 kHz; and special treble loudspeakers for the frequency range from 3 to 16 kHz.

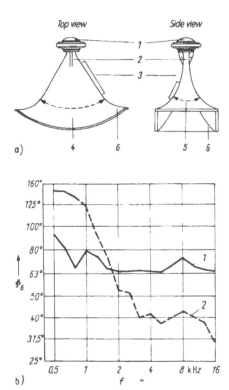

Figure 4.21 Constant-directivity horn. (a) Horn configuration: 1, driver; 2, initial sector with exponential expansion for ensuring low distortion; 3, conical main sector determining the radiating direction; 4, horizontal angle of radiation; 5, vertical angle of radiation; 6, exit opening with wider angle impeding constriction in the medium-frequency range. (b) Radiation angles: 1, horizontal; 2, vertical.

The first type is used mainly as a tweeter in composite units equipped with bass reflex loudspeakers (see Figure 4.13). Apart from diffraction radiators, small CD horns are also often used.

The smaller special treble radiators are mostly of rotationally symmetrical design. Their driving system is of the pressure-chamber type, and consists of a metal or, less frequently, a plastic diaphragm coupled to a very small pressure chamber. As is the case with most pressure-chamber systems, the rear side of the diaphragm is damped by means of a layer of foamed plastic. At the radiating end the pressure chamber is terminated by a phase-correction element, the *phase plug*. This phase plug also constitutes the centre piece of the concentric horn (Figure 4.25). The channels formed by the dome and the walls of the horn are arranged so that they are above the

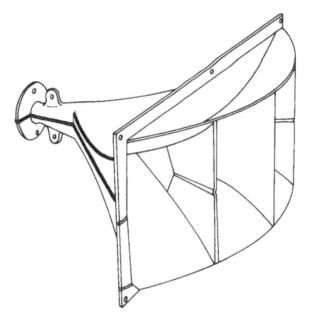

Figure 4.22 Multicellular horn.

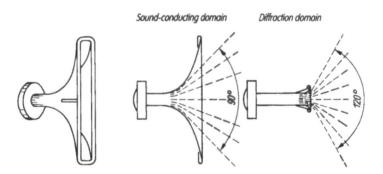

Figure 4.23 Diffraction horn, showing the principle of operation.

peaks (antinodes) occurring with the first self-resonance of the diaphragm. Thereby it is possible to extend the effectiveness of the pressure chamber towards the higher frequency ranges.

The electromechanical transducers used are mainly of the dynamic type, but may also be piezoelectric systems. The necessarily very small systems have a power-handling capacity of up to 50 W. Because of the way in which the nominal power-handling capacity is ascertained (section 4.1.1.2) it must be considered, however, that a tweeter may be operated with only a small

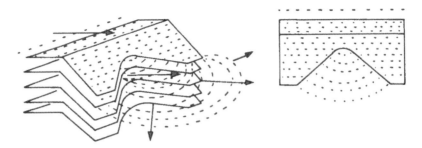

Figure 4.24 Action of an acoustic dispersion lens.

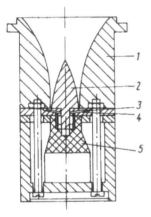

Figure 4.25 Structure of an axially symmetric pressure-chamber loudspeaker (treble horn): 1, horn; 2, phase correction; 3, pressure chamber; 4, diaphragm; 5, magnetic system.

percentage of its nominal power-handling capacity. According to Figure 4.4 (curve 2), a tweeter with a nominal power-handling capacity of $P_N = 50$ W may at 5 kHz be operated with a sinusoidal tone of $\Delta P = 2\% \times P_N = 1$ W.

The directional characteristic, the directivity and consequently also the characteristic sound level are determined by the horn geometry. It is possible to reach characteristic sound levels of up to 108 dB.

If the small compression chamber is subjected to very high loads there is a risk that particle velocity values in the region of supersonic speed are reached in the area of velocity transformation at the horn rim. This may give rise to distortions. If these distortions occur above 8 kHz, the harmonics are no longer audible, but it is possible that lower difference tones may also be generated. For this reason the transmission channels should be limited at 16 kHz or at least at 20 kHz.

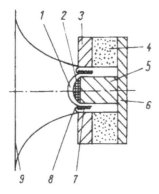

Figure 4.26 Dome-driver horn loudspeaker: 1, dome; 2, damping material; 3, pole plate; 4, magnet; 5, pole body; 6, bottom plate; 7, moving coil; 8, edge suspension, 9, spherical wave horn.

As well as horns with pressure-chamber drive systems, *dome horns* are also used. These consist of a domed diaphragm as driving system, which is provided with a horn for increasing the directivity and thus the characteristic sensitivity (Figure 4.26). Because of their reduced efficiency and directivity, such dome horns have a lower characteristic sensitivity than the corresponding compression-type loudspeakers.

HORN LOUDSPEAKERS FOR INFORMATION SOUND REINFORCEMENT

Information sound systems are generally designed for transmitting speech with high clarity, for which it is often sufficient and even advantageous for intelligibility to emit only the frequency range between 200 Hz and 4 kHz. It is possible to manufacture quite robust folded horns with integrated compression drivers capable of transmitting the whole frequency range concerned. Thanks to their weather-resistant design of sheet metal or various plastics, they are frequently used for outdoor installation. Their pronounced directional characteristic makes them suitable for covering narrow areas.

4.1.3 Fields of application of various radiator types

4.1.3.1 Single loudspeakers

In sound reinforcement engineering, small broadband, single loudspeakers are used only in decentralized systems, for example as ceiling or wall loudspeakers for information systems, or for the reproduction of room effects and of signals serving for enhancing spaciousness and reverberation time. Loudspeakers with especially small dimensions (diaphragm diameter <120 mm) are used also for integration in the backrests of auditorium seats, in a chairperson's desk, or in the balustrades of balconies and the front sides of platforms at close range to the listeners. The power levels are between 1.5 and 3 W, with ceiling loudspeakers also 6–12.5 W.

4.1.3.2 *Large loudspeakers for indoor use*

For the main loudspeaker arrays of large sound reinforcement systems, radiator units are required with high power rating, wide transmission range, and preferably distinct and well-defined directional characteristics as well as high directivity. In general column speakers are used to achieve these characteristics, as well as combinations of bass reflex loudspeakers with horn loudspeakers for the medium and upper frequency range, or – less frequently – combinations of horn loudspeakers alone.

These radiators are frequently arranged in centralized main loudspeaker arrays, which are responsible for the sound level and often also for the timbre of the whole system. With the exception of radiators used for speech transmission only, and for very high-quality music transmissions, the transmission range is normally 80 Hz to 16 kHz. For speech-only systems it is sufficient to have lower and upper limit frequencies of 150–200 Hz and 10–12 kHz, respectively, while for music a transmission range bottom limit of <40 Hz is required. This requirement can usually be realized only by means of additional special loudspeakers having larger dimensions than the normal sound reinforcement loudspeakers (*subwoofers*).

The arrangement and circuitry of the loudspeakers exerts an important influence on the direction from which the acoustic signal is perceived. Still more important, however, is the need to ensure that the sound field built up by the loudspeakers does not cause positive acoustic feedback via the microphones used in the performance.

If a centralized arrangement is not appropriate and a more or less decentralized arrangement has to be chosen, which is generally associated with the use of delay equipment, care has to be taken to ensure that sound is reflected back to the platform as little as possible, because such a 'room response' could give rise to an inadmissible spaciousness or even a disturbing echo within the performance or in the reception area near the platform. Both phenomena are disturbing for the actors and for the audience seated close to the platform. For these reasons the main arrays nearly always consist exclusively of radiators with a cardioid or highly directional characteristic.

For rock and pop concerts, and for systems that have to produce very high sound levels, combinations of special horn loudspeakers are used almost exclusively; for their dimensions alone, these have to be arranged on the platform. The power of the systems used in big halls or in the open may well exceed 100 kW [4.17]. In such cases individual loudspeakers or loudspeaker arrays are assigned to particular instruments, groups of instruments, vocalists or – in the case of purely electronic music – to specific registers [4.18].

4.1.3.3 *Stage loudspeakers*

For monitoring purposes in the performance area (stage or platform) both permanently installed loudspeakers and mobile *stage monitors* are used. Their function is to provide the performer with a signal that is influenced as little as possible by the room or by remote loudspeaker arrays, to enable the performer to coordinate his or her performance with that of other performers or with tape recordings.

The permanent systems may be smaller sound columns arranged in the inner stage proscenium (at the crossbeam as well as at the sides), since this is the point where they have the least travel-time difference with the main loudspeakers.

Mobile stage monitors usually have the form shown in Figure 4.27, so that they may be positioned on the stage between the performers and the audience without unduly hindering the audience's view. Stage monitors should radiate as little sound as possible in the rearward direction, since

Figure 4.27 Stage monitor (wedge cabinet) equipped with one broadband cone loudspeaker and one treble horn (d & b Audiotechnik AG).

otherwise they would have a disturbing effect on the sound pattern in the audience area near the stage. Their necessarily compact design is at the expense of sensitivity and frequency transmission range.

The stage monitors are complemented by *stage effects loudspeakers*, which serve for example in theatres for emitting effects signals that are to be localized in the stage area. Suitable in this respect are permanently installed or mobile large loudspeakers of the type used in the loudspeaker arrays. These loudspeakers or loudspeaker arrays must have a high directivity, as frequency-independent as possible, especially if they are arranged in the backstage area, so as to avoid the majority of their radiated sound being absorbed in the stage house.

A third group of stage loudspeakers may be used to ensure the acoustic localization of weak sound sources (see section 6.5).

The loudspeakers required to this end have to satisfy several – sometimes contradictory – requirements. These include:

- small dimensions, so as not to obstruct the view;
- a wide transmission range corresponding to that of the original source;
- a high maximum sound level to ensure that at a distance the level difference compared with the nearer main loudspeaker radiating the amplified sound is minimized;
- a radiating angle to the audience that is as wide and frequency independent as possible;
- high backward attenuation, so as to avoid positive feedback via microphones in the stage area.

Since it is actually impossible to meet all these requirements simultaneously, one has to decide on the priorities in individual cases. In practice nowadays loudspeakers similar to those of the main loudspeaker arrays are used, but of more compact design. Because of the small dimensions desired, radiation of the lowest transmission range is often abandoned. This is acceptable, however, only if the lowest frequency group transmitted by the system is taken into consideration, since otherwise one runs the risk of a frequency-dependent acoustical localization. It must be said, however, that very low frequencies (<100 Hz) are hardly relevant for localization.

Localizing/reinforcement loudspeakers for lecturers are often built into the lectern or podium [4.19, 4.20].

4.1.3.4 *Loudspeakers for outdoor use*

The mechanical design of loudspeakers for open-air systems has to meet special requirements. Acoustical dimensioning, however, is widely governed by the same conditions as for indoor radiators.

The enclosures have to be especially robust and weather resistant, to protect the radiator against moisture and strong solar radiation. They are generally made of lacquered steel sheet, aluminium, plastic, or less frequently of coated wood.

A special problem is the protection of the paper diaphragms of cone loudspeakers, since it must not impair the radiation conditions. The protection consists mostly of several layers of a fine-meshed gauze behind a protective cover to prevent mechanical damage (see also [4.21]). The protective cover, whose structure is mostly similar to that of a ball protection for indoor loudspeakers, may consist of a metal sheet with more than 50% perforation, a wire mesh, or a rod arrangement. All these have to be fastened in such a way that flutter or rattling noises are avoided. To this end rubber-coated clamping brackets are often used, and in some cases additional rubber-coated pressure strips.

If the diaphragms are made of impregnated paper or plastic, a higher moisture level is permissible.

At temperatures below freezing, the loudspeakers have to be slowly 'run up' to operating level by radiating a low-frequency tone.

The reduction of high-frequency radiation caused by the covering can mostly be kept within permissible limits, or can be compensated by means of electronic equalization. If the radiator is heavily coated with porous material, it is also necessary to ensure that strong solar radiation on the radiating surface does not cause heat accumulation in the enclosure, which may result in blocking of the moving coil due to the different thermal expansion of the coil and of the air gap in the magnetic system.

Compression drive horn loudspeakers of sheet-metal or plastic designs do not require additional weather protection.

4.2 Microphones

The use of microphones in sound reinforcement systems requires the observation of a number of conditions that are additional to normal recording conditions. To avoid positive acoustic feedback (howl round) it is often necessary to keep the microphone closer to the sound source, so that often considerably more microphones have to be used than with normal multi-microphone recordings. Moreover, the live conditions demand very robust microphones.

All these requirements presuppose an exact knowledge of the properties of the microphone types and their connection technique.

4.2.1 Parameters

The microphone data are laid down in European and international standards. In this context we shall consider only those data that are important for sound reinforcement engineering and are required for further calculations, as described in Chapter 5.

The magnitude of the output voltage of a microphone is expressed as a function of the incident sound pressure by the *microphone sensitivity*[1]

$$T_E = \frac{\tilde{u}}{\tilde{p}} \qquad (4.16)$$

in V/Pa or its 20-fold common logarithm, the *sensitivity level*

$$G_E = 20 \log \frac{T_E}{T_0} \text{ dB} \qquad (4.17)$$

The reference sensitivity, T_0, is normally specified for 1 V/Pa.

Depending on the test conditions, we can distinguish the following sensitivities:

- the *pressure open-circuit sensitivity*, T_{Ep}, which is the ratio between the effective output voltage at a certain frequency and the effective sound pressure of a normally (vertically) incident sound source;
- the *(free)-field open-circuit sensitivity*, T_{Ef}, which considers by special measuring conditions the pressure increase caused by the cross-sectional dimensions of the microphone;
- the *diffuse-field sensitivity*, T_{Er}, which mainly reflects the diffuse sound incident on the microphone.

Some manufacturers specify as *sensitivity* the open-circuit sensitivity, the reference value being the square root of the apparent electrical power delivered with rated terminal resistance, and the sound pressure. (This quantity is not standardized.)

The microphone is optimally terminated by means of the *rated terminal impedance*, which is specified by the manufacturer. It is 1 kΩ for the studio microphones generally used with commercial sound reinforcement systems.

The *rated impedance*, Z, characterizes the output impedance. With these microphones it is usually about 200 Ω.

The *transmission range* is the useful frequency range transmissible by a microphone. Apart from the sensitivity level ascertained in the free field at 0° to the main axis of the microphone, the direct-field level resulting for other directions is frequently also taken into consideration. The values measured in the diffuse field, however, are less frequently used.

The *overload sound pressure* is the highest sound pressure that can be handled by a microphone with permissible distortion. For the studio

---

1 In standards the letters B or S are generally used for microphone sensitivity. To achieve uniformity of designations, however, we use here the symbol T, as with the loudspeaker.

microphones generally used with commercial sound reinforcement systems, harmonic distortion of 0.5% (or less frequently of 1%) at 1 kHz is nowadays considered as permissible. Condenser microphones under these conditions are capable of handling sound pressure levels of 120–140 dB.

The effective value of a voltage delivered by a microphone without influence by a sound field or another disturbing field is called the *internal noise voltage*.

The *psophometric voltage* is the voltage delivered under the same conditions and ascertained with an 'ear-filter weighting' (*psophometer weighting*) laid down by CCIR Recommendation 468–4 [4.22] or OIRT Recommendation 71 [4.23] and a quasi-peak value indication. It offers a certain approximation to subjective conditions.

The *equivalent loudness* is the voltage ascertained under these very conditions by means of the A-curve according to IEC 268–1 [4.24, 4.25] (weighting curve of the sound-level meter) and an rms indicator. It corresponds less than the psophometric voltage to the subjective, tone-compensated conditions.

The difference between the voltage given with overload sound pressure and the internal noise voltage including a safety margin produces the *dynamic range* of the microphone. Because of the different weighting curves used in determining the psophometric voltage and the equivalent loudness, the numerical values of one and the same measuring object may show differences of up to 12 dB.

The dependence of the microphone voltage on the direction of incidence of the exciting sound is called the *directional effect*. The following quantities are used for describing this effect:

- the *angular directivity ratio*, $\Gamma(\vartheta)$, which is the ratio between the (free-) field sensitivity, T_{Ed}, for a plane sound wave incident at an angle ϑ to the main microphone axis and the value ascertained at the reference level (incidence angle 0°):

$$\Gamma(\vartheta) = \frac{T_{Ed}(\vartheta)}{T_{Ed}(0)} \tag{4.18}$$

- the *angular directivity gain*, D, which is the 20-fold common logarithm of the angular directivity ratio;
- the *coverage angle*, which is the angular range within which the directivity gain does not drop by more than 3 dB (or 6 dB or 9 dB) against the reference axis.

Apart from the quantities describing the ratio between the sensitivities of the microphone with sound incidence from various directions deviating from the main axis, it is also necessary to describe the relationship between the sensitivities with reception of a plane wave and those with diffuse excitation.

With these quantities it is then possible to ascertain the suppression of the room-sound components against the direct sound of a source to be transmitted. This energy ratio is described by the following parameters:

- the *front-to-random factor*, which is the ratio between the electric power generated by the microphone when excited by a plane wave from the direction of the main axis, and the power generated by the microphone excited in a diffuse field with the same sound level and same exciting signal. If the sensitivity in the direct field is T_{Ed} and in the diffuse field is T_{Er}, the front-to-random factor is

$$\gamma_M = \frac{T_{Ed}^2}{T_{Er}^2} \tag{4.19}$$

- the *front-to-random index*, which is the 10-fold common logarithm of the front-to-random factor.

While the front-to-random factor of an ideal omnidirectional microphone is 1, that of an ideal cardioid microphone is 3. This means that a cardioid microphone picks up only one-third the sound power of a room compared with that picked up by a comparable omnidirectional microphone at the same distance from the source. This implies for instance that with identical proportions of room sound power, the speaking distance for a cardioid microphone may be three times that of an omnidirectional microphone.

4.2.2 Principles of reception

A microphone's reception principle exerts a significant influence on its transmission properties, and especially its directional effect. The most important reception principles used in sound reinforcement engineering are explained below.

4.2.2.1 Pressure microphones

If the pressure variation existing at the microphone location is used to drive the diaphragm and thus to generate a voltage proportional to the pressure variation, an omnidirectional characteristic results because of the scalar quality of pressure.

With higher frequencies, at which the dimensions of the diaphragm are of the same order of magnitude as the wavelength of the sound received, the *pressure increase effect* comes to bear. The sound pressure in front of the diaphragm cannot balance itself, but increases compared with that occurring in the immediate surroundings. Thus sensitivity increases with normal sound incidence as against lateral or diffuse incidence. The

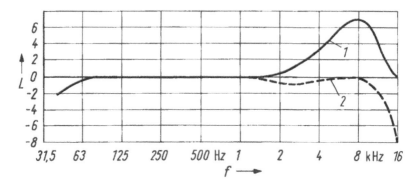

Figure 4.28 Action of the pressure increase effect on a pressure microphone:
1, free field, sound incidence from the front; 2, diffuse field or with
equalization.

directional characteristic varies accordingly, so that a high-frequency
accentuation occurs with normal (vertical) sound incidence in the free field
(Figure 4.28). For this reason such microphones are often equipped with
pressure increase equalization, which compensates the high-frequency
accentuation by means of an electronic network. However, microphones
with such an equalization network show a high-frequency roll-off in the
diffuse field and with lateral sound incidence. Figure 4.29 offers an example
in this respect by comparison with other high- and low-frequency de-
emphasis methods used in microphones.

Boundary microphones (see section 4.2.4.3) are used to extend the
pressure increase, while avoiding lateral sound incidence: that is, a
linearization of the frequency response and an increase of sensitivity.

Pressure microphones are normally used for picking up sound incident
from a large room area without significant local timbre. They allow very
'brilliant' recordings and a clean reproduction of low frequencies.

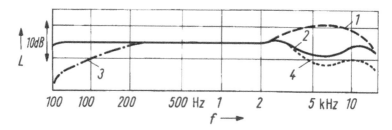

Figure 4.29 Effect of different equalizations on the free-field response level of a
microphone: 1, without equalization when speaking vertically to
the diaphragm; 2, with pressure-increase equalization: 3, with bass
de-emphasis; 4, with treble de-emphasis in addition to pressure-
increase equalization.

However, as they cannot attenuate sound from loudspeakers, they are not often used for sound reinforcement tasks.

4.2.2.2 Pressure-gradient microphones

If the pressure gradient or the velocity between two points is used to drive the diaphragm, one obtains a directional pressure-gradient microphone. The diaphragm must be accessible from both sides for the incident sound wave – perhaps even via an acoustic network. Some pressure-gradient microphones are constructed with two cascaded diaphragms.

Different directional characteristics (cardioid, hypercardioid, supercardioid, bidirectional) can be achieved by influencing the portions of pressure picked up in front of and behind the microphone (Figure 4.30). Taking rotational symmetry for granted, the directivity factors are: cardioid, 3; supercardioid, 3. 7; hypercardioid, 4; bidirectional (figure of 8), 3.

Frequently no distinction is made between hypercardioid and supercardioid characteristics. Such a distinction is of importance, however, because thanks to the reduced bidirectional lobing of the supercardioid, its rearward attenuation surpasses that of the hypercardioid. Thus a sound arriving from the rear is suppressed more effectively by a supercardioid than by a hypercardioid.

Because of their directional effect, pressure-gradient microphones are the type most frequently used in sound reinforcement engineering. They allow appropriate selection of the directional characteristic so as to effectively minimize the positive-feedback potential of a sound reinforcement system.

A drawback of all pressure-gradient microphones is their high sensitivity to air currents. With close-range use, a low-frequency accentuation occurs

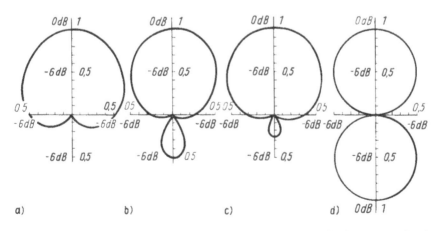

Figure 4.30 Microphone directivity characteristics: (a) cardioid; (b) hypercardioid; (c) supercardioid; (d) bidirectional.

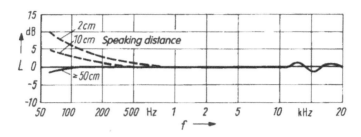

Figure 4.31 Result of the proximity effect on the frequency response of a pressure-gradient (directional) microphone with different speaking distances.

(the *proximity effect*). This is sometimes used deliberately: for example, vocalists use it to add warmth or 'body' to their voice. However, it always reduces the intelligibility of speech. Figure 4.31 shows the impact of the proximity effect on the microphone's frequency response. The effect falls off with increasing distance, and is no longer noticeable beyond about 1 m. To suppress this bass accentuation, pressure-gradient microphones are frequently equipped with switchable or permanently set electronic equalization circuits. To avoid the so-called 'sputter' or 'pop effect' that occurs with bursts of air current, specially formed sound-transparent foamed-plastic caps are placed over the diaphragm access openings. These wind protectors (wind or pop shields) damp the air currents thanks to their flow resistance, and partially neutralize whirls through their dimensions.

4.2.2.3 Interference microphones

If a still higher directional effect is required than can be achieved by means of pressure-gradient microphones, interference microphones have to be used. Currently, the most frequent type is the *directional tube microphone* (the 'shutgun microphone') [4.26]. It consists of a pressure-gradient capsule with a cardioid pattern arranged at one end of a slotted or perforated tube. The openings of this tube have a flow resistance that increases towards the microphone capsule. Thanks to this arrangement the sound waves from the front arrive in phase at the diaphragm, while those incident laterally to the tube are at least partially cancelled (Figure 4.32). The directional effect obtained in this way depends on the length of the tube. Directional tube microphones are used, for example, as tracking microphones for large speaking distances.

Microphone columns, analogous to loudspeaker *columns*, may also be considered as interference microphones. Because of their large dimensions and highly frequency-dependent directivity they are used only for special purposes.

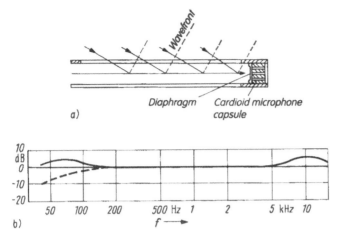

Figure 4.32 Directional tube microphone: mode of operation.

4.2.3 Transducer principles

Various types of transducer can be used to put the different reception principles into practice. Electrodynamic and electrostatic transducers are used almost exclusively for sound reinforcement engineering. Piezoelectric microphones are used only in special designs: for example, as contact microphones, as throat microphones, or in talk–listen combinations for command transmissions.

4.2.3.1 Electrostatic microphones

Electrostatic transduction is effected by variation of the distance between a moving electrode (the diaphragm) and a fixed electrode (Figure 4.33). In the

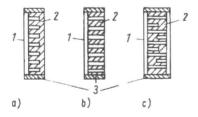

Figure 4.33 Electrode structure of condenser microphones with different directional characteristics: (a) pressure microphone with omni-directional characteristic: (b) pressure-gradient microphone with bidirectional characteristic; (c) double-diaphragm microphone with switchable directional characteristic. 1, Diaphragm; 2, back plate; 3, insulator.

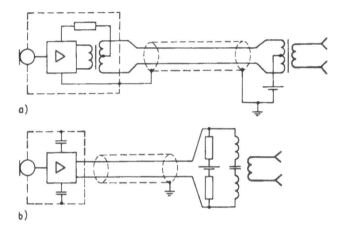

Figure 4.34 Microphone power supply: (a) phantom powered; (b) signal-cable powered.

condenser microphone, the two electrodes are loaded via a high-impedance resistor with a stable direct voltage. The distance variation of the electrodes induces a variation in capacitance. The load voltage varies in proportion to the variation in capacitance, and is tapped off via a high-impedance resistor. This voltage is passed on to the microphone line via an output impedance-converting preamplifier.

Because of the high resonant frequency of the diaphragm (the basic resonance of the system lies above the transmission range), electrostatic microphones have good transmission properties.

The introduction of the field effect transistor has made it possible to dispense with the heating voltage for the electronic tube of the preamplifier and to supply the voltage required for charging the electrodes by *phantom powering* or *direct signal cable powering* (Figure 4.34). Direct signal cable powering is restricted to a certain microphone type, since the signal cable becomes voltage-carrying. Phantom-fed circuits, however, may also be connected to other microphone types, such as dynamic microphones, since the signal cable is potential-free.

The phantom feed voltage is today generally 48 V, or in rare cases still 12 V. The power supply units used are capable of switching the feed voltage automatically according to the impedance of the connected microphone. If a condenser microphone contains two diaphragms, one each on either side of the back-plate electrode (Figure 4.33c), it is possible to change its directional characteristic by applying different polarization voltages.

With the *electret microphone*, polarization of the electrodes is achieved, unlike the condenser microphone, by the electret effect of the diaphragm material. For this reason it does not require any polarization voltage [4.27], but only a supply voltage for the subsequent impedance converter. If only a small dynamic range is to be expected, as for example with microphones in a conference system used exclusively for speech transmission, a supply voltage of 1.5 V is sufficient for operating the impedance converter (microphone amplifier).

Recently, electret microphones have also started to be used for studio purposes, making use of the relatively simple design in the quest for miniaturization. Because of the high dynamic range to be transmitted, however, a higher supply voltage is required. Electret microphones require a manufacturing expenditure similar to that of conventional condenser microphones.

The electrostatic transducer principle, in the shape of the condenser microphone, offers numerous advantages over other transducer principles:

- wide transmission range and balanced frequency response;
- reduced sensitivity to structure-borne sound,
- excellent transient response (thanks to the very low diaphragm mass);
- insensitivity to disturbing magnetic fields.

The possibility of electronic switching of the directional characteristic may be considered as an additional advantage.

These advantages are opposed by the following drawbacks:

- the supply voltage required for the impedance converter;
- the moisture sensitivity of the electrostatic system;
- the relatively high price of high-quality condenser microphones (because of the additional preamplifier).

4.2.3.2 Electrodynamic microphones

The transducer system consists of a conductor moving within a magnetic field with constant flux according to the sound-pressure fluctuations. The electric voltage thus induced in the conductor is proportional in magnitude and frequency to the sound pressure variation.

In its form as the *ribbon microphone*, the metallic conductor consists of a ribbon. Although the very low mass of the ribbon makes possible a balanced frequency response, such microphones are rarely used nowadays because of the large mass of the magnet and the fragility of the ribbon.

Another design variant of the electrodynamic microphone, the *moving-coil microphone*, however, plays an important role in sound reinforcement engineering. With this type of microphone the sound pressure fluctuations move a coil that is firmly fixed to a diaphragm. The coil moves within a magnetic field inducing in it a voltage that is proportional to the motion. The diaphragm–coil system is centred by an elastic suspension (Figure 4.35), and operates like all electrodynamic transducers above the basic resonance. If the back of the system is closed, as shown in Figure 4.35, the result is a pressure microphone, and if it is open, a pressure-gradient microphone.

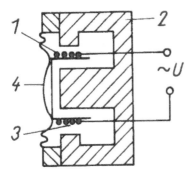

Figure 4.35 Moving-coil microphone (design principle): 1, moving coil; 2, magnet; 3, air gap; 4, diaphragm.

The problem with low-tuned systems consists in the linearization of the basic resonance, which is necessary for obtaining a sufficiently wide and balanced transmission range. With the moving-coil microphone this is achieved either by a corresponding configuration of the air volume behind the diaphragm (coupling of *acoustic labyrinths*) or – as is done with loudspeakers – by combining various complementary and differently tuned systems. In this way it has been possible to manufacture moving-coil microphones of studio quality.

The generally high structure-borne sound sensitivity of the low-tuned moving-coil microphones is adequately reduced by shock-mounting spring systems. This is particularly important with regard to their use as hand microphones. Interference by magnetic fields is largely eliminated by the use of of compensating coils. Thanks to

- their robust design,
- their reduced sensitivity to overload,
- their reduced sensitivity to moisture,
- their simple connection technique and
- their relatively low cost

the dynamic microphones currently available play a dominant role in sound reinforcement engineering, especially as portable microphones for soloists.

4.2.4 *Special microphones*

Apart from the standard designs used as stand, hand and suspension microphones, several special microphone designs are also frequently used in sound reinforcement engineering.

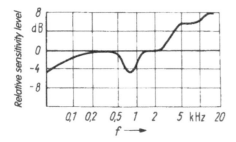

Figure 4.36 Equalization of a Lavalier microphone.

4.2.4.1 Clip-on microphones and Lavalier microphones

To avoid positive feedback with speech and voice amplification of moving persons it is often necessary to locate the microphone very close to the sound source. To maintain nevertheless the best possible freedom of movement, the microphones are carried on the body as clip-on or Lavalier microphones.

Clip-on microphones are very small, and fix to the clothing by means of a clip. The transducer used is often a condenser microphone with omnidirectional characteristics; directional microphones are less frequently used for this purpose because of their more complicated structure and slightly larger dimensions. As the microphone is located below the mouth of the speaker, no proximity effect equalization is required. Because of the very short distance between the speaker and the microphone, as well as the screening effect of the speaker and his or her clothing, it is also possible to use pressure microphones in sound reinforcement systems with a manageable risk of positive feedback. With suitable filtering it is also possible to use clip-on microphones for voice transmissions.

The *Lavalier microphone* is carried by means of a neck cord on the chest of the speaker. A 'structure-borne' component in a relatively narrow frequency range around 700 Hz is transmitted to the microphone through the speaker's breastbone. To suppress this component, such microphones are provided with frequency response equalization (Figure 4.36). Lavalier microphones are almost exclusively realized as dynamic microphones with omnidirectional characteristics. Because of the short speaking distance and the screening effect of the body, these microphones may also be used in sound reinforcement systems without increased risk of positive feedback.

4.2.4.2 Transducers for direct pick-up of low-output instruments

For sound amplification of low-output instruments, especially if they are to play in conjunction with loud instruments, but also if a very direct reproduction is desirable for artistic reasons and crosstalk is to be avoided, special electroacoustic transducers are used. Frequently these are directly operating vibration pick-ups. For guitars, for instance, one uses magnetic devices that pick up directly the oscillations of the steel strings.

A common problem is the need to amplify string instruments and harps against the stronger wind instruments, in order to establish a balance with these and with the louder percussion instruments. Nowadays special clip-on microphones are normally used. They are fixed above the F-holes of the string instruments, and their signals are transmitted to the mixing console via a local preamplifier and possibly a submixer. This pick-up technique gives better results than the former technique of attaching vibration pick-ups to the sound board, but all direct string pick-ups exhibit the drawbacks of mixing coherent sound components: they do not allow the development of a rich, choir-like, dense string sound, and they suffer from comb-filter distortions. To avoid this, an additional incoherent room sound has to be added to the directly picked-up string sound.

4.2.4.3 Boundary microphones

The location of the microphone at normal height (1.20–1.50 m above the floor of the platform) has the disadvantage that the microphone or the stand can obstruct the view, or are generally considered to be disturbing. This is particularly undesirable for theatre performances or television transmissions. If the source is at some distance from the microphone, interference occurs at the microphone location between the direct sound and the sound reflected from the floor. At a certain distance of the microphone from the source, this reflection may give rise to a comb-filter-like distortion of the frequency response, and may be heard as a 'roaring' sound (Figure 4.37); see also section 3.2.2.

To avoid these drawbacks, microphones have been arranged near the floor, with adequate insulation against structure-borne sound (Figure 4.38). But even with these holders there is still a small distance between the microphone and the floor, so the comb-filter distortions are not quite eliminated at higher frequencies. Przybilla and Schmidt [4.28] established that with such microphone arrangements the directional properties are not lost, and moreover the positive feedback threshold is increased thanks to the sensitivity enhancement by 3 dB (because of the arrangement of the microphone in front of a large surface outside the critical distance of the loudspeakers).

Various manufacturers are now producing *boundary microphones*, which are frequently offered under the description *pressure zone microphones* PZM [4.29]. With these the diaphragm is embedded in the area of the reflecting surface (Figures 4.39a and c) or immediately above it (Figure 4.39b). With these microphones it is possible to avoid comb-filter distortion also at the higher frequency ranges. Because the diaphragm has to be arranged flush with the reflecting plane, it is easily possible to produce

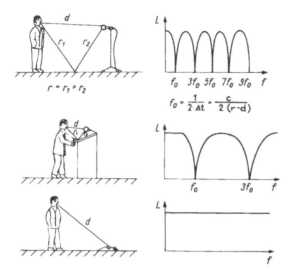

Figure 4.37 Formation of comb-filter distortions with different microphone arrangements: *d*, path length of direct sound; *r*, path length of reflected sound.

Figure 4.38 Structure-borne sound insulation mounts for floor-mounted microphones.

microphones operating on the pressure principle with a hemispherical pick-up characteristic. Several manufacturers also offer microphones with a cardioid characteristic. Figure 4.40 shows a microphone of this kind that is suitable for mounting to the upper edge of a speaker's desk or podium, for example. Strictly speaking, however, these pressure-gradient microphones are not boundary microphones.

Boundary microphones are well suited for picking up small, widely scattered ensembles and soloists. Also, in discussion meetings where the microphones are placed flat on the tables, it is possible to obtain acceptable results. For piano pick-ups it has proved successful to mount the microphone directly on the lid. The flap serves in this case both as reflection surface and as screening, which is especially important for sound reinforcement purposes.

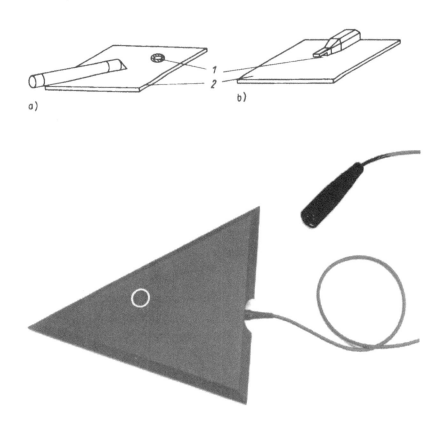

c)

Figure 4.39 Boundary microphone designs: (a) Schoeps; (b) Crown; (c) Neumann.
1, Microphone capsule; 2, reflecting plate.

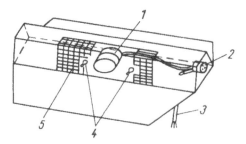

Figure 4.40 Directional boundary microphone designed for mounting on a
speaker's desk: 1, pressure-gradient capsule; 2, connection socket;
3, connecting cable; 4, fixing holes; 5, protective grid.

Boundary microphones are not suited, however, for picking up the sound emitted by large and deeply staggered units such as symphonic orchestras or big choirs. In such cases shadowing of the rear groups occurs, resulting in an insufficient timbre. Summing up, we can establish the following advantages of boundary microphones:

- They avoid comb-filter distortions.
- The frequency response is independent of the pressure increase effect.
- They are visually inconspicuous for the audience and possibly the camera.
- They have a larger positive-feedback distance.
- It is easier to achieve a favourable balance between the parts of an ensemble.

Opposed to these advantages are the following drawbacks:

- There is a greater disturbing effect from structure-borne sound.
- There is a risk of 'shadowing' effect when picking up the sound from deeply staggered ensembles.
- There is a high-frequency drop when the microphone is placed very near the source (passing of the high frequency over the microphone).

4.3 Sound transmission and amplification equipment

The structure of sound reinforcement systems varies widely according to the intended purpose. In its simplest form such a system consists of

- the programme-transmitting section, which includes the signal sources such as microphones, sound recorders, receivers, and preamplifiers;
- the signal distribution systems;
- the control equipment;
- the power amplifiers;
- one or more loudspeakers.

Since the peripheral acoustic transducers such as loudspeakers and microphones have already been treated, this section deals with signal transmission by wire or wireless as well as via intermediate storage, and with signal amplification.

4.3.1 Microphone connection technique

The choice of microphone connections through which the microphone signals are conducted to an amplifier, a mixing console or a microphone distribution amplifier/splitter depends on the operating technology and on the intended purposes of the system. An optimally designed connection technique guarantees for all envisaged kinds of events the most favourable microphone positioning with short and flexible microphone cables.

Large communication centres such as theatres, multi-purpose halls, concert halls, congress centres and conference rooms are for this reason generally equipped with permanently installed microphone cable networks.

The number of *microphone sockets*, N_M, required within the main performance area of a cultural centre is according to [4.30] approximately proportional to its surface H_b (in m^2):

$$N_M = \frac{H_b}{5} \qquad\qquad (4.20)$$

The main performance area comprises in this respect the stage or platform plus the forestage (proscenium) and the orchestra pit.

The microphone sockets are preferably grouped in wall socket panels. Compared with underfloor connections, these offer the advantage that they cannot be made inaccessible by scenery or carpeting; moreover, the risk of ingress of dust and debris is reduced. The sockets are installed mostly on the left and the right of the stage opening and in the rear area of the platform. The number per connection panel ranges between 6 for smaller stages to 30–40 for very large stages. In the orchestra pit one provides, if necessary, 30–40 connecting sockets.

Very large connection panels are mostly realized as collective cable connections. By means of bus (multicore) cables it is possible to run the microphone cables in a clearly arranged form to the connecting points, for instance the position of an orchestra, where they are distributed to the different microphones. For digitized signals, fibre-optic connections are going to play an increasingly important role in this respect. The main collective cable connections are also often additionally provided with a smaller number of indi-vidual connections that are led out in parallel in order to use the connections for events requiring a less sophisticated microphone technique.

Apart from the connections within the main action areas, additional local connection points are required. In this way it is possible for choirs and symphonic ensembles to be picked up by *suspended microphones*, for which connections have to be provided in the ceiling area above the platform. In larger concert halls microphone hoist systems may also be used to enable correction of the microphone position from the central control console. Further microphone connections are arranged as *room microphones* above the auditorium.

Of particular importance for the internal communication of a house are the *monitoring microphones*. These are mostly arranged in pairs above the stage area, the orchestra pit and the rear auditorium, and supply the accompanying sound for the image of a remote observation system. This system is normally coupled with the stage-manager system (show relay).

Congress centres are often equipped with a discussion system. In smaller conference rooms one frequently uses to this effect *automatic mixers*: that is, mixer channels that are opened by a threshold switch. The opening level can be adjusted so that disturbing noises are not amplified during intervals [4.31].

With the extensive microphone networks of large centres one has to take into account the permissible line length. To avoid attenuation of the higher frequencies of the transmitted signal, a line length of 250 m should not be exceeded with normal line cable materials and a source resistance of 200 Ω.

In order to minimize the number of microphones used for cooperative productions of various media, larger systems are equipped with *distribution amplifiers* to which the microphone signal is supplied for distribution via (frequently) four outputs to four individual internal and external users. Apart from reducing the number of microphones it is thus also possible to extend the microphone transmission line, for instance to a mobile transmission unit (O.B. van).

4.3.2 Wireless transmission

To allow sound pick-up without the restrictions of cables, *radio microphones* were developed. In these the cable is usually replaced by a high-frequency transmission link. Infrared equipment is nowadays also used in sound reinforcement engineering.

4.3.2.1 High-frequency transmission

Since the transmission quality of high-frequency links is, under normal conditions, equal to that of wire lines nowadays, these links are frequently used for mobile sources. The transmitters are often combined with the microphones. The transducers used are either directional microphones (mostly of the pressure-gradient type with cardioid characteristic) or microphones with an omnidirectional characteristic. The aerial is usually either a wire leading out below the handle housing the transmitter, or of the electronic type and incorporated in the handle. Apart from transmitter microphones, *body pack transmitters* are also quite common. These are of the size of a wallet, and are carried on the body with the aerial hidden in the garment. It is possible to connect to these transmitters not only hand-held microphones but also clip-on microphones: these have the advantage that one does not see the actor to be equipped with a microphone.

With a view to reducing sensitivity to interference, *frequency modulation* has been internationally adopted for stage transmission techniques. The carrier frequencies used in general are in VHF (150–250 MHz) and UHF (450–950 MHz) bands. Note that these frequencies vary from country to country. For example, in the USA (900 MHz) and in the UK (860 MHz). Since for the highest quality it is necessary for adjacent channels to have a

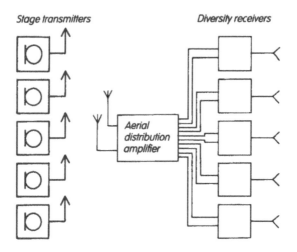

Figure 4.41 Diversity receiver system for radio microphones.

minimum separation of 300 kHz, this frequency band can accommodate no more than three broadband channels of studio quality. In order to allow the highest-quality transmission of a greater number of channels (up to 88), the use of locally unoccupied VHF bands (174–230 MHz) and 470–790 MHz are often employed. In the USA the carrier frequencies are in the range 169–218 MHz VHF and 780–900 MHz UHF.[1]

In most countries the transmission equipment has to be licensed by the appropriate authorities.

Since stage radio transmission is often used in halls whose building structure contains numerous metallic reflection surfaces, one must reckon with heavy interference, which may cause significant disturbances, especially when the transmitter is moved. For this reason high-quality systems employ several receiving aerials distributed round the room and operating with *diversity reception* (Figure 4.41).

High-frequency radio transmission is also used in sound reinforcement engineering for sound signal transmission to mobile systems with active loudspeakers equipped with receivers and power amplifiers. The procedures used are the same as those used for stage transmission, except that one transmitter usually supplies a whole range of loudspeakers, which are coupled with HF receivers.

---

1 Authorized radio frequencies vary from country to country, but generally lie in the bands quoted.

4.3.2.2 Infrared transmission

Apart from HF radio techniques, infrared transmission is also increasingly being used in sound reinforcement engineering. This section describes the technique that is currently used [4.32, 4.33].

As *transmitter* one uses GaAs light emitting diodes, whose wavelength is about 930 nm. Because the radiation is incoherent it is possible to use any number of diodes in order to increase the radiated power. In practice one uses 12–120 diodes per transmitter unit.

As *receiver* one uses a photodiode provided with a frogeye-shaped collecting lens, enabling it to receive as many direct and reflected rays as possible. The visible light that would be received by the diode as a disturbing quantity is absorbed by a black filter. Frequently there are also several reception diodes arranged side by side. For transmission, a carrier frequency of 95 kHz is modulated with the audio signal. If several channels are to be handled by the same transmitter, several auxiliary carriers have to be used. To avoid mutual interaction it is necessary for the frequency modulation to be very low in harmonics. Figure 4.42 shows some designs of transmitters and receivers.

a)

Figure 4.42 Infrared transmission technique: (a), (b) receivers (Sennheiser) (c) principle – 1, diode-equipped high-power radiator; 2, modulator and amplifier; 3, receiver.

b)

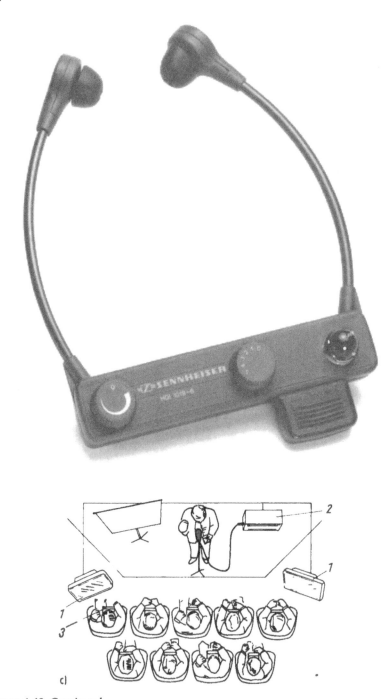

c)

Figure 4.42 Continued.

Because of the large dimensions of the transmitters, and their pronounced directivity, the infrared transmission technique is currently not applicable for stage microphones. However, it is frequently used for interpretation systems, in which the transmitters are arranged above the audience in the congress hall, and the participants are equipped with receiving sets. By choosing the appropriate channel it is possible to receive the required language. In contrast to the RF and audio-frequency induction transmission techniques, the infrared process is safe from undesired monitoring outside the room. Another field of application for the infrared transmission technique is *guide systems* in museums, galleries and exhibitions. Here one makes use of the restricted propagation of the infrared light to provide each exhibit with separate information in various languages, which can only be received in the immediate area [4.34] (see section 6.6.5).

4.3.3 Sound storage devices

Sound storage devices are of great importance for sound reinforcement systems. Some of their most frequent applications are:

- playing of background music via information systems;
- playback and partial playback reproduction of music in show and music programmes (for example, playing a continuous rhythm – a 'click track' – to establish synchronism between artists and music, or insertion of musical accompaniment for soloists);
- playing of a wide range of sound effects to establish an acoustical atmosphere for theatre productions;
- reproduction of specially prepared magnetic tapes as part of a modern composition in the concert hall or the opera house.

4.3.3.1 Analogue sound storage devices

The programmes required for playback in a room are today normally recorded on two-channel studio magnetic tape recorders. Contrary to the stereo recording technique for disk and radio it is common practice to record different groups, such as voices and accompaniment or string and brass instruments, on separate tracks so as to be able to correct the balance in the reproduction, or to assign groups to different loudspeakers.

For special purposes, multi-channel sound reinforcement systems also use devices with eight or more tracks. For the production of records, but also for reproduction via multi-channel loudspeaker systems, one uses to an increasing extent also 4-, 8- and 16-channel magnetic tape machines.

One drawback of the analogue magnetic-tape recording procedure consists in the reduced signal-to-noise ratio against the remaining sound channel. With $\frac{1}{4}$ in stereo machines this ratio is no more than 60 dB, and reduces still further as the number of recording channels increases. Quarter-inch four-channel stereo magnetic-tape machines, for instance, have a signal-to-noise ratio of only 35–40 dB. To decrease the tape noise nowadays highly developed noise suppression systems such as Dolby SR [4.35] or Telcom C4 [4.36] are used; these enable a maximum improvement of the signal-to-noise ratio by 30 dB. The systems are based on a variable frequency-dependent compression (dynamic contraction) during recording and a corresponding expansion during reproduction. Such a combination of compressor and expander is called a *compander*.

Apart from the magnetic-tape studio machines, cassette recorders are acquiring increasing importance. The devices used are mostly stereo recorders with equalization for different tape types according to IEC I (Fe_2O_3), II (CrO_2), III (double-layer tape) and IV, as well as with different noise suppression systems (Dolby B, Dolby C, dbx, etc.). Thanks to their simple operation and the widespread utilization of the systems, cassettes are frequently used by soloists for their musical accompaniment. At congresses they serve for recording the proceedings (see also [4.37]).

In addition to the use of analogue magnetic-tape storage devices, any larger sound studio should be able to reproduce vinyl gramophone records as well as their modern version, the CD.

4.3.3.2 Digital sound storage devices

Compared with analogue units, digital storage devices provide entirely new possibilities. We can distinguish between magnetic, optical, and electronic memories. All offer almost unlimited copying without quality deterioration, fault correction, and extensive freedom from distortion.

All digital procedures used in sound engineering and thus also in sound reinforcement engineering require sampling of the signals with a clock rate of at least 30 kHz, in order to enable transmission of 15kHz. Because of the finite slope steepness of the limiting low-pass filters, and because the control signals have also to be transmitted, a clock rate of as much as 44 kHz is desirable. The usual rates are 44.1 kHz for home devices and 48 kHz for professional equipment. Some *magnetic-tape storage systems* (R-DAT) allow the clock rate to be switched from 44 kHz to 48 kHz [4.38]. As digital magnetic-tape storage systems are currently still very expensive, and moreover lack proper standardization, they have been used so far almost exclusively in recording studios. In contrast to the S-DAT equipment (stationary magnetic head with several parallel tracks), however, cassette recorders of the R-DAT type (rotating magnetic head for recording and reproducing) have acquired a greater importance.

One of the most important optical information carriers is the *compact disc* (CD) [4.39], on which the information is contained in tiny cavities (pits), which are scanned by means of a laser. Nowadays erasable and re-recordable optical disks are increasingly being used. To this effect magneto-optical as well as amorpho-crystalline procedures have been developed. Compared with magnetic tape, these procedures offer the advantage of contactless, and hence wear-free, scanning, a feature that can be considered as an essential advantage for frequent use [4.40]. Further features are quick access, programmability, perfect marking, and adjustment.

Electronic storage by means of RAMs and EPROMs is already being used in sound reinforcement engineering, where it offers the advantage of enabling the recording of 'takes' of ever increasing duration as an audio signal and of recalling them quickly – under computer control, for example – in different combinations. Such devices are used, for instance, in information systems to generate 'synthetic' announcements (see section 6.1).

LEVEL CONTROL

In contrast to the analogue technique, in which peak modulation is usually limited by the recording characteristic, with the digital technique overload of the A/D converters will cause sudden, strong distortions. Signals that are to be digitally recorded must therefore be monitored by means of a genuine peak-value level indicator.

The reproduction of digital recordings is therefore richer in peaks than that of analogue ones, and this has to be taken into account with digital playback. The reproduction of a digital recording requires a greater power reserve than that of an analogue recording to generate the same (r.m.s.-value determined) sound level. This is why a 'headroom' of at least 10 dB is internationally required. A further restriction of the dynamic range is caused by the digitizing noise, which requires a 'footroom' of 16–20 dB. These restrictions mean that the reproduction qualities of good companded analogue recordings are still today similar to those of normal digital recordings (see also section 5.2.3). Thus the superiority of the digital technique can be expected only with the full introduction of the 24-bit technique.

4.3.4 Amplifiers

In every sound reinforcement system the electrical power produced by the various sources has to be amplified to a value that will allow operation of the loudspeakers. In simple terms, amplification is effected in two stages: preamplification and power amplification. If summation, volume regulation or processing of the audio signal are required, the voltage and power losses caused must be compensated for by additional amplification.

4.3.4.1 Voltage amplifiers

The relatively low voltage produced by the signal sources of a sound reinforcement system (Table 4.2) has within the scope of preamplification to be amplified to a value that will allow mixing, filtering, distribution, etc., without the signal-to-noise ratio being essentially reduced by outside fields

Table 4.2 Input conditions at the preamplifier for different sound sources (according to DIN 45500)

Source	Source voltage	Source resistance	Minimum load resistance	Amplifier input
Dynamic microphone	0.5–10 mV	200 Ω	>500 Ω	Balanced earth-free
Condenser microphone	5–100 mV	200 Ω	>500 Ω	Balanced earth-free
Mixing console (studio equipment)	0.775–1.55 V	>20 kΩ	>600 Ω	Balanced earth-free
Studio tape recorder	0.775–1.55 V	>20 kΩ	>600 Ω	Balanced earth-free
Magnetic pick-up	5–15 mV		≈50 kΩ	RC input with equalization
Domestic tape recorder	⩽800 mV	⩽50 kΩ	⩾300 kΩ	RC input, linear
Cassette recorder	⩽500 mV	⩽50 kΩ	⩾500 kΩ	RC input, linear
Radio tuner	⩽300 mV to 2 V	⩽50 kΩ	⩾500 kΩ	RC input, linear

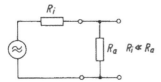

Figure 4.43 Adaptation of a voltage amplifier.

on account of inadmissibly low levels. The voltage that can be achieved by preamplification is generally 0.775 V, which is the internationally usual reference level. An r.m.s. voltage of 0.775 V applied to a reference resistance of 600 Ω allows a power of 1 mW to become effective. The level referred to this voltage (corresponding to the reference quantity of 1 mW) is therefore indicated in dBm. For studio equipment one uses as a reference value twice this voltage:

$$U_{\text{eff}} = 1.55 \text{ V} \equiv 6 \text{ dBm}$$

For this reason the transmission lines for this voltage are called *6 dB lines*.

Within the scope of voltage amplifiers (that is, preamplifiers and booster amplifiers), a wattless interconnection is aimed for in order to avoid frequency-dependent attenuations. The external resistance, R_a, connected to such an amplifier (Figure 4.43) should therefore have a value 25 times that of the amplifier output resistance. By means of parallel connections it is possible to reduce this value to $\frac{1}{5}$, i.e. $R_a : R_i = 5 : 1$, before noticeable disturbances occur [4.41]. When connecting a line with a capacity of C_L, the capacitance $1/(\omega C_L)$ must be greater than $5R_i$. These conditions have to be considered, for instance, when connecting various power amplifiers to one mixing-console output. Under the conditions described a level drop of 1.6 dB occurs. However, if the value of the load resistance is only twice that of the internal resistance, the level drop amounts under the same conditions to 3.5 dB and may thus be quite disturbing.

If a greater number of devices have to be hooked up in parallel to one amplifier, or if the load resistance becomes too small for other reasons, it is necessary to insert a *buffer amplifier* to act as an impedance converter.

4.3.4.2 Power amplifiers

Power amplifiers can be coupled directly to the preamplifiers. By adding a mixing unit and connecting peripheral equipment (microphone, sound recorder, loudspeaker) they may form the smallest possible sound reinforcement system and be operated independently as so-called *powered mixers*. In larger installations the power amplifiers are either assigned to the loudspeakers or located in separate racks in the sound reinforcement control room.

Power amplifiers are offered in two main forms. The first type is designed for loading by a relatively high-impedance network with an output voltage that is kept constant by strong negative feedback up to the total wattage rating. The maximum value of this voltage in Europe is 100 V. (In some countries it is intended to lower it to 50 V in order to reduce the potential hazard in case of accidental contact; the maximum value in the USA is 70 V.) The second type of power amplifier is designed to be permanently assigned to the loudspeakers. Since the latter are connected directly via a relatively short, low-impedance line, accurate power balancing is required. The maximum power-handling capacity of the power amplifiers is determined by the permissible distortion. In this respect harmonic distortion coefficients of up to 1% are frequently permitted.

Nowadays power amplifiers with a rating of less than 50 W are rarely used in sound reinforcement engineering. In the 100 V technique one normally uses units of 100–300 W, whereas the low-impedance technique also works with larger power units of 500–1000 W.

HIGH-IMPEDANCE TECHNIQUE (100 V TECHNIQUE)

Thanks to the reduced internal resistance of the power amplifiers, this technique allows the connection of different types of loudspeaker up to the rated power-handling capacity of the amplifier, without influencing its output voltage. To this effect it is necessary for the loudspeakers to be adapted according to the required connecting load capacity, maximally up to their rated power load capacity, to the output capacity of the power amplifier (100 V line). Thus it is possible to operate widely distributed loudspeaker networks non-reactively with varying loads (Figure 4.44).

To avoid overloading by mismatching or faulty connections, before a system is put into operation it is recommended that the impedance of the loudspeaker network is checked, usually at a frequency of 1 kHz. The impedance value must not fall below

$$Z_L = \frac{U_S^2}{P_v} \tag{4.21}$$

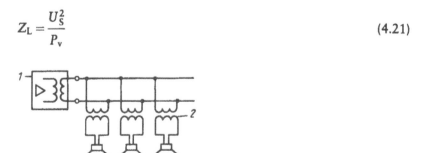

Figure 4.44 Structure of a 100 V network: 1, power amplifier with output transformer; 2, matching transformer for loudspeakers of different power ratings.

where U_S is the system voltage (e.g. 100 V), and P_v is the rated power of the power amplifier.

The driving-point impedance of the loudspeaker is

$$Z_S = \frac{U_S^2}{P_L} \tag{4.22}$$

where P_L is the load of the loudspeaker (maximally the rated power).

Apart from facilitating uncomplicated connection, the 100 V technique offers the advantage of enabling transmission of high powers by cables of relatively small cross-section. For this reason it is always preferred for widely distributed loudspeaker networks. When connecting high-quality loudspeakers the line resistance should not exceed a maximum of 5% of the rated impedance of the connected loudspeakers, in order to avoid excessive power loss. A drawback of this connection technique is the matching transformers that are required: their inductance and capacitance may cause a restriction of the transmission frequency range.

LOW-IMPEDANCE TECHNIQUE

In this case the power amplifiers and the loudspeakers function with direct matching. This means that the maximum power of the amplifier is supplied directly to the loudspeaker, if the terminal resistance – that is, the impedance of the loudspeaker including the supply line – is equal to the internal resistance of the amplifier system. By avoiding transformers it is thus possible to bring directly to bear the high transmission quality of high-standard loudspeaker systems. The output impedances are, like the rated impedances of loudspeakers, 4, 6, 12, or 16 Ω, the preferred value being 8 Ω.

From these conditions it follows that the resistance of the loudspeaker feed lines must be kept as low as possible. It must not exceed 10% of the rated impedance of the connected loudspeaker: in other words, no greater than 0.8 Ω for a rated impedance of 8 Ω.

The required line cross-section, q, can be calculated from the following equation:

$$q = \frac{2LpP(100 - n)}{U^2 n} \tag{4.23}$$

where q is the cross-section in mm^2; L is the length of the cable between the amplifier and loudspeaker in m; P is the power absorbed by the loudspeaker in W; n is the permissible voltage loss in the cable, expressed as a percentage; and ρ is the specific resistance of the cable ($\rho_{cu} = 1.78 \times 10^{-2}$ Ω mm^2/m).

> **Example:** Given a distance between amplifier and loudspeaker of 40 m, a maximum amplifier output voltage of 28.3 V (at 8 Ω), a power rating of the loudspeaker of 100 W, a copper line having a specific resistance of 1.78×10^{-2} Ω mm^2/m and a maximally permissible voltage loss of $n = 10\%$, there results a line cross-section of
>
> $$q = 1.6 \text{ mm}^2$$

Apart from the resistance of the loudspeaker feed line one has also to consider the maximum current density in the cable, since the low impedance of the loudspeaker implies a high current. To avoid excessive heating of the cable, the current density in it should not exceed $G = 5\text{--}10$ A/mm^2. For thicker cables it is convenient to choose the lower value, since a cable of larger diameter has a less efficient heat dissipation. Taking into account the maximum current density G, the resulting minimum wire cross-section is:

$$q = \frac{100P}{GU(100 - n)} \tag{4.24}$$

For the example given, for a maximum current density of $G = 5$ A/mm^2, the resulting minimum cross-section is

$$q = 0.8 \text{ mm}^2$$

Thus the cross-section calculated according to equation (4.23) is sufficient [4.42].

These conditions imply that the low-impedance technique requires the amplifiers to be permanently assigned to the loudspeakers (Figure 4.45 shows an example of this). As a rule, manufacturers specify the maximum length and the minimum cross-section of the feed line. In most cases this means that the power amplifiers cannot be centrally installed, but (with larger systems) have to be located near the loudspeakers.

OVERLOAD PROTECTION

As a rule, modern power amplifiers are fully transistorized. In contrast to valve amplifiers, which feature a curved characteristic where distortion sets in slowly along with the amplitude peaks, transistor amplifiers have a characteristic that bends sharply, resulting in a sharp limitation of the transmitted signal, called *clipping* (Figure 4.46a). Limitations of this kind

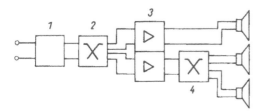

Figure 4.45 Power connection of a low-resistance amplifier to a two-way loudspeaker combination: 1, limiter; 2, active crossover; 3, power amplifier; 4, passive crossover.

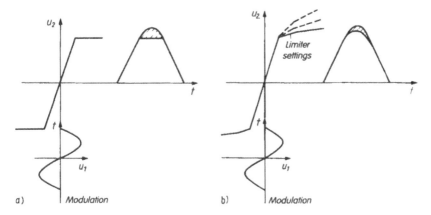

Figure 4.46 Limitation of overmodulated signals (hatched): (a) effect of a 'clipping' characteristic; (b) effect of a limiter.

cause very disturbing distortion, which because of its sudden onset is difficult to conceal, and may destroy the tweeter systems. For this reason, systems with transistor power amplifiers for high-quality transmissions are designed to be capable of briefly processing overmodulation peaks of up to 10 dB, for which purpose they require a corresponding headroom [4.43]. This implies that when sizing a system one has to calculate the tenfold power to obtain the desired level. To reduce this high expenditure, valve characteristics are often approximated by an appropriate circuit design. Since these circuits are frequently realized as an integrated circuit in miniaturized hybrid design, amplifiers of this kind are also called *hybrid amplifiers*.

Other designs use additional regulating amplifiers, which impede a sudden onset of distortion. These amplifiers, which reduce amplification gently when overload peaks occur, are called *limiters* (Figure 4.46b). Their response time is relatively short; regulating amplifiers with longer time constants are generally used for volume compression.

In large power units, in addition to the limiters, *overload safety switches* are also used; if overmodulation occurs, these reduce the output power significantly in order to avoid destruction of the loudspeakers. The use of limiters can reduce the overmodulation risk for amplifier and loudspeaker by about 10 dB; for this reason they are widely used in large transportable systems for rock and pop music.

PROCESSOR AMPLIFIERS (CONTROLLERS)

Yet another type of overload protection is achieved by means of processors integrated in the power amplifiers. As shown in Figure 4.47, a measurement is taken at a specific point – for instance in the amplifier, at the amplifier output, at the moving coil, or at the diaphragm of the loudspeaker – fed into the processor and compared with the input or reference value. The parameter measured may be the level, the moving-coil temperature increase, the diaphragm movement, or a distortion quantity. If the measured parameter shows an inadmissible deviation from a predetermined maximum value, the level or the frequency response of the output signal is corrected accordingly until the correct operating value is restored.

These processors are best used for direct coupling between power amplifier and loudspeaker. When using active crossovers – that is, crossovers inserted before the power amplifier – the processors may also effect a redistribution of the energy between the different frequency ranges.

Such systems make it possible to achieve relatively high rates of amplifier utilization. However, this may be accompanied by a compression of the output signal, and this is not always artistically acceptable.

It is increasingly becoming possible to control the processor amplifiers by means of external computers (PC control), so that preprogramming becomes feasible [4.44].

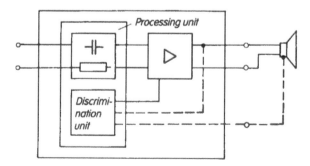

Figure 4.47 Structure of a controller.

CROSSOVERS

To avoid undesired interference and overloading of the tweeter systems in the overlap region between component loudspeakers, filters are inserted between the loudspeakers. These have a slope steepness of 6, 12, 18 or 24 dB/octave. Good loudspeaker combinations require at least 12 dB/octave.

If the appropriate power amplifiers are assigned to the loudspeakers intended for the different frequency ranges, the crossovers must be arranged before the power amplifiers. In this way it is also possible to adapt the level distribution between the component loudspeakers to the requirements, and the loudspeaker impedance has no influence on the crossover. In this case one speaks of *active crossovers*, which also make it possible to use an optimum limiter as well as the optimized amplifier and overload protection for every range. Subdividing the frequency ranges in this way, and with different response times and threshold levels, offers the advantage of making the control processes less conspicuous for the listener.

Passive crossovers are used especially for loudspeaker combinations housed in one common cabinet. In this case the crossovers are an integral part of the loudspeakers, enabling optimization to be carried out by the manufacturer.

4.4 Technical equipment for sound control systems

An essential task of modern sound reinforcement engineering is the acoustic coverage of large performance centres. These include cultural establishments such as theatres, opera houses and concert halls, but also multi-purpose facilities such as sports and congress halls, which are also used for cultural purposes. Each of these facilities requires specific equipment of its own, which also depends on the size and importance of the establishment. The range of technical devices used in this respect – shown schematically in their combined operation in Figure 4.48 – are explained in the following sections.

4.4.1 Input signal distributor

An input signal distributor (patch panel) is required for handling the vast number of lines coming from the signal sources available in large sound systems, since a permanent assignment to the mixing console inputs is neither economically acceptable nor easy to monitor or control. The sources are connected to the inputs of the mixing console or consoles via this distribution panel before every event and in some cases also during an event. Large systems must therefore often handle more than 100 microphone lines [4.45] in addition to even more high-level lines. Only a small portion of these connections can be coupled directly to the mixing consoles.

The rack containing the input distribution panel often also contains the power supply units for the condenser microphones, if appropriate.

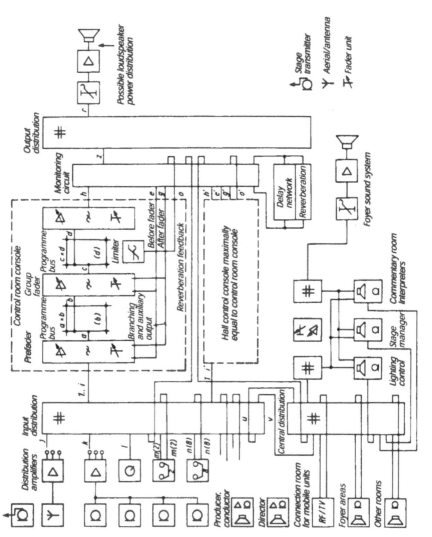

Figure 4.48 Schematic plan of the technical equipment of a large sound reinforcement system.

If the microphones are also likely to be used by radio and television during performances, or if the microphone signals will also be required at other parts of the house, *distribution amplifiers* are inserted. These distribute the low-level signal to, in most cases, a maximum of four co-users. The number of microphone distribution amplifiers, and in some cases also distribution transformers, depends on the number of microphones normally used in a performance. It is generally between 20 and 30, but may also be as high as 60.

Apart from the microphone lines, the outputs of all other transmission lines and sound recording devices are also routed to the input distribution panel. In certain cases (for mixing consoles with switchable inputs), some studio tape recorders and frequently used microphones are directly linked to the mixing consoles. The input distribution (patch panel) is often complemented by some freely available *microphone filters* or equalizers, enabling suppression of certain microphone-specific timbres or positive feedback frequencies before the input to the mixing console, so that in the case of a redistribution during the performance these changes of the input filters in the mixing console do not have to be considered.

Because of the large number of sources to be considered, the input distribution panels are often realized as patch panels with patch cord connections or, in modern installations, as multi-stage, computer-aided switching and routing circuits. BUS groups and parallel patch connections or matrix outputs are also used, especially in theatres, for switching frequently used sources.

4.4.2 Sound-mixing consoles

The input distribution panel is followed by one or more mixing consoles. One of these mixing consoles is installed in the central control room (the *monitoring room*). It serves, among other things, if further mixing consoles are available, for premixing, for operating the sound recording devices including the level matching of recorded signals, and for recording performances. A second mixing console is often arranged at a point in the hall that is acoustically representative for the auditorium area (hall control). The final balance of the output signal is monitored here, since an assessment of the acoustic condition and of the overall sound quality is possible only from within the hall. (Tests aimed at controlling the output signal by means of one or several dummy heads [4.46] did not produce the desired effect.)

The hall mixing console is particularly necessary in those cases where performances are mounted with little rehearsal time, heavy use of microphones is made, and extensive loudspeaker arrangements are employed. Such conditions usually arise in multi-purpose halls. In small halls, with only a few sound technicians, a whole evening's performance may need to be controlled exclusively from the hall mixing console.

More recently there has been a requirement for a *mobile sound-mixing console* in theatres for the rehearsal of drama productions or for use with musical performances, with corresponding location facilities in the auditorium. Similar conditions apply in concert halls, where it must be possible in special cases, as in the performance of works of electronic music for example, to locate such a console without having to run additional cables [4.47, 4.48].

With very large performances a special *monitoring console* is also used in the stage area. Bands and groups often also require a mixing console near the platform for ensuring the appropriate balance within the group.

The sound mixing consoles used in small and medium-sized multi-purpose halls for the sound control room and the hall sound control usually contain 24 or up to 32 inputs, which in some cases are switchable for high- and low-impedance sources. The input paths are equipped with preamplifiers, level controllers, and often also with filters for signal matching.

Mixing consoles usually provide two cascaded mixing facilities with level controllers, filters and booster amplifiers, so that adjustment of the individual sources and group mixing is possible before setting the overall sound balance. To this end eight group buses and four main outputs are often used. The group buses should be switchable between output and auxiliary paths, in order to provide partial programmes, for effects or for monitoring loudspeaker groups, for example. Loops (effects returns), used for reverberation or effects processing for example, must also be provided. Apart from the group outputs it should also be possible to connect individual paths to an output directly after preamplification, so as to create for example paths allowing individual level control.

Whereas in multi-purpose halls and show theatres three-stage mixing consoles are generally preferred, two-stage mixing consoles with many outputs are usually adequate for sound effects use in theatres.

The accumulation of cascaded amplifier stages, which occurs for instance when several multi-stage mixing consoles are connected in series, carries the risk that the signal-to-noise ratio at the output of the system may become inadmissibly low. This drawback is avoided with modern automated mixing consoles by the fact that although the levels of the individual stages are adjustable by individual controllers, the overall level adjustment is effected only via the control in the input stage of the mixing console. To achieve this, the input stage is realized as a voltage-controlled amplifier (VCA), a digital level controller [4.47], or an optoelectronically controlled resistor. These mixing consoles have to be equipped with special circuits to create the required group outputs.

Such mixing consoles (some of which are now fully digital) offer the advantage of programmable control. This makes it possible, for instance, to preset the required attenuations and – provided the corresponding distribution devices and controllable filters are used – the required paths and frequency characteristics. In this context we can distinguish between *static* control and *dynamic* control.

Static control makes it possible to preset all control processes, which can be operated by a single command (pressing of a button, insertion of a memory card, or a disk). Dynamic control allows continuous changing of the settings during the performance; with static control this has to be done by hand.

Static 'automation' control offers great advantages, especially for theatres in which one production and thus the sequence of settings of the sound system is often maintained unchanged for a long time, and relieves the sound engineer of many routine jobs. With a live production, it offers simultaneously the possibility of interfering in the sound reproduction according to the behaviour of the actor and to leave it unchanged, to interrupt it or to change it. If required for reutilization, it is possible to store the changes effected in the settings.

Dynamic automation control, however, requires a determined chronological sequence of the settings programmed for the sound channel, as it becomes necessary in sound reinforcement engineering, for instance, with given sound-panning configurations or the reproduction of electronic music. But with live productions such a system may give rise to problems, if contingency plans have not been made for possible programme changes. Dynamic automation control proves to be advantageous especially for the production of multi-track recordings, where a number of otherwise necessary input and group buses can be dispensed with.

4.4.3 Output signal distribution

An output distribution panel is connected after the mixing consoles. It accommodates all main, group and individual branch outputs of the mixing consoles in the system. The paths between the mixing consoles, as well as certain paths frequently used for reverberation and special effects, are usually connected directly, and not via the output distribution panel.

POWER AMPLIFIER PATHS

Distribution of the lines to the power amplifiers supplying the individual loudspeaker groups and subgroups is normally effected via distribution amplifiers or parallel patch or matrix connections. If delay devices are used, they are connected after the output or (for multi-channel delay), after outputs of the mixing console. After the outputs of these devices, distribution to the power amplifiers is done via a distribution matrix, which is either permanently interconnected or has to be interconnected by hand. If a multi-channel delay system is used, such as the Deltastereophony System [4.48], this matrix is realized as a mixing matrix.

Filters for equalizing the loudspeakers are arranged in the power amplifier paths after the output distributor. In larger systems, before the power amplifiers or groups of amplifiers there are remotely controllable volume regulators, which are adjustable – for example in 3 dB steps – from the hall sound control console, and often also from the monitoring control room. They permit a separate level adjustment for each individual loudspeaker group without using the output level controllers of a mixing console (which would not be possible at all when using a delay system). This is also important for establishing the loudness balance between sound system and effects loudspeakers. In the region of output distribution before the power amplifiers it is possible to install remotely controlled changeover switches that can distribute the signals between the power amplifiers and thus between the loudspeakers, thereby producing with distributed sound reinforcement systems acoustic travelling effects (panning facility). To avoid switching noise, and also to permit coupling with other techniques, such as automatically controlled lighting regulation systems, VCAs, digital level controllers or optoelectronic couplers are used as switches [4.49]. Corresponding switches may also be used on the power side (see below).

POWER DISTRIBUTORS

After the power amplifiers there is no additional distribution facility for the main loudspeaker groups and the loudspeaker groups of a room sound reinforcement system (except the insertion facility for replacement amplifiers). Stage loudspeaker groups and, under certain conditions, effects loudspeaker groups or individual loudspeakers can, however, be switched between different power amplifiers. This switch may serve to activate rarely used lines for connecting mobile loudspeakers or for supplying particular loudspeakers with special signal components originating, for example, from a delay system. Power distribution also includes the *panning switches*, which are arranged on the power side and enable individual loudspeakers installed in the room to be switched out of a larger loudspeaker group and simultaneously onto a special power amplifier. This happens in a certain sequence, until the power amplifier designated for the panorama group is fully loaded. During the switching sequence, each time the first loudspeaker is switched off and re-connected to the original power amplifier, before a further loudspeaker is connected. This process is controlled by a special element assigned to the hall mixing console or, in theatres, to the monitoring control console [4.45].

OUTPUTS FOR OTHER USERS

Apart from the connections for the loudspeaker groups used for covering the hall, the output distribution panel serves also for connecting the lines to recording devices and to cross-connections for further users. These include the stage management system, the system for the hard of hearing, and – if required – a connection room for the mobile units of radio and television networks.

4.4.4 *Systems for sound reinforcement in theatres*

A theatre sound reinforcement system is an example of a large system that is specially designed for artistic effects. Because of the wide subdivision of the permanent and mobile loudspeaker groups, the control room has to handle a large number of independent sound channels (see section 6.6.1). Nowadays medium-size theatres frequently use two-stage mixing consoles with at least 24 inputs, switchable to various sources. To enable single sources to be individually processed, these inputs are connected to between 4 and 24 group outputs. At least eight premixed programmes should be available via busbars at an equal number of outputs. These outputs are fed to matrix distributors, which should be controllable remotely. The 'final stage' connected after the output distribution panel consists primarily of remotely controllable attenuation regulators and room adaptation filters (see section 4.6.5.3). The power amplifiers used are frequently 100 W units using the 100 V technique, but amplifiers of higher rating – mainly using the low-impedance technique – are now increasingly been used.

The main loudspeaker groups and the frequently used stage loudspeakers are permanently assigned to these power amplifiers. The remaining stage loudspeakers, as well as all other mobile loudspeakers, are often connected to the power amplifiers via a power distributor located in the stage area.

The effects loudspeakers arranged at the walls, and in many cases also at the ceiling, are operated through a switching device – the panning switch described above – but allow additionally still other interconnections. If required, delay devices are inserted between the mixing consoles and the output distributor.

4.5 Sound reinforcement with remote or satellite control rooms

There is frequently a requirement for coverage of larger complexes, such as: sports centres with extensive facilities; exhibition centres with multiple halls and open-air spaces; hotels with halls, foyers and restaurants; and railway stations with reception halls, platforms and enterprises. In such cases it is necessary to reproduce a central programme throughout the whole complex as well as individual programmes in the different areas. The central programme may also be transmitted from one of the remote control rooms.

Such installations normally consist of a central studio and several satellite stations, which may also operate independently of one another.The satellite stations can also be activated from the central control room by means of an override circuit, and often remote-control circuits as well. Only rarely is the whole programme, including the decentralized tasks, handled by the central control room.

4.5.1 Central sound studio equipment

At the heart of the system is a mixer. For small systems this may consist of a mixing amplifier or a powered mixer; medium-sized systems require rack-type equipment, and larger systems require a small or medium-sized mixing console.

The programme sources are in most cases permanently connected to the mixer. Very large systems with decentralized programme components handled by the central studio require an input distributor, usually equipped with a key-operated switch-in device. The programme transmitters that can be connected to the mixing or switch-in equipment include:

● one or more paging stations, which are either switched on by the central control room or are capable of switching with priority into the programme;
● a programme transmitter unit equipped with one or more mono playback units (mostly cassette recorders or endless-loop recorders) and a radio tuner;
● in case of need, an *alarm system* that has to cover individual areas as well as the whole complex. It is switched into the loudspeaker system by means of an *override circuit* (see section 4.5.3), with priority over all other programmes and any volume regulators that may be inserted in the lines. The alarm equipment usually comprises storage units (cassette or digital memories) with various announcements (some of them as an endless loop). Alarm equipment is subject to special maintenance and supervision regulations. It is often equipped with batteries to provide an *emergency power supply*, which can sustain operation for a predetermined time – e.g. 15–30 minutes – after power failure.

The central mixer is usually interconnected via an output distributor to the power amplifiers and the linking paths to the satellite stations.

4.5.2 Satellite station equipment

The satellite station equipment is usually simple. It often consists of only one or two paging stations for the area concerned. In hotels and restaurants there are also cassette players for providing music programmes. The equipment may be somewhat more extensive if the coverage area of the satellite station includes a small hall or a music platform. In this case it is possible for the satellite station to put together its own programme, which is transferred to the central studio and distributed as a central programme.

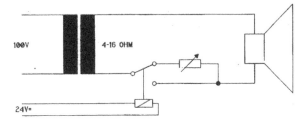

Figure 4.49 Structure of a compulsory override circuit.

To this end the satellite station must have its own mixing console or other mixing facility.

On the reproduction side the satellite stations are equipped with the power amplifiers and loudspeakers required for their coverage area.

4.5.3 Compulsory override circuits

In distributed sound reinforcement systems there are individually adjustable volume controllers and disconnection devices for the different areas. Thus the playback level can be adjusted for the actual attendance, and to cope with local extraneous noise. These circuits are often located on the power side of the amplifiers. To make important announcements or alarm signals fully audible under these conditions, the attenuation controllers must be bridged by means of a compulsory override circuit.

As can be seen in Figure 4.49, the loudspeaker is supplied in normal operation through a two-pole circuit. On the secondary side of the transformer a potentiometer makes it possible to reduce the volume or mute the loudspeaker.

When the compulsory override circuit is activated, the regulator on the loudspeaker side is bridged by means of a changeover relay. On the installation side this requires two additional wires in the supply line. As compared with former three-wire systems, the advantages of such a circuit are safe attenuation or muting without residual noise with normal operation, and the fact that the volume of the override signal is not affected by the actual position of the loudspeaker volume regulator.

4.6 Sound-processing equipment

Various sound-processing devices used in studio recording are also important for sound reinforcement engineering, where they are used for such things as improving intelligibility, reducing the risk of positive feedback, and adjusting or modifying sound balance.

4.6.1 Delay equipment

Frequently it is delay devices that make successful sound reinforcement possible by delaying electrical signals so that echo disturbances do not occur, even between widely spaced loudspeakers. A range of solutions for sound reinforcement problems made possible by means of the delay technique will be explained in Chapter 6; however, in this section we shall concentrate on the technical design of the devices.

The first delay devices prolonged the sound travel time by means of narrow tubes [4.50], but the resulting timbre change set narrow limits to transmission quality. Better quality was later achieved by means of *endless-loop recorders*, in which the spacing between the recording head and one or more replay heads is used for obtaining various delay times. It was possible to obtain practically any required delay time by suitable choice of tape speed and head spacing. Disadvantages were the relatively low signal-to-noise ratio, the abrasion of tape and heads, and in particular the breakdown of the whole device if the tape broke. To avoid this main drawback, special devices were developed for automatic changeover to a bypass or a standby unit.

Because of their high maintenance cost and poor reliability these devices were not widely used. They were used mainly in temporary installations for large events.

Electronic devices for sound signal delay were developed at the beginning of the 1970s. First came MOS flip-flop devices [4.51], and then *digital delay devices* [4.52]. The signal was digitized in the input section. Then, in the early digital devices, the digitized signal was fed into a clock-pulse-operated shift register. The delay time was determined by the retention time in this register. After recall, the signal was converted back to analogue and then transferred to the output. The clock frequency (sampling frequency) determined the smallest delay time. The longest delay time produced by such a device resulted from the clock frequency and the register length. Figure 4.50 shows the structure of such a device.

Modern studio-quality delay devices rely extensively on computational processing of the digitized signal. These devices permit automated control of the delay times as well as noise-free changeover during operation thanks to subdivision of the delay times into very small quantization steps. Figure 4.51 depicts the basic structure of such a delay device.

4.6.2 Effects devices

By bit completion or bit omission, and by changing the clock frequency, the procedures used in the digital delay technique may be employed for creating a wide range of audio effects.

Decreasing or increasing the clock frequency may produce a lower or a higher pitch respectively. According to [4.41] *pitch variations* of up to

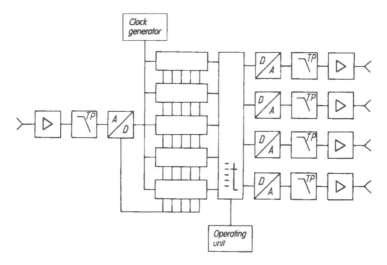

Figure 4.50 General circuit diagram of a delay unit with shift register.

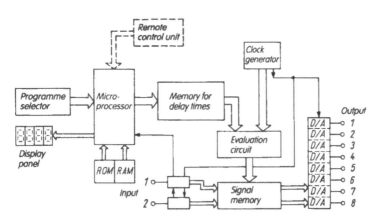

Figure 4.51 General circuit diagram of a delay unit with computerized signal
processing.

3 octaves (+1 to −2) can be achieved in this way. Figure 4.52 shows the
basic structure of such a device, called a *harmonizer*. As can be seen, it is
possible to mix the delayed and the undelayed signal in suitable
proportions. It is also possible to feed back a portion of the signal for
signal iterations and echo effects.

By means of such a device it is also possible to influence the frequency
and the time expansion of a sound signal independently from each other.

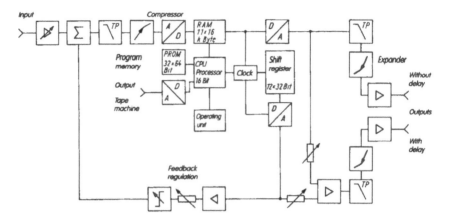

Figure 4.52 Schematic circuit diagram of a harmonizer (Eventide 949).

To do this, the signal is recorded and reproduced with different scanning frequencies. An important application of this procedure is *pitch stabilization* with sound storage devices that operate with variable speed. This is required for playback as accompaniment for live performances with varying speed.

Small pitch shifts were used in the so-called *feedback suppressors* for reducing the propensity to positive feedback in sound reinforcement systems [4.4]. Such devices are rarely used nowadays because of the resulting timbre changes.

However, these processing devices are used especially in electronic music or producing the *phasing, flanger* and *Leslie effects*.

For a *phasing effect* part of the signal is tapped off and routed through all-pass networks, where it is subjected to variable phase shifts. After recombination of the component signals non-harmonic nulls (cancellations) occur. The number of nulls depends on the number of all-pass networks [4.41]. The effect thus produced is a beating phase vibrato.

The *flanging effect* is achieved by varying the delay of one signal component relative to the undelayed component. This gives rise to comb-filter distortions: that is, nulls in the harmonic ratio. In practice the delay times amount to between 2 and 20 ms. With shorter delay times the effect diminishes. With this signal one also uses iterations – feedback and reprocessing of the composite signal – whereby the effect is further amplified. The audible result is a 'turning' or 'rotating' timbre change.

The *Leslie effect* is based on the Doppler effect, which may also be created by rotating loudspeakers.

All the mentioned effects can nowadays be realized by means of digital effect devices similar to those that have just been described.

4.6.3 Reverberation equipment

Unlike a delay device, which merely repeats a signal with delay but without changing the amplitude and phase, a reverberation device is expected to add to the original signal additional and delayed repetitions, which moreover are incoherent with the original signal and grow denser as well as increasingly attenuated over time. The duration of the decay process thus produced should if possible be variable. The same applies to the frequency dependence: a faster decay is normally expected for the higher frequencies than for the lower ones.

The length of the audible decay process is subjectively perceived as the *reverberation duration*. The time lapse required for a signal to decay from its original level to a value 60 dB lower is by definition called the reverberation time, even if the decay process has already been swamped by noise. Reverberation equipment should allow variation of the reverberation time of a signal in a room from 1 s to 5 s.

To avoid timbre changes, a minimum density and uniform distribution of the eigenfrequencies are required. The human ear is most sensitive to timbre changes between 800 Hz and 1 kHz. According to [4.15] there should be at least three eigenfrequencies per hertz bandwidth in this frequency range; at higher and lower frequencies the density may be less.

In sound reinforcement engineering, reverberation equipment is also needed for generating a so-called *reverberation tail*. Another application consists in adding reverberation to signals that have been picked up directly, as required in sound reinforcement to avoid positive feedbacks, so that they can be perceived in an aesthetically satisfactory fashion.

The various types of reverberation equipment are described below.

4.6.3.1 Reverberation rooms (echo chambers)

The first professional means used for artificial reverberation of sound signals were relatively small rooms with predominantly hard boundaries, into which the signals to be reverberated were introduced by loudspeakers and from which the reverberated signals were taken by microphones. To obtain a favourable distribution and sufficient density of the eigenfrequencies, the volume must be greater than 50 m^3, and the dimensions (length : height : width) must be in an optimum ratio within the so-called 'Bolt region' [4.53]. Smaller volumes produce an insufficient eigenfrequency density in the lower frequency range, but this effect can be mitigated by attenuating the low-frequency sound-channel range. Volumes of more than 200 m^3 are unsuitable because the reverberation time then becomes too dependent on temperature and air moisture.

One advantage of natural reverberation rooms is that it is relatively easy to obtain various incoherent sound signals: all that is required is for the microphones required for each channel to be spaced more than 2 m apart.

One drawback, apart from the inherent immobility, is the poor variability of the reverberation time, which can be altered only by means of different acoustic attenuation materials.

4.6.3.2 Two-dimensional reverberation equipment

While the density of eigenfrequencies increases in rooms by the third power of the frequency, it increases only by the square with two-dimensional reverberation devices, as represented by the reverberation plate [4.54] developed in the 1950s, and later on by the smaller reverberation foil [4.55]. This implies that, provided the values are equal in the medium frequency range, the resulting density of eigenfrequencies is higher in the lower frequency range and lower in the higher frequency range, than in rooms [4.15].

The *reverberation plate*, which is frequently still used today, is an elastically and stress-free suspended plate of high-quality steel about 2 m^2 square, which is excited by an electrodynamic system. It oscillates with its eigenfrequencies, and the vibrations are picked up by contact transducers. The reverberation time can be varied fairly easily by moving a sound-absorbing plate close and parallel to the oscillating plate: the smaller the distance, the more energy is taken away from the plate and the shorter becomes the reverberation time. With little damping, the plate has a longer reverberation time in the lower frequency range, which diminishes towards the higher frequencies. This is very similar to the behaviour of natural rooms. With increasing damping the frequency response becomes linearized (Figure 4.53).

To reduce the dimensions of the vibrating medium, the *reverberation foil* was developed. This is a 18 μm thick gold foil of about 300 mm × 300 mm,

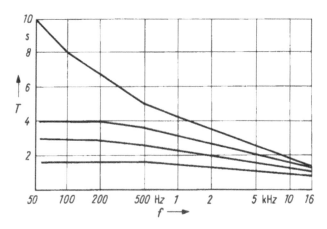

Figure 4.53 Frequency response of a reverberation plate with different amounts of damping.

which because of its sensitivity has to be well protected against outside influences. For this reason it is doubly encapsulated, and provided with multiple insulation against structure-borne sound. It may even be located in the sound control room adjacent to the monitoring loudspeakers, without positive acoustic feedback occurring.

Damping of the reverberation foil is effected as for the reverberation plate by the proximity of an absorbent layer. However, unlike the reverberation plate, only a slight increase in reverberation time is obtained in the lower frequency range compared with that at medium frequency. This effect is often wanted, because timbre changes are thus avoided in this region [4.55].

Incoherent output signals can be realized with the reverberation plate as well as the reverberation foil by means of various pick-ups arranged at correspondingly varying locations.

4.6.3.3 One-dimensional reverberation equipment

A reverberation device that was widely used for a long time consists of one or more *helical springs* excited by electrodynamic systems into performing torsional oscillations. A proportionable feedback loop provides the possibility of undamping the spring and thus prolonging the decay time. Given that the eigenfrequency density of the springs increases only linearly with the frequency, a sufficient eigenfrequency density can be obtained only by either exciting several springs of different length simultaneously or by providing the springs with arbitrary imperfections, which allow the resulting sections to oscillate independently of one another [4.56, 4.57]. In the interest of maintaining a natural timbre one has in both cases to take care that an integer ratio does not occur between the spring lengths or the sections between the imperfections. The metallic sound that would result is due to the comb-filter effect.

Because of material fatigue, eigenfrequency shifts may occur over time, and give rise to timbre variations of the reverberated signal. Incoherent output signals are realized by means of additional and simultaneously excited spring systems.

4.6.3.4 Digital reverberation equipment

As for delay and effects devices, the structure of digital reverberation devices is based on methods of digital signal delay. Here the signal is returned to the input of a delay line via a proportionately damped feedback loop. Given that a single feedback loop has only harmonic eigenfrequencies, like a one-dimensional reverberation element, to avoid comb-filter distortions it is necessary to combine various loops of different delay times. But since satisfactory results similar to natural reverberation cannot be achieved even by means of a larger number of such loops and by additionally inserting phase-rotating all-pass filters before the coupling

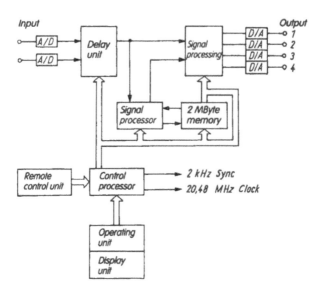

Figure 4.54 Schematic circuit diagram of a computer-aided digital reverberation unit (room simulator).

points in order to disrupt the coherence of the individual signal
components, an additional processor control has to be used to avoid a
repetition of similar amplitude and phase sequences over longer periods of
time. This requires a very fast-operating processor capable of redetermin-
ing constantly, according to control programs, the damping of the loops as
well as their coupling [4.58]. Frequently several incoherent outputs are
supplied.

More modern equipment (so-called *room simulators*) makes it possible
not only to influence the frequency response of the reverberation time, but
also to delay the reverberation and insert various differently damped *single
reflections* between the triggering sound signal and the reverberation
process (Figure 4.54). Some devices allow processor control not only of the
decay process, but also of the overall room simulation, coupled in many
cases with an effects device. They thus make possible effects that cannot be
created by means of natural rooms or mechanical reverberation devices.
These include extremely long and frequency-independent reverberation
times. So-called *freeze effects* allow constant maintenance of a sound with
superposition of newly added sounds. Moreover it is possible to create
effects such as decreasing echo loops and stereo phasing.

By choosing the appropriate software, modern computer-controlled
equipment, unlike the first fixed reverberation programmes, makes it
possible to achieve very natural and adaptable settings, which correspond to
all requirements.

4.6.4 Analogue processor for improving intelligibility

If one overmodulates a signal, then filters out from the totality of harmonics produced the harmonics corresponding to the fundamental wave of the original signal, and reintegrates them proportionately into the input signal, one obtains a composite signal that is essentially richer in harmonics and thus distinguishes itself definitely from the original signal as regards brilliance, clearness and intelligibility.

A processor operating along these lines was developed for the US Navy in order to improve speech intelligibility in noisy environments without having to increase the level of the useful signal [4.18]. As Figure 4.55 shows, the proportion of correctly understood words, compared with an untreated signal, could be noticeably increased in this way. The untreated signal required a level 6–8 dB higher than the treated signal.

Apart from the economic importance of this processor (the same degree of intelligibility can be obtained at a considerably lower power), it also offers a host of acoustical advantages. One of these consists in a reduction of the problems of positive acoustic feedback in sound reinforcement systems. Moreover it makes it possible to improve considerably the quality of old recordings whose brilliance has faded owing to an obsolete recording technique or abrasion of the magnetic tapes. This effect could not be achieved by means of a filtering process.

Such devices, which are offered by the Aphex Company under the name of *Aural-exciter*, have frequently been used in sound reinforcement systems as

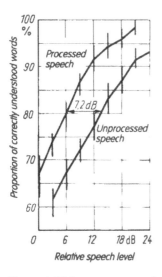

Figure 4.55 Improvement of intelligibility with high background noise by increasing the harmonic content of the signal. Background noise: noise band limited to between 100 and 500 Hz.

well as in recording and radio studios. Variants of the device differ mainly with respect to the technique used for damping the harmonics [4.59].

4.6.5 Filters

Filters for influencing the amplitude–frequency characteristic of transmitted sound signals are among the classical means of sound processing used in sound reinforcement engineering. Two main fields of application have to be distinguished:

- optimization of timbre within the reception area concerned;
- suppression of positive acoustic feedback frequencies.

Optimization of timbre depends on the particular field of application of the system. For example, a balanced frequency response over the whole frequency spectrum may be desirable for high-quality systems designed for music transmissions (see Chapter 7). However, to improve intelligibility in speech-only transmission systems, a reduction in the lower frequency range and an enhancement of certain formats in the range of about 2 kHz is appropriate [4.60]. Quite different requirements may apply when optimizing the timbre of stage monitoring. In larger sound reinforcement systems filters are used at various points. For controlling the microphone frequency response, and usually also for suppressing the main positive feedback frequencies, they are normally located in the input channel of the mixing console. However, the basic adjustment of the loudspeaker frequency response needed to compensate for the linear distortions caused by the installation and by the acoustics of the auditorium is carried out using so-called *room adaptation filters* connected before the power amplifiers.

The use of filters is nearly always at the expense of the maximum realizable sound level. It is therefore necessary to consider corresponding power reserves when designing the system.

There are two basic types of filter: passive filters and active filters. *Passive filters* operate without an additional power supply. They thus do not offer any possibility of amplification (level enhancement), nor any reduction of the signal-to-noise ratio. *Active filters*, which are more widely applicable, smaller, and cheaper, are nowadays used almost exclusively in studio equipment.

Another distinguishing characteristic of filters is the influence of damping on the behaviour of the filter curve, and here we can distinguish between filters of constant q and filters of constant bandwidth (Figure 4.56) [4.61].

For reasons of expenditure, available equipment and ease of operation a range of different practical designs are used. The most important ones used in sound reinforcement engineering will be briefly described below.

Under the name of *feedback controller*, microprocessor units have recently become available that trigger frequency-dependent attenuation in

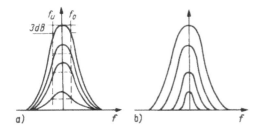

Figure 4.56 Attenuation behaviour of filters of (a) constant bandwidth and (b) constant *q*.

response to incipient positive feedback phenomena such as timbre changes, reverberation effects, and fluctuations in level. These automatic filters are assigned to the individual microphone channels, and are capable of enhancing the positive feedback limit of a system by up to 15 dB [4.62].

4.6.5.1 Treble and bass correctors (shelf filters)

To raise or lower the frequency response in the lower and/or upper section of the transmission channel, relatively simple RC or RL combinations are used, together with a resistor-type voltage divider. They are often directly integrated in the power amplifiers or used in simple mixing consoles. Figure 4.57 shows that their slope steepness depends on the actual setting. The maximum achievable level enhancement depends on the basic attenuation, which for a maximum enhancement of the higher and lower frequencies of 15 dB must be about 34 dB.

Filters of this kind offer only limited possibilities for frequency response correction, since they are not sufficiently adaptable to suppress, for example, frequency response peaks or individual positive feedback frequencies.

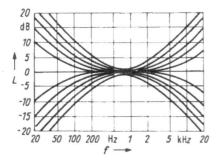

Figure 4.57 Attenuation curves of a shelf filter.

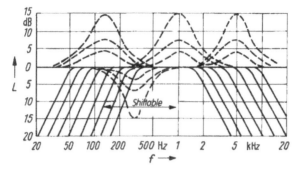

Figure 4.58 Attenuation curves of a multiple parametric filter.

4.6.5.2 Parametric filters (channel filters)

The filters used in larger mixing consoles and assigned to the individual channels apart from providing boost and cut of the treble and bass frequencies also provide narrow band boost and cut filtering of the medium frequency range (typically around 1 kHz). In many cases the filter is also tunable over a given frequency range. The steepness of the filter slope is generally related to the degree of cut or boost applied (Figure 4.58).

Given a sufficiently steep slope, the cut filters permit the effective suppression of individual positive feedback frequencies without essentially impairing the timbre, since narrow-band notches are only rarely noticeable in the fundamental-tone range, thanks to the masking effect. Wider presence adjustments are suitable for enhancing certain harmonics (e.g. around 2 kHz, see [4.60]), which improve considerably the intelligibility of unaccentuated speech.

4.6.5.3 Muliti-bandpass filters (equalizers)

By lining up a larger number (7 to 30) of individual bandpass filters (width usually 1, $\frac{1}{2}$ or $\frac{1}{3}$ octave, or less frequently different relative bandwidth) allowing variable attenuation, it is possible to form almost any desired linear transmission function. By arranging the control elements (sliding faders) appropriately it is possible to obtain a visual representation of the adjustment of the transmission function. Multi-bandpass filters of this design are for this reason also called *graphic equalizers.*

By attenuating individual bandpass elements it is possible to suppress successively several positive feedback frequencies: the narrower the individual filters and the steeper their slopes, the less the timbre of the transmitted signal will be affected. Moreover, the filters are well suited for equalizing room or loudspeaker timbre changes. This normally requires various adjacent bandpasses with different attenuation.

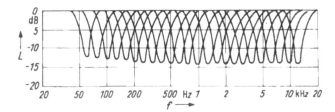

Figure 4.59 Filtering curves of a multi-bandpass filter with attenuation function only.

Multi-bandpass filters are also produced in attenuation-only form, as Figure 4.59 shows by means of an example. Such filters have proved to be widely 'foolproof', since neither overmodulation nor frequency peak emphases with associated heavy masking effects can be caused by misadjustment.

A special type of multi-bandpass filter is the *transverse equalizer*, in which the actual frequency response is synthesized after a Fourier transform of the input signal in the frequency domain. Thereafter the signal is retransformed. In this way it is possible to obtain smooth and linear phase frequency responses [4.63].

4.6.5.4 *Preset filters*

Apart from the variable filters, which are of great importance for sound reinforcement engineering, simple preset equalizers are also used for pressure-increase equalization, for bass de-emphasis of microphones, or as footfall-sound filters.

References

4.1 DIN 45570 *Lautsprecher; Begriffe, Formelzeichen, Einheiten* (Loudspeakers; Concepts, formula symbols, units) (November 1979).
4.2 DIN 45573 Part 2, replaced by IEC 268, Part 5.
4.3 *Sound system equipment Part 5: Loudspeaker, simulated program signal* IEC Publication 268–5 (1972).
4.4 Ahnert, W. and Reichardt, W., *Grundlagen der Beschallungstechnik* (Foundations of sound reinforcement engineering) (Berlin: Verlag Technik, 1981).
4.5 Zwicker, E. and Zollner, M., *Elektroakustik* (Electroacoustics), 2nd edn (Berlin: Springer, 1987).
4.6 Wöhle, W. In: *Taschenbuch Akustik 'Schallabstrahlung'* (Sound irradiation). (Vade mecum of acoustics), (Berlin: Verlag Technik, 1984), pp 32–53.
4.7 *Technische Parameter von Studio-Abhöreinrichtungen* (Technical parameters of studio monitoring equipment) OIRT Recommendation 55/1 (Matinkylä, 1985).

4.8 Keibs, L., 'Die physikalischen Bedingungen für optimale Baß-reflexgehäuse' (The physical conditions for optimum bass reflex boxes), *Technische Mitteilungen BRF* (1957), 11–19.

4.9 Thiele, A.N., 'Loudspeakers in vented boxes', *Journal of the Audio Engineering Society,* **19** (1971) 5, 382–392; 6, 471–483.

4.10 Small, R.H., 'Vented-box loudspeaker systems', *Journal of the Audio Engineering Society,* **21** (1973): 'Part I, Small-signal analysis', 5, 363–372; 'Part II, Large-signal analysis', 6, 438–444; 'Part III, Synthesis', 7, 549–554; 'Part IV, Appendices', 8, 635–639.

4.11 Kammerer, E., 'Lautsprechergruppen in Zeilenform ohne Höhenverluste und weitgehend frequenzunabhängiger Richtcharakteristik' (Horizontal loudspeaker lines without treble losses and with largely frequency-independent directivity ratio), *Siemens Zeitscrift,* **42** (1968) 2, 131–134.

4.12 Möser, M., 'Amplituden- und phasengesteuerte akustische Sendezeilen mit gleichmäßiger horizontaler Richtwirkung' (Amplitude and phase controlled acoustic loudspeaker lines with uniform horizontal directivity), *Acustica,* **60** (1987) 2, 91–104.

4.13 Orth, D., 'Tonsäulen für Innen- und Außenbeschallung' (Sound columns for indoor and outdoor sound reinforcement), *Technische Mitteilungen RFZ,* **20** (1976) 4, 73–77.

4.14 Kuttruff, H. and Fasbender, J., 'Zur Beschallung von Sälen mit laufzeit-kompensierten Lautsprecherzeilen' (On the sound coverage of halls by means of travel-time compensated loudspeaker lines), *Acustica,* **54** (1983) 1, 39–40.

4.15 Kuhl, W., 'Notwendige Eigenfrequenzdichte zur Vermeidung von Klang-färbung von Nachhall' (Eigenfrequency density required for avoiding timbre changes in reverberation), International Congress on Acoustics, Tokyo (1968), Paper K-2-8.

4.16 *Sound in action; professional music '88,* Company information, Electro-Voice, Buchanan, MI, USA.

4.17 Chaix, P., "Le Zenith" Parc de la Villette Paris', *Architektur, Innenarchitektur* (Architecture, interior design), *Tech. Ausbau* (1986) 1/2, 16–19.

4.18 Davis, D. and Davis, C., *Sound System Engineering,* 2nd edn (Indianopolis: Howard W. Sams, 1987).

4.19 Meyer, E. and Kuttruff, H., 'Zur Raumakustik einer großen Festhalle' (On the room acoustics of a large festival hall), *Acustica,* **14** (1964), 138.

4.20 Steffen, F., 'Akustische Probleme bei der elektroakustischen Beschallungsan-lage des Großen Saales des PdR' (Acoustic problems with the electroacoustical sound reinforcement system of the great hall of the Palast der Republik), *Technische Mitteilungen RFZ,* **21** (1977) 3, 51–55.

4.21 Orth, D., 'Tonsäule mit veränderbarer Richtcharakteristik' (Sound column with variable directivity ratio), *Technische Mitteilungen RFZ,* **29** (1985) 2, 33–36.

4.22 CCIR Rec 468–4, *Measurement of audio frequency noise in broadcasting, in sound-recording systems and on sound programme circuits,* CCIR Recommendation 468–4 (Dubrovnik, 1986).

4.23 *Geräuschspannungsmessung in Tonkanälen für Rundfunk und Fernsehen* (Psofometric voltage measurement in sound channels for radio and television), OIRT Recommendation 71 (Kazanlyk 1986).

4.24 *Sound-system equipment, Part 4, Microphones.* IEC Publication 268–1 (Geneve, 1972).

4.25 *Specification for precision sound level-meters,* IEC 179 (1973).

4.26 Griese, H.-J., 'Ein neues Fernsehmikrofon' (A new type of television microphone), *Internationale Elektronische Rundschau,* (1965) 2, 68–70.

4.27 Werner, E., 'Kondensatormikrofon mit Elektretmembran' (Condenser microphone with electret diaphragm), *Funkschau* (1973) 287 ff.

4.28 Przybilla, T. and Schmidt, W., 'Erfahrungsbericht über den Einsatz von herkömmlichen Studiomikrofonen als Boden- oder Grenzflächenmikrofone' (Report on experiences gathered with using conventional studio microphones as floor or boundary microphones), 13th Tonmeistertag (München: Bildungswerk des Verdandes deutscher Tonmeister, (1984), pp. 208–221.

4.29 Lipschitz, S.P. and Vandercoy, J., 'The acoustical behaviour of pressure responding microphones position on rigid boundaries – a review and critique', 71st Convention of the Audio Engineering Society, Los Angeles (1981), preprint no. 1976.

4.30 Ahnert, W. and Schmidt, W., *Akustik in Kulturbauten* (Acoustics in cultural buildings) (Berlin: Instut für Kulturbauten, 1987), p. 53.

4.31 *AMS microphones with directional control of output*, Company information, Shure, Evanston, IL (1982).

4.32 Veit, I., Lips, H. and Mayer-Fasold, S., *Beschallungstechnik* (Sound reinforcement engineering) (Sindelfingen: Expert Verlag 1986), pp. 114–121.

4.33 *Infrarottechnik in Theorie und Praxis* (Infrared technique in theory and practice), Company information, Sennheiser (Wedemark, 1980).

4.34 *Anordnung für ein audiovisuelles Informationssystem über Ausstellungsobjekte* (Set-up for an audiovisual information system on exhibition objects), German patent no. DD 258697A1.

4.35 Kaminski, P., 'Rauschunterdrückungssystem Dolby SR' (The noise suppression system Dolby SR), *dB Magazin für Studiotechnik* (1988) März/Apr., 46.

4.36 Wermuth, J., 'Dynamikerweiterung durch neuartige Studiokompander' (Dynamic range extension by novel studio companders), *Funkschau* (1975) 18, 571.

4.37 Mühlstädt, G., 'Über das Rauschen von Magnettonbändern' (On the noise of magnetic sound tapes), *Funkschau*, 43 (1971), 215 ff.

4.38 Thewes, M. and Eller, W., 'R-DAT ein neuer Digital-Standard?' (R-DAT a new digital standard?), *dB Magazin für Studiotechnik* (1986) Jul/Aug, 50–53.

4.39 Baggen, C.P.M.I., 'Compakt Disk: Grundlagen und Systeme' (Compact disc: foundations and systems), *Tagungs-Bericht DAGA*, Darmstadt (1984), 19–29.

4.40 Redlich, H. and Joschko, G., 'Direct Metal Mastering Technologie Ein Schritt zur rationellen Herstellung der Compakt Disc' (Direct metal mastering technology – a step towards efficient production of the compact disc), *dB Magazin für Studiotechnik* (1987) März/Apr., 17–24.

4.41 Webers, J., *Tonstudiotechnik* (Sound studio equipment), 4th edn (München: Franzis, 1985), p. 278.

4.42 Herrman, U.F., *Handbuch der Elektroakustik* (Handbook of electroacoustics), 2nd edn (Heidelberg: Hüthig, 1983).

4.43 Boye, G., 'Freiluft-Beschallungsanlagen für Großveranstaltungen und Sportstadien' (Open-air sound reinforcement systems for mass events and sports stadiums), *Fernseh- und Kino-Technik*, 33 (1977) 1, 13–17.

4.44 Müller, M., 'PC-gesteuerte Großbeschallungsanlage von Stage Accompany' (PC-controlled large sound reinforcement system by Stage Accompany), *dB Magazin für Studiotechnik* (1988) Jul/Aug., 33–34.

4.45 Steffen, F., Fels, P. and Schwarzinger, W., 'Die tontechnischen Einrichtungen der Semperoper in Dresden' (The sound-technical facilities of the Semperoper in Dresden), *Technische Mitteilungen RFZ* (1986) 3, 49–56.

4.46 Keller, W. and Widman, M, 'Kommunikationssysteme im ICC Berlin' (Communication systems in the ICC Berlin), *Funkschau*, 51 (1979) 15, 858–864.

4.47 Steffen, F., 'Die elektroakustischen Einrichtungen im Neuen Gewandhaus Leipzig' (The electroacoustical facilities in the Neues Gewandhaus Leipzig), *Technische Mitteilungen RFZ* (1983) 2, 19–20.

4.48 Hoeg, W. *et. al.*, 'Ein Schallübertragungssystem zur richtungs-und entfernungsgetreuen Beschallung großer Auditorien' (A sound transmission system for the direction-true and distance-true sound coverage of large auditoriums), Lecture, 6th Acoustics Convention, Budapest (1976); also *Technische Mitteilungen RFZ*, 20 (1976) 25–27.

4.49 Müller, R. and Pennewitz, V., 'Computersteuerung für Showeffekte' (Computer control for show effects), *Kulturbauten* (1986) 1, 10–20.

4.50 Kammerer, E., *Technische Elektroakustik* (Technical electroacoustics) (Berlin: Siemens, 1975), pp. 118–120.

4.51 Hollmann, J. and Burth, A.D., 'Verzögerung von NF-Signalen mit MOS-Eimerketten' (Delay of LF signals by means of MOS bucket brigade devices), *Funkschau*, 47 (1973), 967–1009.

4.52 Frese, S., 'Konzeption eines digitalen Zeitverzögerungsgerätes' (Conception of a digital time delay device), *Fernseh- und Kino-Technik*, 33 (1979) 8, 293–295.

4.53 Bolt, R.H., 'Influence of room proportions on normal frequency spacing', *Journal of the Acoustical Society of America*, 17 (1945) 1, 101.

4.54 Kuhl, W., 'Über die akustischen und technischen Eigenschaften der Nachhallplatte' (On the acoustical and technical properties of the reverberation plate), *Rundfunktechnische Mitteilungen* (1958) 3, 111.

4.55 Kuhl, W., 'Eine kleine Nachhallplatte ohne Klangfärbung' (A small reverberation plate without timbre change), *Proceedings of the Seventh International Congress on Acoustics*, Budapest (1971).

4.56 Fidi, W., 'Erfahrungen mit dem Studionachhallgerät "BX 20"') (Experiences with the studio reverberation unit "BX 20"), *Fernseh- und Kino-Technik*, 27 (1973) 9, 342 ff.

4.57 Indlin, Ju.A., 'Issledovanie i technologiceskaya razrabotka pruzinych reverberatorov' (Investigation and technological elaboration of spring-type reverberation devices), Dissertation, Moscow State University (1967).

4.58 Bäder, K.O. and Blesser, B., 'Digitaltechnik im Studio, ein elektronisches Nachhallgerät' (Digital equipment in the studio, an electronic reverberation unit), *Fernseh- und Kino-Technik*, 31 (1977) 12, 443–445.

4.59 'Aphex C Aural Exciter billiger und besser' (Aphex C Aural Eciter – cheaper and better), *Studio* (1986) 9, 26–27.

4.60 Tool, F.E., 'Loudspeaker measurements and their relationship to listener preferences', *Journal of the Audio Engineering Society*, 34 (1986): 4, 227–238; 5, 323–348.

4.61 Bohn, D.A., 'Operator-adjustable equalizers: an overview', 6th AES Conference on Sound Reinforcement, Nashville (1988), Paper 8.B.

4.62 'Feedback Controller FC 100', *Studio Magazin* April (1990) 135, 20.

4.63 'Equalizer networks and methods of developing scaling coefficients therefrom', US Patent no. 4,566,119 (21 January 1986).

5 Calculations

As already described in section 2.1, any sound reinforcement system is nowadays confronted by many general demands, which vary according to the planned use. However, certain basic requirements have to be met in all cases, as follows.

The *sound level* produced by the system within the audience area of the hall or in the open air must be adequate. The following quantities are relevant here:

- loudness (expected natural level, adaptation to original sources);
- sound energy density (efficiency of the sound reinforcement system);
- ratio between disturbing sound and useful sound, etc. (signal-to-noise ratio, dynamic range of reproduction).

The *sound level distribution*, which makes it possible to assess the spatial distribution of the loudness, must be sufficiently uniform. It depends on:

- the arrangement of the loudspeakers;
- the directivity characteristics of the loudspeakers;
- the diffusivity of the room in which the reproduction takes place.

The *clarity* of reproduction must be up to the planned use. It depends on:

- definition, clarity (D/R ratio);
- masking of the reproduction signal;
- freedom from echoes.

It is necessary to obtain a sufficiently natural transmission. Contributing factors in this respect are:

- the timbre, which depends on the transmission frequency range and the frequency response of the signal transmitted;
- the frequency response;
- freedom from distortions.

Sound reinforcement systems must be sufficiently insensitive to *positive acoustic feedback*. This implies constraints that include:

- the level of loop amplification;
- the sound level conditions around the microphone;
- the directivity characteristics of the microphones and loudspeakers.

The physical bases and resulting technical conclusions are briefly expounded below.

5.1 Important acoustic parameters

5.1.1 *Sound energy density: sound level*

According to [5.1], the *acoustic power* of a sound source (such as a loudspeaker) in the free sound field is

$$P_{ak} = \int J_d \, dS'$$

Since sound intensity results from the squared sound pressure related to the acoustic wave resistance, $J_d = p^2/\rho c$, if we take account of directional radiation by means of the *angular directivity ratio*, $\Gamma(\vartheta) = p(\vartheta)/p_0$ (cf. equation (4.12)), we obtain:

$$P_{ak} = \frac{p_0^2}{\rho c} \int \Gamma(\vartheta) dS' = \frac{p_0^2}{\rho c} \frac{4\pi r^2}{\gamma_L}$$

(cf. [5.2], pp. 120/121).

After further transformations, and taking into account the fact that $J_d = w_d c$, we obtain the *sound energy density*:

$$w_d = \frac{P_{ak}}{c} \frac{\gamma_L \Gamma_L^2(\vartheta)}{4\pi r^2} \tag{5.1}$$

where the subscript d indicates direct source.

If the loudspeaker (or source in general) radiates into a solid angle other than $\Omega = 4\pi$, one ought in equation (5.1) to replace 4π by, for instance, 2π (hemisphere), π (quarter sphere), etc. For spherical loudspeakers ($\gamma_L = 1$), equation (5.1) simplifies to

$$w_d = \frac{P_{ak}}{c} \frac{1}{4\pi r^2} \tag{5.1a}$$

In this case the sound pressure is

$$\tilde{p}_d = \sqrt{\left(\rho c P_{ak} \frac{1}{4\pi r^2}\right)} \tag{5.1b}$$

Thus the sound energy density decreases proportionally to the square of the distance and the sound pressure decreases proportionally to the distance.

In level notation one obtains from equation (5.1) the *sound pressure level*:

$$L_d = 109 \text{ dB} + 10 \log P_{ak} \text{ dB} + 10 \log \gamma \text{ dB} + 20 \log \Gamma(\vartheta) \text{ dB}$$
$$- 20 \log r \text{ dB} \tag{5.2}$$
$$L_d = L_W + 10 \log \gamma \text{ dB} + 20 \log \Gamma(\vartheta) \text{ dB} - 20 \log r \text{ dB} - 11 \text{ dB}$$

where P_{ak} is in W, and r is in m (cf. equation (5.20) with the inclusion of equation (5.21)). (The sound power level, $L_W = 10 \log(P_{ak}/P_0)$ dB; $P_0 = 10^{-12}$ W.)

For a spherical loudspeaker one obtains:

$$L_d = 109 \text{ dB} + 10 \log P_{ak} \text{ dB} - 20 \log r \text{ dB}$$
$$L_d = L_W - 20 \log r \text{ dB} - 11 \text{ dB} \tag{5.2a}$$

We can see that the direct sound level decreases by 6 dB when the distance is doubled (the so-called $1/r$ law).

According to equation (3.3) the sound energy density in the *diffuse* field of a room having the equivalent absorption area A is:

$$w_r = \frac{4P_{ak}}{cA} \tag{5.3}$$

where the subscript r indicates reverberance, enhanced spaciousness, and spatial impression.

In terms of levels:

$$L_r = 126 \text{ dB} + 10 \log P_{ak} \text{ dB} - 10 \log A \text{ dB}$$
$$= L_W - 10 \log A \text{ dB} + 6 \text{ dB} \tag{5.4}$$

where P_{ak} is in W, and A is in m^2 (see also equation (5.24a)).

If the loudspeakers are directed at highly absorptive areas that are not uniformly distributed in the room, such as the audience area in a hall, much

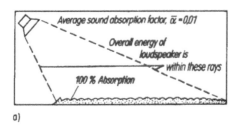

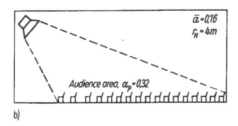

Figure 5.1 Explanation of the equivalent front-to-random factor, γ_{PL}. (a) Borderline case: $\gamma_{PL} = (1 - 0.01)/(1 - 1) = \infty$.   (b) Practical case: $\gamma_{PL} = (1 - 0.16)/(1 - 0.32) = 1.24$. $r_R = \sqrt{\gamma_p} \times r_H \approx \sqrt{1.24} \times 4\text{ m} \approx 4.5$ m

of the radiated energy is absorbed immediately, and in the steady-state conditions the result is a lower sound energy density than for omni-directional and uniform diffuse reflection in the room. This fact can be expressed by the *effective front-to-random factor*, γ_{PL} [5.3, 5.4]:

$$\gamma_{PL} = \frac{1 - \overline{\alpha}}{1 - \alpha_p} \tag{5.5}$$

where $\overline{\alpha}$ is the mean sound absorption coefficient of the room, and α_p is the sound absorption coefficient of the audience area.

Figure 5.1 explains the relationship. Under these conditions, the sound energy density in the diffuse field is, according to equation (5.3):

$$w_r = \frac{4P_{ak}}{cA\gamma_{PL}} \tag{5.6}$$

The diffuse sound level according to equation (5.4) is thus reduced to the value of

$$L_r = L_W + 6\text{ dB} - 10\log A\text{ dB} - 10\log \gamma_{PL}\text{ dB} \tag{5.6a}$$

where A is in m^2.

5.1.2 Critical distance and equivalent acoustic distance

An important quantity in sound reinforcement systems is the distance from the loudspeaker at which the direct sound and the diffuse sound show the same energy density, $w_d = w_r$ (or, for a microphone, the distance at which the two sound-field components produce the same electric voltage). With directional transducers this distance is called the *critical distance*, r_R, whereas with omnidirectional transducers it is called *reverberation radius*, r_H. By setting equations (5.1) and (5.3) equal, we obtain for this distance

$$r_R = \Gamma_L(\vartheta)\sqrt{(A/16\pi)}\sqrt{\gamma_L}$$
$$\approx \Gamma_L(\vartheta)\sqrt{\gamma_L}\sqrt{(A/50)} \tag{5.7}$$

(For microphones we obtain analogously $r_R = \Gamma_M(\vartheta)\sqrt{\gamma_M}\sqrt{(A/50)}$.)

For microphones with an omnidirectional characteristic the reverberation radius is thus the boundary between the free-field and the diffuse-field influence:

$$r_H = \sqrt{(A/16\pi)}$$
$$\approx \sqrt{(A/50)} \tag{5.8}$$

Consequently, this amount is a room constant (cf. equation (3.5)), and so can also be written in the form

$$r_R = \Gamma_L(\vartheta)\sqrt{\gamma_L}r_H = r_H^* \tag{5.8a}$$

This means that because of the directional effect, the reverberation radius is increased by the root of the directivity factor (cf. equation (4.14)). The *angle-dependent directivity factor*, Q, is thus in general terms

$$Q(\vartheta) = (r_R/r_H)^2 = (r_H^*/r_H)^2 = \gamma * \Gamma^2(\vartheta) \qquad \text{(cf. equation (4.14))}$$

If the source (the loudspeaker) radiates directly into the audience, the sound energy density, w_r, is determined by means of equation (5.6). The resulting critical distance is thus

$$r_R = \Gamma_L(\vartheta)\sqrt{(\gamma_L\gamma_{PL})}\sqrt{(A/50)} \tag{5.9}$$

The effective front-to-random factor of the loudspeaker used is thus apparently increased by γ_{PL} through the direct sound coverage of the absorbing area.

If in a room there are n loudspeakers of the same type operating and all exciting the diffuse sound field uniformly, but only one of which is directed to the receiving position concerned, after several transformations and considering the fact that

$$P_{\text{acou, total}} = \sum_n P_i$$

has to be entered for the diffuse-field excitation, we obtain

$$r_R = \Gamma_L(\vartheta)\sqrt{\gamma_L} \sqrt{\frac{P_L}{\sum_n P_i}} \cdot r_H \tag{5.10}$$

If the radiated powers $P_{\text{ak}} = P_L = P_i$ of the loudspeakers are equal, the result is

$$r_R = \Gamma_L(\vartheta)\sqrt{\gamma_L/n} \cdot r_H \tag{5.10a}$$

That is, because of the number of loudspeakers the effective critical distance is diminished by the factor $1/\sqrt{n}$. However, this holds true only if the wavefronts of the individual loudspeakers incident at the listener's position are not pushed together by a corresponding arrangement or a delay system in such a way that they arrive within about 30 ms. In this special case all loudspeakers contribute to the enhancement of definition and may in this way also be 'direct-sound supporting'.

If there is a microphone in the hall for sound reinforcement, its directional characteristics have to be considered when ascertaining the positive feedback conditions. By means of equations (5.1) and (5.3) we obtain the sound energy density (cf. [5.2], pp. 121–122), from which it is possible to calculate the *coupling critical distance*, r_{LMR}:

$$r_{\text{LMR}} = \Gamma_M(\vartheta_M)\Gamma_L\sqrt{(\gamma_M\gamma_L)} \cdot r_H \tag{5.11}$$

(See also Figure 5.3.)

The *coupling directivity value*, $Q_{\text{LM}}(\vartheta_L, \vartheta_M)$, is formed analogously to the directivity factor:

$$Q_{\text{LM}}(\vartheta_L, \vartheta_M) = \frac{r_{\text{LMR}}^2}{r_H^2} = \gamma_L\Gamma_L^2(\vartheta_L)\gamma_M\Gamma_M^2(\vartheta_M) \tag{5.12}$$

This factor should be made as small as possible to avoid positive feedback from the loudspeaker to the microphone. However, the *coupling directivity factor* between the original source and the microphone,

$$Q_{\text{SM}}(\vartheta_S, \vartheta_M) = \gamma_M\Gamma_M^2(\vartheta_S)\gamma_S\Gamma_S^2(\vartheta_{SM}) \tag{5.12a}$$

must be as large as possible to guarantee good sound transmission from source to microphone. For example, if the microphone is directed towards a

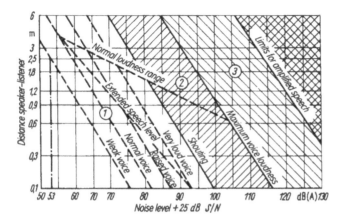

Figure 5.2 Determination of the equivalent distance, r_a: 1, direct communication possible with normal level; 2, direct communication difficult; 3, direct communication impossible.

speaker so that it is directly excited by it, the two directivity ratios $\Gamma_M(\vartheta_S)$ and $\Gamma_S(\vartheta_{SM})$ assume the maximum value 1. The coupling factor then corresponds to the product of the two front-to-random factors $\gamma_M \gamma_S > 1$ (in practice the values are between 10 and 20).

If a sound reinforcement system is installed in a room, it is necessary to check what level of maximum amplification can be achieved without causing positive acoustic feedback. A source lacking electroacoustic amplification can be perceived distinctly and intelligibly only up to the *equivalent acoustic distance*, r_a, which depends on the noise level and on the sound-power level of the source (Figure 5.2). Moreover it is taken for granted that a sound level L_{ra} is required for achieving audibility [5.4].

Apart from the critical distance and the reverberation radius, the *equivalent acoustic distance*, r_a, serves for assessing the so-called *necessary amplification* when using sound reinforcement systems (see section 5.4).

An example will help in understanding Figure 5.2.

> If the prevailing noise level in a room is 35 dB(A), then a desired signal-to-noise ratio of 25 dB requires a sound level of $L_{ra} = (35 + 25)$ dB(A) $= 60$ dB(A). For a sound source with enhanced loudness, an equivalent acoustic distance of $r_a \approx 4$ m should not then be exceeded.

5.1.3 Sound energy density: sound level in the room

In the sound field of a room the sound pressure components stemming from the direct field and the diffuse field are additive, producing the *resulting energy density:*

$$w = w_d + w_r \tag{5.13}$$

According to equations (5.1) and (5.3):

$$w = \frac{P_{ak}}{c} \left(\frac{\gamma_L \Gamma_L^2(\vartheta)}{4\pi r^2} + \frac{4}{A} \right)$$

In level notation this becomes

$$L = L_W + 10 \log \left(\frac{\gamma_L \Gamma_L^2(\vartheta)}{4\pi r^2} + \frac{4}{A} \right) \text{ dB} \tag{5.14}$$

where L_W is the sound power level of the loudspeaker.

If we consider the reduction of the radiated energy when directing the loudspeakers to an absorbing audience area, a fact that has to be taken into account when applying equation (5.6), equation (5.14) changes to

$$L = L_W + 10 \log \left(\frac{\gamma_L \Gamma_L^2(\vartheta)}{4\pi r^2} + \frac{4}{\gamma_{PL} A} \right) \text{ dB} \tag{5.14a}$$

The sound energy density at a listener's position H at a distance r_{LH} from the exciting loudspeaker L in a room is calculated from

$$w_{HL} = \frac{P_L}{c} \left(\frac{\gamma_L \Gamma_L^2(\vartheta_H)}{4\pi r_{LH}^2} + \frac{4}{A} \right) \tag{5.15}$$

or, in level notation,

$$L_{HL} = L_{WL} + 10 \log \left(\frac{\gamma_L \Gamma_L^2(\vartheta_H)}{4\pi r_{LH}^2} + \frac{4}{A} \right) \text{ dB} \tag{5.15a}$$

Corresponding relationships can also be established for the energy density at the listener's position when produced by an original source:

$$w_{HO} = \frac{P_S}{c} \left(\frac{\gamma_S \Gamma_S^2(\vartheta_{SH})}{4\pi r_{SH}^2} + \frac{4}{A} \right) \tag{5.16}$$

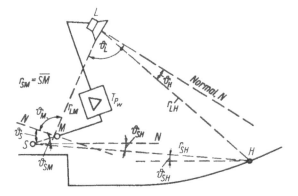

Figure 5.3 Use of a sound reinforcement system chain. Position of source S,
microphone M, loudspeaker L and listener H plus associated angles.

By taking the directional properties of the microphone into account it is
possible to calculate for microphone location M fictive sound energy
densities for different sources:

(a) arriving from the loudspeaker L:

$$w_{ML} = \frac{P_L}{c} \left(\frac{\gamma_L \Gamma_L^2(\vartheta_L)\Gamma_M^2(\vartheta_M)}{4\pi r_{LM}^2} + \frac{4}{\gamma_M A} \right) \tag{5.17}$$

(b) arriving from the original source S:

$$w_{MS} = \frac{P_S}{c} \left(\frac{\gamma_S \Gamma_S^2(\vartheta_{SM})\Gamma_M^2(\vartheta_S)}{4\pi r_{SM}^2} + \frac{4}{\gamma_M A} \right) \tag{5.18}$$

These equations are starting points for the considerations in section 5.4.
The angular relationships to be considered in this respect may be gathered
from Figure 5.3.

5.2 Sound level and dynamic range of the radiated signal

The initial considerations for the design and specification of sound
reinforcement systems require the following information to be ascertained:

- the noise level at or affecting the listener;
- the anticipated sound level;
- the required signal-to-noise ratio;

- the power of the sources operating in conjunction with the system;
- the audio power to be installed to obtain the required sound level at the listener's position.

5.2.1 Required sound level

Leaving aside considerations of the level of extraneous sound, the sound level required for a system depends on the listener's expectations. For speech this level is about 70–75 dB, which can be easily achieved at short distances from a speaker. For music the value is considerably higher, because of the wider dynamic range. Values of 95 dB are rarely exceeded in the reverberant field of a hall even by large symphonic orchestras playing in fortissimo, but with pop music values of 105–110 dB(A) have been measured, even at larger distances from the stage.

5.2.2 Useful level; noise level

When specifying a sound reinforcement system, we have to start from the noise level that has to be overcome. In the open (that is, without the surround sound produced by reverberation) the average sound level achievable at the listener's seat must be at least 10 dB above the noise level in order to obtain sufficient intelligibility. (Strictly speaking, this condition must be fulfilled for each critical band.)

Figure 5.4 shows how the intelligibility in enclosures depends on the signal-to-noise ratio. Peutz [5.5] and Klein [5.6] found that, depending on the reverberation time, T, an improvement of intelligibility can be achieved by increasing the signal-to-noise ratio up to a value of 35 dB (see section 3.1.2.3).

In rooms we also have to consider the disturbing effect of the reflected sound arriving at the listener's seat later than 50 ms (speech) or 80 ms (music) after the direct sound. If this delayed sound is caused by the system itself, the disturbing effect that may be produced cannot be overcome by merely increasing the sound level, but only by a better selection and arrangement of the loudspeakers (see Chapter 6). This holds true especially for large reverberant rooms. Therefore one must take care to ensure in these cases that the direct sound level that is subject to free-field conditions does not, in the reception area, fall more than 6–9 dB below that of the diffuse (reverberant) field. This means that the listener's distance should, if possible, not exceed twice the critical distance (see equation (5.7)). In this connection it is possible to check the intelligibility to be expected according to Peutz [5.5] and Klein [5.6].

When calculating the power, apart from the sound level resulting from signal-to-noise ratio and dynamic range considerations, we also have to allow an overload reserve for the distortion-free transmission of occasional

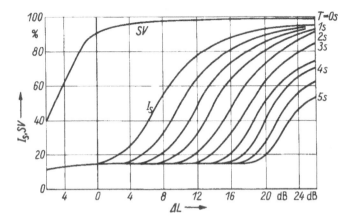

Figure 5.4 Intelligibility of phrases, SV, and of syllables, I_s, as a function of the signal-to-noise ratio, ΔL. Parameter: reverberation time, T.

signal peaks, which therefore need not be included, for instance, in the thermal calculation of the system.

The resulting sound level is often called the *calculation level* of the system (effectively usable level including overload reserve), and taken as a basis for the ensuing calculations. Table 5.1 contains typical calculation levels, which may serve as guide values if more exact values are not available. The values apply to the far field or the diffuse field for the frequency range from 1000 to 2000 Hz. From these values it is then possible to determine the required sound power approximately according to Figure 5.5. A-weighted sound level values may also be used for calculation.

Table 5.1 Typical calculation levels

Calculation level (dB)	Coverage area
86	Speech in areas with low noise level (conference rooms, foyers, churches, sports grounds, open-air swimming pools)
92	Speech and entertainment music in areas with elevated noise level (department stores, waiting rooms, booking halls, indoor swimming pools, traffic facilities, mass rallies)
96	Speech and music in areas with high noise level (sports halls, large stadiums, open-air theatres)
104	Concert transmissions, performances of electronically amplified music, theatre effects (theatres, opera houses, concert halls, multi-purpose halls)

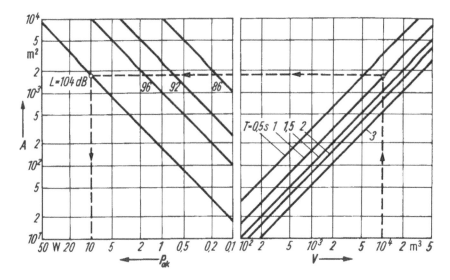

Figure 5.5 Sound power, P_{ak}, required to obtain a predetermined sound level, L, in the diffuse field of a room, as a function of the equivalent absorption area, A, and the reverberation time, T. Example: room volume $V = 10\,000$ m³ and $T = 1$ s results in $A \approx 1600$ m². For $L = 104$ dB one requires in this case $P_{ak} = 10$ W.

5.2.3 Dynamic range

Any signal requires a certain range of levels for its transmission. Comprising the range between noise and maximum modulation, this range is described in sound engineering and acoustics by the term *dynamic range*, which originates from music theory. Dynamic range in this connection is understood as a *subjective* concept for loudness conditions, and as an *objective* concept for physical quantities such as sound intensity relations.

We must distinguish between

- *original dynamic range* – property of the original signal of the source;
- *programme range* – dynamic range of the sound signal in the transmission channel;
- *reproduction range* – dynamic range and reproduction conditions at the listener's position.

A sound reinforcement system has to be dimensioned for the dynamic range of the original sound, for which one has to take into account

- the ratio between the maximum and minimum sound levels, measurable as the weighted sound pressure difference (objective concept); and

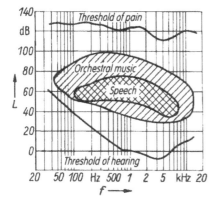

Figure 5.6 Dynamic range of natural speech and orchestral signals as a function of frequency.

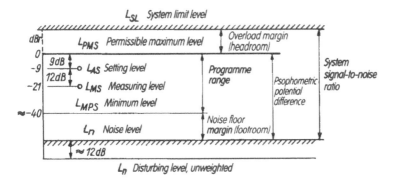

Figure 5.7 Gain structure of a transmission channel [5.7].

- the ratio between the maximum and minimum sound perception on hearing natural sound events (subjective concept).

Figure 5.6 shows the dynamic ranges of speech and orchestral signals. We can see that the *original dynamic range* may lie between 30 and 100 dB.

The gain structure of the transmission equipment (Figure 5.7) serves to explain the *programme range*, and shows that it is the difference between signal-to-noise ratio and overload reserve (headroom) as well as noise (safety) margin (footroom). This implies that, on the one hand, we have to consider a certain margin or 'footroom' between the signal level and the inherent noise of the system (the equivalent loudness of a good microphone is about 20 dB) and, on the other hand, we have also to reserve a level margin as headroom for undistorted reproduction of individual programme peaks, which may exceed the effective value of the useful signal by up to 20 dB (especially if digital storage and transmission techniques are envisaged). For sound reinforcement systems, a footroom of 10 dB and a headroom of likewise 10 dB are generally aimed at. If we wanted under these conditions to transmit the original dynamic range of 80 dB without any restriction, the internal signal-to-noise ratio of the system would have to be 100 dB. This cannot be achieved, however, even by excellent digital systems, and is in any case not necessary, as the quality of radio transmissions shows.

The *reproduction range* that can be produced by electroacoustic systems varies between 40 dB with conventional broadcasting and maximally 60 dB with digital sound recordings. Thus the maximum original dynamic range is usually wider than the reproduction range.

Since, in addition to the programme range, the background noise in the reproduction room has also to be considered, circumstances may cause the signal-to-noise ratio to drop (Figure 5.7), bringing about a reduction of the reproduction range. Experience shows that sound reinforcement systems are often rather loud in fortissimo, but that the pianissimo components of a signal are often eclipsed by the inherent noise of the system. Also, clipping of the signal (overmodulation limitation) may give rise to unacceptable distortions at the upper modulation limit. This can be counteracted by a deliberate redirection of the programme range, which nowadays is often achieved using limiters (see section 4.3.4.2), which also serve as a protection for the high-quality loudspeakers. The use of limiters and other AGC amplifiers may, however, interfere significantly with optimal signal level adjustment, a fact that should be taken into account for the transmission of classical music.

Table 5.2 shows the signal-to-noise ratios (programme range, inclusive of footroom) of electroacoustic systems and devices. By using noise suppression systems such as companders (see section 4.3.3) it is possible to achieve

Table 5.2 Signal-to-noise ratios of electroacoustical devices and facilities

Device	Signal-to-noise ratio (dB)
Analogue studio tape machine	48–55
Analogue studio tape machine with compander (Dolby SR, Tel-Com C4)	60–72
Cassette recorder	35–45
Cassette recorder with compander (Dolby B, High-Com)	40–50
Gramophone record	40–50
Gramophone record with compander (CX, UC)	50–60
Microphone	70–75
Compact disc (CD)	80–90
Digital cassette recorder (R-DAT)	70–90

With digital recording procedures it is necessary, in order to avoid significant distortion, to provide a minimum 'footroom' (noise floor margin) of 20 dB (quantization noise) and a headroom of 10 dB (bit faults).

Table 5.3 Acoustic power of selected sources

Source	Acoustic power (W)
Conversational speech (average value)	0.000 007
Human voice, maximum value	0.002
Violin, fortissimo	0.001
Clarinet, piccolo flute, horn, triangle	0.05
Piano, bass tuba	0.2
Grand piano, bass saxophone, trumpet	0.3
Orchestra (15 musicians)	2
Organ, drum, bass drum, fortissimo	5
Orchestra (75 musicians)	15

signal-to-noise ratios better than 70 dB (see the useful synopsis given by Krause [5.8]).

5.2.4 Adaptation to the original sources

When the sound reinforcement system is required to operate in conjunction with an orchestra or other original sources (for example, in the case of half-playback – taped music with a live singer or soloist), the system must be capable of generating at least the same acoustic power as the original source that is to be accompanied.

The sound powers of different sources are listed in Table 5.3; they can be converted by means of equation (5.6a) into level values for the diffuse field. Here it is also necessary to consider the dynamic range required in each individual case.

5.3 Achievable sound level and audio power

5.3.1 Sound level calculation

5.3.1.1 Free field (direct field of the loudspeaker)

To calculate the sound level in the free field we have to use the loudspeaker parameters ascertained in the free field. These include the characteristic sensitivity, E_K, according to equation (4.6) and the characteristic sound level, L_K, according to equation (4.7), as well as the directivity ratio, $\Gamma_L(\vartheta)$, according to equation (4.12) and the directivity gain, D, according to equation (4.12a).

The sound pressure in the direct field of a loudspeaker at a given location at a distance r_{LH} from the loudspeaker and an angle ϑ from the reference axis is

$$\tilde{p}_d = E_K \sqrt{P_{el}} \, \frac{r_0}{r_{LH}} \, \Gamma_L(\vartheta) \tag{5.19}$$

and the corresponding direct sound level (with $r_0 = 1$ m) is

$$L_d = L_K + 10 \log P_{el} \, \text{dB} - 20 \log r_{LH} \, \text{dB} + 20 \log \Gamma_L(\vartheta_H) \, \text{dB} \tag{5.20}$$

where P_{el} is in W, and r_{LH} is in m.

The 1 m, 1 W level is then $L_{d, 1\,m, 1\,w} = L_K$.

For larger distances (above 40 m) one has still to consider the atmospheric propagation loss $D_r = D_{LH}$ according to Figure 3.20. Equation (5.20) is then transformed to

$$L_d = L_K + 10 \log P_{el} \, \text{dB} - 20 \log r_{LH} \, \text{dB} + 20 \log \Gamma_L(\vartheta_H) \, \text{dB} - D_{LH} \tag{5.20a}$$

where r_{LH} is the distance from loudspeaker to listener location in m; $D_{LH} = D_r$ is the additional level attenuation according to Figure 3.20; L_K is the characteristic sound level of the loudspeaker in dB; and P_{el} is the installed power of the loudspeaker in W.

The characteristic sensitivity, the directivity ratio and the propagation loss are frequency dependent. To ascertain the characteristic parameters of broadband loudspeakers in practice one therefore generally uses an averaged value of the range between 200 Hz and 4 kHz. (This range is generally recommended in the ISO and DIN standards for ascertaining the characteristic sensitivity of broadband loudspeakers.)

The above-mentioned method for calculating the sound level to be expected in the free field is generally applicable to outdoor systems. But the free-field propagation is also of interest for indoor systems: not only for information about the direct-field component, but also if for instance the loudspeakers of effects sound systems in theatres have to be installed in the rear stage area. In this case we must reckon that only the directly irradiated sound components passing the first stage opening (and in some cases also the second one) will become effective in the auditorium, whereas all room reflections are absorbed in the stage house itself or become noticeable only as long-delayed detrimental room sound. An example will illustrate the importance of the characteristic sound level for the free-field propagation.

In an open-air theatre the aim is to produce a sound pressure level of $L_d = 85$ dB at a distance of 20 m from a loudspeaker having an assumed characteristic sound level of $L_K = 97$ dB. What is the required installed power, P_{el}?

To find the solution we start with equation (5.20). At a distance of 20 m additional attenuations, D_{LH}, may still be neglected ($D_{LH} = 0$ dB).

Moreover we consider the main radiating direction of the loudspeaker $(20 \log \Gamma_{\mathrm{L}}(\vartheta_{\mathrm{H}})\ \mathrm{dB} = 0\ \mathrm{dB})$.

After transforming equation (5.20) we obtain

$$10 \log P_{\mathrm{el}}\ \mathrm{dB} = L_{\mathrm{d}} - L_{\mathrm{K}} + 20 \log r_{\mathrm{LH}}\ \mathrm{dB} = (85 - 97 + 26)\ \mathrm{dB}$$
$$= 14\ \mathrm{dB}$$

The resulting value of P_{el} is 25 W ($\rightarrow 10^{14/10} = 25$ W).

If a 10 dB higher level ($L_{\mathrm{d}} = 95$ dB) is to be produced, a loudspeaker with a rated power capacity of no less than 250 W will, however, be required. This can hardly be achieved by means of broadband loudspeakers. In this case it is better to resort to a more sensitive and more efficient loudspeaker type.

If in special cases we know only the sound power level, L_{W}, of the loudspeaker and the front-to-random index, $C = 10 \log \gamma_{\mathrm{L}}\ \mathrm{dB}$, it is possible to calculate the required characteristic sound level, L_{K}, according to equation (4.7a):

$$L_{\mathrm{K}} = L_{\mathrm{W}} + C - 10 \log P_{\mathrm{el}}\ \mathrm{dB} - 11\ \mathrm{dB} \tag{5.21}$$

or by means of $L_{\mathrm{W}} = 10 \log(P_{\mathrm{ak}}/P_0)\ \mathrm{dB} = 10 \log(P_{\mathrm{L}}/P_0)\ \mathrm{dB}$:

$$L_{\mathrm{K}} = 109\ \mathrm{dB} + 10 \log \eta\ \mathrm{dB} + C \tag{5.21a}$$

η is the efficiency of the loudspeaker ($P_{\mathrm{ak}}/P_{\mathrm{el}}$ or $P_{\mathrm{L}}/P_{\mathrm{n}}$). If the characteristic sensitivity and the front-to-random factor are known, η can be ascertained according to equation (4.9), or from the characteristic impedance according to equation (4.10).

Let us assume that the front-to-random factor, γ_{L}, in our example is 8. With $L_{\mathrm{d}} = 85$ dB and by means of equation (5.21) we obtain $L_{\mathrm{W}} = 113$ dB ($C = 9$ dB). With this acoustic power $P_{\mathrm{ak}} = 0.2$ W we then obtain an efficiency of $\eta = 0.8\%$.

However, when using a loudspeaker with the same directivity behaviour at 1000 Hz, but with a higher characteristic sound level of $L_{\mathrm{K}} = 104$ dB, we obtain with equation (5.20) an installed power of only 50 W (and not 250 W as above) required to produce a sound level of $L_{\mathrm{d}} = 95$ dB at a distance of 20 m. From equation (5.21) with $C = 9$ dB the resulting sound-power level, $L_{\mathrm{W}} = 123$ dB: that is, a sound power of $P_{\mathrm{ak}} = 2$ W and thus an efficiency of $\eta = 4\%$. This means that the required efficiency has risen by a factor of 5. Loudspeakers of such high efficiency, however, have only a limited frequency range or a heavily harmonics-prone reproduction. If such a loudspeaker were to be used with a front-to-random index 3 dB higher

assumed: $L_K = 104$ dB; $C = 9$ dB. This gives

$$10 \log P_{el} \text{ dB} = (104 - 104 + 32 + 9 - 17) \text{ dB} = 24 \text{ dB}.$$

From this we obtain the sum of the powers of all loudspeakers installed with $\Sigma_n P_{el} \approx 250$ W (see also Figure 5.10).

5.3.1.3 Real rooms

In enclosures the sound field stems from the diffuse sound as well as the direct sound of the loudspeakers. Starting from equation (5.15a), entering the attenuation coefficient, D_{LH}, and using equation (5.21), the sound level to be expected in the room at a distance r_{LH} is

$$L = L_K + 10 \log P_{el} \text{ dB} - C + 11 \text{ dB} + 10 \log \left(\frac{\gamma_L \Gamma_L^2(\vartheta_H)}{4\pi r_{LH}^2} + \frac{4}{A} \right) \text{ dB}$$

or

$$L = L_K + 10 \log P_{el} \text{ dB} + 10 \log \left(\frac{\Gamma_L^2(\vartheta_H)}{r_{LH}^2} + \frac{16\pi}{\gamma_L A} \right) \text{ dB} \tag{5.25}$$

where P_{el} is in W, r_{LH} is in m, and A is in m^2.

These equations can only be applied, however, if only one loudspeaker is used, or if the loudspeakers are arranged in a very concentrated form, as, for instance with a single cluster. Here the directivity ratio, Γ_{tot}, and the front-to-random factor, γ_{tot}, of the array must be known, however.

The transition from the spatially limited free-field behaviour to the diffuse-field behaviour of the loudspeakers is characterized by the critical distance (see section 5.1.2). This critical distance may be reduced for several loudspeakers arranged at greater distance from each other (see equation (5.10)).

5.3.2 Determining the required audio power

When designing a sound reinforcement system one first ascertains the sound power required to achieve the necessary sound level. The starting point here is equation (5.14), wherein

$$L_W = 10 \log \frac{P}{P_0} \text{ dB}$$

is to be entered. Putting $r_0 = 1$ m, we obtain

$$10^{L/10 \text{ dB}} = \frac{P}{J_0} \cdot \left(\frac{\gamma_L \Gamma_L^2(\vartheta)}{4\pi r^2} + \frac{4}{A} \right)$$

or

$$P = \frac{10^{L/10 \text{ dB}} J_0 A \cdot 4\pi r^2}{16\pi r^2 + A\gamma_L \Gamma_L^2(\vartheta)} \qquad (5.26)$$

where $J_0 = 10^{-12}$ W/m^2 is the reference sound intensity.

As has already been mentioned in section 5.3.1.2, it is in many cases sufficient to consider only the diffuse field, for example when it is sufficient to precalculate the sound level required as a minimum. Under these preconditions (in equation (5.26) $4\pi r^2 \blacktriangleright A\gamma_L \Gamma_L^2(\vartheta)$) we can use the equation

$$P_r = \frac{10^{L/10 \text{ dB}} J_0 A}{4} \qquad (5.27)$$

The required sound power as a function of equivalent absorption area, reverberation time and volume of a room can be taken from Figure 5.5. The installed power then results by multiplying by the reciprocal value of the loudspeaker efficiency. If only the free field is to be considered – for outdoor systems or decentralized systems in heavily damped flat rooms for example – the necessary sound power, with a required sound level, L, at a given distance r from the loudspeaker is obtained as

$$P_d = \frac{4\pi r^2 \cdot 10^{L \; 10 \text{ dB}} J_0}{\gamma_L \Gamma_L^2(\vartheta)} \qquad (5.28)$$

The dependence of the sound power, P_d, on the distance from source to listener is plotted in Figure 5.8. The parameter is the sound level, L, to be achieved. The power to be installed is also ascertained via the efficiency of the loudspeaker.

As described in section 5.3.1.1, the characteristic sensitivity or the characteristic sound level may serve for determining the audio signal power (see equation (4.9)). Below we summarize the corresponding dimensioning equations for calculating the installed audio power (the starting point is equation (5.25)).

In the complex sound field, i.e. when a required sound level, L, is to be achieved at a given point in the room, the necessary *installed audio power of the loudspeaker* is ascertained by means of the following

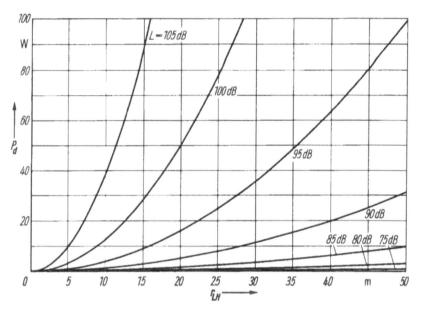

Figure 5.8 Required sound power, P_d, as a function of the distance from source to listener, r_{LH}. Parameter: desired sound level, L, at the listener's location in the open.

equation:

$$P_{el} = \frac{10^{L/10 \text{ dB}} \gamma_L A \rho c r^2 J_0}{E_K^2 r_0^2 (16\pi r^2 + A \gamma_L \Gamma_L^2(\vartheta))}$$

In level notation this reads as follows:

$$P_{el} = 10^{(L - L_K - 10 \log(16\pi/A\gamma_L + \Gamma_L^2(\vartheta)/r^2 \text{ dB}))/10 \text{ dB}} \qquad (5.29)$$

If only the free field is considered, the level-oriented equation for the required installed power is

$$P_{el_d} = 10^{(L - L_K - 20 \log \Gamma_L(\vartheta) \text{ dB} + 20 \log r)/10 \text{ dB}} \qquad (5.30)$$

With outdoor sound reinforcement systems in which there is a great distance between the loudspeaker and the listener, the propagation loss has also to be considered additionally (further term, $-D_r$, in the numerator of the exponent of equation (5.30)).

In the diffuse field the installed power required for generating a sound level L is, in level notation,

$$P_{el_r} = 10^{(L - 17\,dB - L_K + 10\,\log A\,dB + C)/10} \tag{5.31}$$

Since the characteristic sensitivity (characteristic sound level, L_K) and the front-to-random index, C, are frequency-dependent and loudspeaker-related quantities, the correct selection and arrangement of the loudspeaker determine whether, in a room with given absorption area A or in the open at distance r, the desired sound pressure levels are produced by the required installed powers, or the loudspeaker with the required power capacity, P_n, is selected. For the necessary overload margin, see section 5.2.3.

The dependence of the required audio power, P_{el}, on the distance r (free-field conditions) and on the absorption area A (diffuse field) is depicted in Figures 5.9 and 5.10. Parameters of both families of curves are the level difference $L - L_K$ and the front-to-random index, C (putting $\Gamma_L(\vartheta) = 1$).

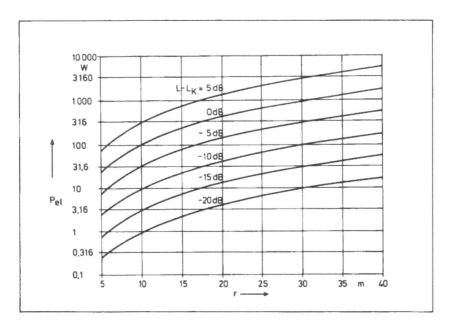

Figure 5.9 Required audio power, P_{el}, as a function of the distance, r, from the source. Parameter: level difference, $L - L_K$.

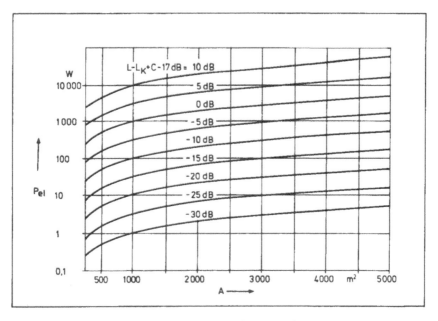

Figure 5.10 Required audio power, P_{el}, as a function of the equivalent absorption area, A. Parameter: level difference, $L - L_K + C - 17$ dB.

5.4 Acoustic gain

5.4.1 General principles

In the reproduction of sound events in rooms and in the open, for many people the listening expectation is strongly determined by previous experience. It is well known, for instance, that a reverberant room impression results in large enclosures with acoustically hard boundaries. Sound transmissions in the open, however, are perceived as 'dry', since reflections are largely missing here.

This may differ with the use of electroacoustic systems. These make possible the realization of sound patterns that are contradictory to expectations. This may be employed as an intentional means of artistic treatment: for example, in the presentation of (virtual) sound images. To help avoid this unnatural effect with 'normal' sound reinforcement tasks, we are going to show the interrelations involved in sound reinforcement. The following requirements are to be met:

- realization of the required sound levels;
- safeguarding of the frequency response that corresponds to the uses of the system and the expectations of the listener;

- unconditional avoidance of positive acoustic feedback and its corollaries (timbre changes, reverberant impression, disturbing noises).

For illustration, a simple sound reinforcement system consisting of microphone, amplifier and loudspeaker is shown in Figure 5.3. At the listener's location, H, the direct sound from the original source arrives along path r_{SH}, and the direct sound from the loudspeaker arrives along path r_{LH}. For effective sound reinforcement to take place, the sound of the loudspeaker arriving at the listener must be louder than the original sound. Although a loudspeaker is directional, it radiates not only towards the area to be covered, but also into the whole room, so that the amplified sound radiated by the loudspeaker may quickly return via path r_{LM} to the microphone, thus giving rise to positive feedback.

In order to enable a quantitative assessment of the interrelations between achievable sound reinforcement and incipient positive feedback, the parameters explained in section 5.1 are used. The sound energy density, and the sound level derived from it, are especially suited for establishing these relationships.

In a sound reinforcement situation like that in Figure 5.3, the sound energy density at the listener's location H, originating from the source S (see also equation (5.16)) is

$$
w_{HO} = \frac{P_S}{c} \left[\frac{\gamma_S \Gamma_S^2(\vartheta_{SH}) 10^{-D_{SH}/10\ dB}}{4\pi r_{SH}^2} + \frac{4}{A} \right] \tag{5.32}
$$

where γ_S is the front-to-random factor of source S; P_S is the acoustic power of source S; Γ_S is the angular directivity ratio of source S; r_{SH} is the distance from source to listener; c is the speed of sound; ϑ_{SH} is the angular deviation between the main radiation direction and the line from source to listener's location; A is the equivalent sound absorption area of the room; and D_{SH} is the attenuation of propagation in air to be considered in the direct field of the source (see Figure 3.20).

To simplify equation (5.32) the *directivity factor*, $Q_S(\vartheta_{SH}) = \gamma_S \Gamma_S^2(\vartheta_{SH})$ (cf. equation (4.14)), is introduced. Thus equation (5.32) becomes

$$
w_{HO} = \frac{4P_S}{cA} \left[Q_S(\vartheta_{SH}) \left(\frac{r_H}{r_{SH}} \right)^2 \times 10^{-D_{SH}\ 10\ dB} + 1 \right] \tag{5.32a}
$$

where $r_H = \sqrt{(A/16\pi)}$ is the reverberation radius (see equation (3.5)).

If the bracketed value in equation (5.32a), which is a dimensionless module that characterizes the transmission conditions between original source S and listener H, is termed the sound *transmission factor*, q_{SH}, equation (5.32a) becomes

$$w_{HO} = \frac{4P_S}{cA} q_{SH} \tag{5.33}$$

If the signal source is not the original source, but a loudspeaker L, we obtain by analogy

$$w_{HL} = \frac{4P_L}{cA} q_{LH} \tag{5.34}$$

where P_L is the sound power of the loudspeaker, and q_{LH} is the sound transmission factor from loudspeaker to listener.

This allows us to calculate the sound amplification at the listener's location, H:

$$v_L = \frac{w_{HL}}{w_{HO}} = \frac{P_L}{P_S} \cdot \frac{q_{LH}}{q_{SH}} \tag{5.35}$$

The radiated sound power, P_L, is the multiplication product of the sound energy density, w_M, existing at the microphone location, and the transmission coefficient, T_{Pw}, of the amplification channel (cf. Figure 5.3). Since this sound energy density consists of components from the original source and from the loudspeaker, it follows that

$$P_L = T_{Pw}(w_{MS} + w_{ML}) = T_{Pw}w_M \tag{5.36}$$

The second term in the brackets is responsible for the positive feedback susceptibility of the system. If the loop amplification of the system

$$v_S^2 = \frac{w_{ML}}{w_M} \tag{5.37}$$

tends →1, the amplification channel becomes unstable; the resulting undamped oscillations make themselves felt first by timbre change or increased reverberation and later on by whistling and howling.

The loudspeaker sound is then prevailing at the microphone location, whereas that coming from the original source becomes negligible.

If equation (5.37) is entered in equation (5.36), we obtain

$$P_L = T_{Pw} \frac{w_{MS}}{1 - v_S^2} \tag{5.36a}$$

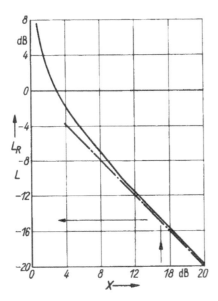

Figure 5.11 Feedback index, L_R, as a function of the level value, X (separation of the average value of the loop gain from the positive feedback threshold).

The ratio between the sound energy density components attributable at the microphone location to the loudspeaker and to the original source is called the *feedback factor*, $R(X)$ (X being the ratio between the average value of loop amplification and the positive feedback threshold in dB):

$$R(X) = \frac{w_{ML}}{w_{MS}} = \frac{v_S^2}{1 - v_S^2} \tag{5.38}$$

with $v_S^2 = 10^{-X/10 \text{ dB}}$.
The *feedback index*

$$L_R = 10 \log R \text{ dB} \tag{5.38a}$$

is plotted in Figure 5.11 as a function of the ratio X. We can see that L_R behaves for $X > 10$ dB approximately like $-X$ (see also Figure 5.18).

By using equations (5.36a), (5.37) and (5.38) we obtain from equation (5.35), after several transformations, the maximum sound amplification

$$v_L = R(X) \frac{q_{SM} q_{LH}}{q_{LM} q_{SH}} \tag{5.39}$$

Thus we obtain the following transmission factors:

- Between source S and microphone M:

$$q_{SM} = Q_{SM}(\vartheta_{SM}, \vartheta_S)\left(\frac{r_H}{r_{SM}}\right)^2 \times 10^{-D_{SM}/10 \text{ dB}} + 1 \qquad (5.33a)$$

(Because $D_{SM} \approx 0$ dB, the partial factor attributable to it can mostly be neglected.)

- Between loudspeaker L and listener H:

$$q_{LH} = Q_L(\vartheta_H)\left(\frac{r_H}{r_{LH}}\right)^2 \times 10^{-D_{LH}/10 \text{ dB}} + 1 \qquad (5.33b)$$

- Between loudspeaker L and microphone M:

$$q_{LM} = Q_{LM}(\vartheta_L, \vartheta_M)\left(\frac{r_H}{r_{LM}}\right)^2 \times 10^{-D_{LM}/10 \text{ dB}} + 1 \qquad (5.33c)$$

(For distances and angles see Figure 5.3.)

The corresponding directivity and beaming properties are summarized in the directivity factors (see equations (4.14) and (5.5)):

$$Q_L(\vartheta_H) = \gamma_L \Gamma_L^2(\vartheta_H)(\gamma_{PL}) \qquad \text{loudspeaker towards listener}$$
$$Q_L(\vartheta_H) = \gamma_L \Gamma_L^2(\vartheta_L)(\gamma_{PL}) \qquad \text{loudspeaker towards microphone}$$
$$Q_S(\vartheta_{SM}) = \gamma_S \Gamma_S^2(\vartheta_{SM}) \qquad \text{source towards microphone}$$
$$Q_M(\vartheta_S) = \gamma_M \Gamma_M^2(\vartheta_S) \qquad \text{microphone towards source}$$
$$Q_M(\vartheta_M) = \gamma_M \Gamma_M^2(\vartheta_M) \qquad \text{microphone towards loudspeaker}$$

and in the coupling factors (see equation 5.12)):

$$Q_{SM}(\vartheta_{SM}, \vartheta_S) = Q_S(\vartheta_{SM})Q_M(\vartheta_S) \qquad \text{source and microphone}$$
$$Q_{LM}(\vartheta_L, \vartheta_M) = Q_L(\vartheta_L)Q_M(\vartheta_M) \qquad \text{loudspeaker and microphone}$$

If the sound is directly incident on absorbing surfaces, the equivalent front-to-random factor γ_{PL} should be used, but in most applications this is usually neglected.

If the loudspeaker is aimed mainly at the audience area, in order to reduce excitation of the internal reverberation of the room (see section 5.1.1), the angle-dependent directivity factor of the loudspeaker is increased by the equivalent beaming factor, γ_{PL}.

In level notation one obtains the *acoustic gain index*

$$VE = 10 \log v_L \text{ dB}$$

or

$$VE = L_R + L_{SM} + L_{LH} - L_{LM} - L_{SH} \tag{5.40}$$

where $L_R = 10 \log R$ dB is the feedback index (see equation (5.38a)), and $L_{XY} = 10 \log q_{XY}$ dB is the transmission measure between the quantities X and Y, which have to be entered each time according to current data.

In section 5.1.2 we introduced the equivalence distance r_a, establishing that a sound level L_{ra} given at this distance is still perceived as sufficient. This level is now also to prevail at a far greater distance, so that amplification is required. This *required acoustic gain* is

$$v_{req} = \frac{q_{ra}}{q_{SH}}$$

or, in the form of level notation with $VN = 10 \log v_{not}$ dB,

$$VN = L_{ra} - L_{SH} \tag{5.41}$$

If the required gain VN (equation (5.41)) is set equal to the achievable gain VE (equation (5.40)) and provided freedom from positive feedback is given (loudspeaker–microphone distance, $r_{LM} \gg$ reverberation radius, r_H), i.e. $L_{LM} = 0$ dB, the equivalent sound level is

$$L_{ra} = L_{SM} + L_{LH} + L_R$$

(cf. equation (5.44))

From this it is possible to derive for transducers with omnidirectional characteristics the minimum distance from source to microphone, r_{SM}, required to obtain this sound level:

$$r_{SM} = r_a \frac{r_H}{r_{LH}} \sqrt{[R(X)]} \tag{5.42}$$

If a directional loudspeaker and a directional microphone are used – which is usually the case – equation (5.42) extends approximately to

$$r_{SM} = r_a \frac{r_R}{r_{LH}} \sqrt{[R(X)]\gamma_M} \tag{5.42a}$$

(valid for $r_{LH} < \sqrt{\gamma_L r_H}$; for $r_{LH} > r_R = \sqrt{\gamma_L r_H}$, r_{LH} has to be replaced by r_R.)

For open-air systems one obtains by analogy to equation (5.42)

$$r_{SM} \approx r_a \frac{r_{LM}}{r_{LH}} \sqrt{[R(X)]} \tag{5.42b}$$

If there is more than one amplification channel in the room, sound amplification can be ascertained by an analogous calculation (see also [5.2], pp. 132, 135):

$$v_L = \frac{R(X)}{q_{SH}} \cdot \sum_{i=1}^{n} \frac{q_{SMi}/q_{LHi}}{\sum\limits_{j=1}^{n} k_{ji} q_{LMji}}$$

This equation relates the radiated sound power of the *i*th channel via the factor k_{ji} to that of the *j*th channel ($P_{Lj} = k_{ji} P_{Lj}$; $0 \leqslant k_{ji} < \infty$).

With level notation we obtain by analogy to equation (5.40) the *acoustic gain index*:

$$VE = L_R - L_{SH} + 10 \log \sum_{i=1}^{n} \frac{q_{SMi}/q_{LHi}}{\sum\limits_{j=1}^{n} k_{ji} q_{LMji}} \text{ dB} \tag{5.43}$$

5.4.2 Deductions in practice

5.4.2.1 Indoors

ONE AMPLIFICATION CHANNEL

In a room it is possible to neglect in equations (5.33) to (5.33c) the additional level attenuations D_{SH}, D_{LH} etc. (i.e. $D_i = 0$ dB). Since with a centralized sound reinforcement system the distances from listener to source, r_{SH}, and from loudspeaker to microphone, r_{LM}, are greater than the critical distance, $r_R = \sqrt{\gamma_L r_H}$, prevailing in the room, equation (5.40) simplifies to

$$VE = L_R + L_{SM} + L_{LH} \tag{5.44}$$

For the feedback index, L_R, the values to be expected are according to Figure 5.11 between -6 and -15 dB, depending on the degree of equalization of the sound reinforcement system (the recommended value is $L_R = -9$ dB).

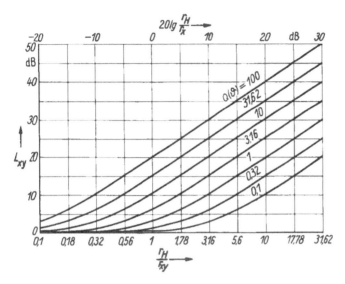

Figure 5.12 Sound transmission index, L_{XY}, as a function of the distance ratio, r_H/r_{XY}. Parameter: directivity factor $Q(\vartheta)$ or coupling directivity factor $Q(\vartheta, \varphi)$.

The values for the sound transmission measures L_{SM} and L_{LH} can be gathered from Figure 5.12, which shows the dependences of the sound transmission measures, L_{XY}, in general terms as a function of the ratio r_H/r_{XY}. Parameters are the respective directivity factors or the coupling factors $Q(\vartheta)$.
Approximately there is

$$
L_{SM} \approx 10 \log(\gamma_S \gamma_M) + 20 \log \left(\frac{r_H}{r_{SM}} \right) \text{ dB}
$$

which implies that in Figure 5.12 $Q(\vartheta) = \gamma_S \gamma_M$ and r_{SM} should be entered for r_{XY}.
For the sound transmission measure L_{LH} we obtain by analogy

$$
L_{LH} \approx 10 \log(\gamma_L \gamma_{PL}) + 20 \log \left(\frac{r_H}{r_{LH}} \right) \text{ dB}
$$

This can be more easily explained by the following example.

In a room with a reverberation radius $r_H = 5.7$ m a loudspeaker with front-to-random factor $\gamma_L = 7$ at a loudspeaker–listener distance $r_{LH} = 10$ m and a microphone with front-to-random factor $\gamma_M = 3$ at a source–microphone distance $r_{SM} = 0.5$ m are installed.

Further data: surface of the room, $S = 7400$ m^2; with the equivalent absorption area $A = 1630$ m^2 ($r_H = 5.7$ m) the resulting average sound absorption coefficient $\alpha = 0.22$. The area occupied by the audience amounts to about 1000 m^2, and an absorption coefficient of $\alpha_P = 0.75$ has been determined. By means of equation (5.5) we obtain a directivity increase to $\gamma_{PL} = 3.1$ as a consequence of the directional radiation towards the absorbent audience.

For the sake of simplification let us assume that the front-to-random factor of the original source is $\gamma_S = 1$. Figure 5.12 shows that the sound transmission measures result as $L_{SM} \approx 26$ dB ($r_H/r_{SM} = 11.4$; $Q(\vartheta) = 3$) and $L_{LH} \approx 8.4$ dB ($r_H/r_{LH} = 0.57$; $Q(\vartheta) = 7 \times 3.1$). With a feedback index $L_R = -9$ dB the achievable reinforcement is $VE = 25.4$ dB.

Further examples are compiled in Table 5.4.

SEVERAL AMPLIFICATION CHANNELS

For the practical case we assume that all source–listener distances, r_{SM}, and loudspeaker–microphone(s) distances, r_{LMji}, are much greater than the applicable critical distances. Thus equation (5.43) simplifies to

$$VE = L_R + 10\log \sum_{i=1}^{n} \frac{q_{SMi}/q_{LHi}}{\sum_{j=1}^{n} k_{ji}} \tag{5.45}$$

Moreover we take it for granted that all loudspeakers of the sound reinforcement system radiate with equal intensity into the room so as to

Table 5.4 Example cases of sound reinforcement

Case	γ_L	γ_M	r_{SM}	r_H/r_{SM}	r_{LH}	r_H/r_{LH}	VE (dB)
1	5	1	3 m	2	12 m	0.5	≈ 0
2	5	1	0.5 m	12	12 m	0.5	7
3	5	3	3 m	2	12 m	0.5	≈ 0
4	5	3	0.5 m	12	12 m	0.5	12
5	5	3	10 cm	60	12 m	0.5	26
6	10	3	10 cm	60	24 m	0.25	23
7	10	5	5 cm	120	24 m	0.25	31
8	10	5	5 cm	120	6 m	1	43

$R(X) = 3 \times 10^{-2}$ $\gamma_S = 1$ $r_H = 6$ m

influence equally the overall loudness of the diffuse field in the room (i.e. $P_{Li} = P_{Lj}$, and hence $k_{ji} = 1$). Thus we obtain the acoustical gain index

$$VE = L_R - \log n\,\text{dB} + 10 \log \sum_{i=1}^{n} q_{SMi}/q_{LHi}\,\text{dB}. \tag{5.45a}$$

Furthermore we assume that there is only *one* microphone whose signal is supplied to all loudspeakers distributed in the room, and that at the listener's location there is effective approximately only the loudspeaker showing the least distance from loudspeaker to listener, r_{LHmin}. Under these conditions equation (5.45a) simplifies to

$$VE \approx L_R - 10 \log n\,\text{dB} + L_{LHmin} + L_{SM}. \tag{5.46}$$

Thus equation (5.46) differs from equation (5.44) only by the term $-10 \log n$ dB.

With n amplification channels, individual loudspeakers appropriately distributed in the hall allow for many room configurations to achieve a better sound level distribution than a centralized loudspeaker arrangement, such as a monocluster in the centre of the room or above the platform. If positive feedback across the different channels can be avoided by means of balanced equalization of the amplification channels, the term $-10 \log n$ dB may theoretically be neglected or at least be entered in a reduced form (e.g. $-10 \log \sqrt{n}$). Under favourable conditions it is thus possible to obtain the same amplification as with only one channel, but with a more balanced sound level distribution.

5.4.2.2 Outdoor areas

ONE AMPLIFICATION CHANNEL

Let us assume that the directivity factor of the source is $Q_S = 1$ (which means a source with omnidirectional characteristics). According to Figure 3.20 the additional level attenuations may also be $D_i = 0$ dB for distances $r_i < 40$ m. In this case equation (5.39) is simplified by using equations (5.33) to (5.33c) (the 1 term is inapplicable in the open) to

$$v_L = R(X) \frac{Q_M(\vartheta_S) Q_L(\vartheta_H)}{Q_L(\vartheta_L) Q_M(\vartheta_M)} \left(\frac{r_{LM} r_{SH}}{r_{SM} r_{LH}}\right)^2 \tag{5.47}$$

Since in practice it is often possible to make the distances r_{SH} and r_{LH}

approximately equal, and since the microphone is directed towards the source and the loudspeaker towards the listener, equation (5.47) is further simplified so as to produce an acoustic gain index of

$$VE \approx L_R + 10 \log Q(\vartheta) \, dB + 20 \log \left(\frac{r_{LM}}{r_{SM}} \right) dB \qquad (5.48)$$

The coupling factor is

$$Q(\vartheta) = \frac{1}{\Gamma_L^2(\vartheta_L) \Gamma_M^2(\vartheta_M)}$$

For transducers with omnidirectional characteristic, $Q(\vartheta) = 1$; for directional transducers the coupling factors can be $Q(\vartheta) = 50–150$.

When for instance $Q(\vartheta) = 100$ and $r_{LM}/r_{SM} = 10$, the acoustic gain index according to Figure 5.12 results as $VE = 31$ dB with a feedback index $L_R = -9$ dB. With a ratio between the loudspeaker–microphone and source–microphone distances $r_{LM}/r_{SM} = 100$, it would even be possible to achieve an acoustic gain index of $VE = 51$ dB without positive feedback occurring.

SEVERAL AMPLIFICATION CHANNELS

It is also possible to calculate the acoustic gain index for a greater number of loudspeaker channels according to equation (5.43). If a loudspeaker arrangement with omnidirectional characteristics is used for the original sound emission ($Q_S = 1$), the acoustic gain index for centralized radiation ($r_{LHi} \approx r_{LH} \approx r_{SH}$) and equality of all partial powers ($k_{ji} = 1$) is

$$VE = L_R + 10 \log Q(\vartheta) \, dB + 10 \log \sum_{i=1}^{n} \frac{\Gamma_{Li}^2(\vartheta_{Hi}) \cdot r_{SMi}^{-2}}{\sum_{j=1}^{n} r_{LMji}^{-2}} \, dB \qquad (5.49)$$

Therefore

$$Q(\vartheta) = \frac{1}{\Gamma_{Lj}^2(\vartheta_{Lji}) \Gamma_{Mi}^2(\vartheta_{Mij})} \approx \frac{1}{\Gamma_L^2(\vartheta_L) \Gamma_M^2(\vartheta_M)}$$

If the distances between the *i*th microphone and the *j*th loudspeaker are equal, the resulting acoustic gain index is

$$VE = L_R + 10\log\frac{Q(\vartheta)}{n}\,dB + 10\log\sum_{i=1}^{n}\Gamma_{Li}^2(\vartheta_{Hi})\left(\frac{r_{LMii}}{r_{SMi}}\right)^2 dB$$

If all distances r_{LMii} and r_{SMi} are also equal, the result is

$$VE = L_R + 10\log\frac{Q(\vartheta)}{n} + 20\log\frac{r_{LM}}{r_{SM}} + 10\log\sum_{i=1}^{n}\Gamma_{Li}^2(\vartheta_{Hi})\,dB \qquad (5.50)$$

It is only if additionally the loudspeakers are aimed mainly at the listener's location, and the cross-couplings between the channels are negligible ($-10\log n$ dB $\rightarrow 0$ dB), that the resulting conditions are as usually specified for one reinforcement channel (see equation (5.48)).

If the individual distances r_{LMii} are different, only the smallest distance $r_{LMjimin}$ needs to be used in equation (5.49) as approximate value for calculation. In this case the acoustic gain index is

$$VE = L_R + 10\log Q(\vartheta)\,dB + 10\log\sum_{i=1}^{n}\Gamma_{Li}^2(\vartheta_{Hi})\,\frac{r_{LMji}^2}{r_{SMi}^2}\,dB\,\Bigg|_{r_{LMij} = r_{LMmin}}$$

For omnidirectional loudspeakers, and when using only one microphone for all loudspeakers, this equation simplifies to

$$VE = L_R + 10\log\frac{\displaystyle\sum_{i=1}^{n}r_{LM_{ji}}^2}{r_{SM}^2}\,dB$$

With approximately equal distances between the loudspeakers and the microphone, i.e. with $r_{LM_n} \approx r_{LM}$, the resulting acoustic gain index is

$$VE = L_R + 10\log n\,dB + 20\log\frac{r_{LM}}{r_{SM}}\,dB \qquad (5.51)$$

Under these conditions one can thus expect an n-fold increase of the achievable gain, as compared with the use of only one loudspeaker.

5.4.3 Conclusions

For rough calculations of the achievable sound reinforcement in enclosures and in the open it is sufficient to consider only one reinforcement channel:

that is, the one whose loop amplification is nearest to the feedback threshold. Below the feedback threshold of the critical channel, the resulting sound level values are not normally higher than with n reinforcement channels. The sound level distribution will, of course, usually be better than with only one channel or with a centralized loudspeaker arrangement in the middle of the room or in the middle of an open-air auditorium.

The procedure for ascertaining the sound level is as follows:

1. Determine the distance relations to be taken into account (r_H/r_{SM}, r_H/r_{LH} or r_{LM}/r_{SM}, etc.), where the subscripts mean: S, source; M, microphone; L, loudspeaker; H, listener.
2. Determine the actual directivity factor or the coupling factor, $Q(\vartheta)$.
3. Read the actual sound transmission measure, L_{XY}, from Figure 5.12.
4. Apply equations (5.40) or (5.44), (5.46) or (5.48), etc.

A more exact calculation of the sound level values can only be performed by means of a computer simulation program that excludes approximations and considers exactly the interactions that exist between the different operating quantities. For practical purposes, however, the above algorithm will suffice.

5.5 Required frequency transmission range and timbre

5.5.1 Causes of timbre change

In the recording and reproduction of a sound signal, as well as in the processing and transmission of the tone signal derived from it, a host of unintentional or undesirable linear distortions may occur that influence the timbre of the output signal. Among the unintentional timbre changes there are those that occur during the recording and reproduction as a result of the design and arrangement of the acoustic transducers. These include, on the recording side:

● the bass emphasis that results from close-range speaking into pressure-gradient microphones (the proximity effect);
● the treble emphasis caused with pressure microphones by the so-called pressure increase in the direct field of the source;
● comb-filter distortions produced by interference of the direct sound and the first reflections;
● the influence of the frequency-dependent equivalent absorption area of the recording room with greater microphone distances from the source;
● incorrect aiming of highly directional microphones;
● incorrect selection of microphone equalization.

Transducer-conditioned timbre changes are more likely to occur during reproduction than during recording. This is because microphone diaphragms are small in comparison to the wavelength of the sound signal to be transmitted across nearly the whole transmission range, which is not the case with loudspeakers. Nevertheless, a linear frequency response of direct sound reproduction has largely been achieved in the main radiating direction. However, this has been at the expense of the frequency linearity of the radiated sound power, especially for single systems.

The following factors may give rise to timbre changes in sound radiation:

- insufficient transmission range (bandwidth) of the radiating loudspeakers;
- peaks and dips in the response curve of the loudspeaker;
- radiating angles that are too small for the auditorium to be covered by a loudspeaker arrangement;
- cancellations caused by interference due to close-range reflections or by interaction between neighbouring loudspeakers;
- treble attenuation caused by high frequency propagation loss when covering large distances between loudspeaker and listener;
- bass emphasis due to a concentrated arrangement of loudspeakers radiating in phase and the consequent increase of the radiation resistance in the lower frequency range;
- selection of an ear-weighted monitoring loudness.

Unlike the timbre changes that occur during recording and reproduction, such changes rarely occur in the course of electronic transmission, since modern analogue and digital sound transmission systems generally have not only a sufficient transmission range (bandwidth), but also sufficient linearity.

A still frequent exception are the reverberation devices (see section 4.4.3). With the mechanical types, timbre changes often occur because of insufficient eigenfrequency density [5.9] or insufficient irregularity of eigenfrequency distribution. In the first case isolated frequency peaks occur, and in the second, comb-filter distortions. It is true, however, that sometimes well-designed devices are rejected because their reproduction behaviour is perceived as being too 'sterile'.

Some timbre changes are intentionally generated in sound processing. They may be used as an artistic means for deliberate modification of signals, as in the flanger effect and Leslie effect, which utilize the comb-filter effect in different ways (see section 4.4.2). Alternatively, it is possible to emphasize specific colorations or peaks, as for example in intelligibility-enhancing effects used with speech transmission. This includes the level reductions effected in the lower frequency range to enable sufficient clarity of reproduction in the case of high loudness levels.

Several of these problems have already been treated in Chapters 3 and 4. In the following we are going to explain further questions that are of special importance for the mechanism of timbre changes.

5.5.2 Transmission range (bandwidth)

A universally applicable electroacoustic system should be capable of transmitting or reproducing all occurring audible signals and sounds with true level and frequency response. With traditional music this frequency range extends from 40 Hz to 18 kHz (see Figure 5.6); with electronically generated music it may be still larger. Modern analogue and digital sound transmission equipment easily allows transmission ranges of between 20 Hz and 20 or 25 kHz, the limits of the transmission range being essentially a question of the transducers used. But it is not always sensible to transmit this maximum frequency range. In the lower frequency range it may well be that

- the clarity of the signal is impaired because of prolonged transient decay characteristics of the room as well as of the loudspeakers, or that
- frequency peaks occur as a result of interference between the radiated and the reflected signals, and an insufficient eigenfrequency density of the excited room.

An excessive expansion of the upper frequency range may bring about the following disadvantages:

- audibility of otherwise inaudible distortions;
- occurrence of differential tones in the required transmission range;
- with decentralized sound reinforcement systems functioning with delay equipment, mislocalizations caused by excessive transmission of high-frequency components at larger distances from the source (precedence effect).

Questions of expenditure, especially concerning the selection of the loudspeakers, and in the lower frequency range also the installation conditions (dimensions that are too large), are arguments that preclude the selection of too wide a transmission range. In general the transmission ranges given in Table 5.5 ought to be sufficient.

5.5.3 Timbre changes of the transmitted signal

A timbre change in sound transmission affects the definition of speech and the quality of music reproduction. It normally has a negative effect, but in certain cases it may also be positive (brilliance enhancement of music, improvement of articulation of speech by response peaks and dips due to interferences, etc.).

Table 5.5 Examples of transmission ranges of typical information systems

Minimum frequency range	Coverage area
400 Hz to 3 kHz	Public address systems in noisy rooms
200 Hz to 4 kHz	Speech transmissions at sports events, rallies, open-air meetings
100 Hz to 8 kHz	Transmission of information and entertainment programmes in foyers, department stores, waiting rooms, booking halls
80 Hz to 10 kHz	Transmission of entertainment music in restaurants, culture rooms
63 Hz to 15 kHz	Transmission of music and sound effects in theatres, concert halls, multi-purpose halls, etc.
31 Hz to 15 kHz	Direct music transmission from electronic instruments

According to section 5.3.1 the sound level of a loudspeaker is

$$L = L_K + 10 \log P_{el} \, dB + 10 \log \left(\frac{16\pi}{A\gamma_L} + \frac{\Gamma_L^2(\vartheta)}{r^2} \right) \, dB \qquad (5.52)$$

(cf. equation (5.25)).

The influence on the frequency response of the electrical system is contained in P_{el}, and that on the loudspeaker in the main radiating direction in the characteristic sound level, L_K (see section 4.1.1.2). The influence of the absorption area and front-to-random factor become most evident in the diffuse field:

$$L_r = 17 \, dB + L_K + 10 \log P_{el} \, dB - 10 \log(A\gamma_L) \, dB \qquad (5.52a)$$

(cf. equation (5.24b).)

If the characteristic sound level, the installed power and the equivalent absorption area are frequency independent, the frequency response of the loudspeaker reproduction is proportional to the front-to-random index, $C = 10 \log \gamma_L \, dB$ (see also [5.10]).

With sound reinforcement systems in the open, where the diffuse field is of little importance, a similar treble drop may take place as a result of the air absorption's increasing with the frequency (see Figure 3.20). The sound level may among other things be heavily influenced by the frequency dependence of directional characteristic and of propagation loss:

$$L_d = L_K + 10 \log P_{el} \, dB + 20 \log \Gamma_L(\vartheta) \, dB - 20 \log r \, dB - D_r \qquad (5.52b)$$

(cf. equation (5.20a)).

The design of the loudspeaker arrangement may cause either an eclipse of certain frequency ranges with directional loudspeakers, or bass emphasis due to the increasing radiation resistance with concentration of in-phase loudspeakers, as in multiclusters for example.

Impairments of signal transmission due to hum, non-linear distortions, switching noises, etc., which in the end may also result in treble changes, are treated in Chapters 3, 4 and 7. In the following we shall, however, discuss tonal colorations (linear distortions): all factors in equation (5.52) will be examined from this point of view. Tonal colorations make themselves felt at the listener's location in the form of:

- frequency-response limiting;
- bass emphasis;
- treble emphasis;
- narrow frequency range peaks;
- residual tone formation by positive feedback and comb-filter effects.

5.5.3.1 Influence of changes in the amplitude–frequency characteristic

If one link of the sound transmission chain does not transmit the whole signal applied to its input, but limits at the upper and lower ends of its spectrum, an appreciable information loss may result, making itself noticeable mainly by timbre changes.

Such frequency response limiting is often caused by the insufficient transmission range of the loudspeakers (the weakest link of the chain). Because of the difference between the sound pressure (free field) and sound power (diffuse field) frequency responses, as well as the increasing directivity with the ensuing variation of the directional characteristic, it is in such cases often not possible to define a clear frequency limit. With high-quality systems the frequency band limits are therefore defined by a level drop ⩾3 dB or ⩾6 dB compared with a reference value in the middle of the frequency response curve (see e.g. section 7.2.2.2).

Given that the loudspeakers produce a higher power output and a wider directional characteristic in the lower frequency range, it is often convenient to radiate the low-frequency component of the signal only from near the stage, while the higher-frequency components are radiated by additional loudspeakers, mostly delay operated, at close range to the audience (e.g. below and above the balcony). Experience has shown that the same applies to the radiation of the highest-frequency components ⩾12 kHz. On the other hand any independently usable sound reinforcement system should be capable of transmitting the whole possible range.

The question of how to adjust the frequency response of a sound reinforcement system audible at the listener's location cannot easily be answered. For the near range of the operational area one should certainly aim for an essentially linear frequency response, since here the original sources are also heard almost linearly. At greater distances within a room this no longer holds true, since the original sources also show a strong increase of directivity, resulting in a reduction of the diffuse sound energy (see Chapter 7).

Large-distance radiation is influenced not only by the propagation loss, D_r (cf. equation (5.52b)), but also by its frequency dependence (see Figure 3.21). As in air-conditioned enclosures one may reckon with air moisture values of $\varphi = 60\%$, additional attenuations have thus to be taken into consideration only from 3–4 kHz upwards. With bad conditions in the open, however, this may already be the case at low frequencies. This also explains the varying propagation conditions in open-air theatres.

According to Toole [5.11], narrow-band notches in the amplitude response are much less disturbing than shallow but broad dips. This can be explained by the masking mechanism of the human ear (see Chapter 3). Everybody also knows from experience that no appreciable timbre changes are perceived in spite of considerable frequency irregularities in the room.

Apart from the width and depth of a dip, its position within the frequency response is of importance. Dips in the range between 600 Hz and 2 kHz (so-called mid-band dips) are perceived to be less disturbing, whereas between 200 Hz and 600 Hz and especially between 2 kHz and 6 kHz they are quite disturbing. According to Zwicker and Feldtkeller [5.4] the frequency range around 2 kHz is of particular importance in this respect, because it is decisive for the intelligibility of consonants. Experience of many years also shows that the brilliance of music is greatly reduced by level reductions at 2 kHz, whereas it is enhanced by boosts in this region.

Toole [5.11] explains the insensitivity of the human ear to mid-band dips at around 1 kHz by the fact that in this region the ear shows a 'physiological crossover': whereas below this region one perceives mainly the fundamental modulation ('carrier frequency'), above it one hears the envelope with a certain pitch. It is also known that below this frequency range acoustic localization is effected by inter-aural sound arrival time differences between the ears, while above this range it is by intensity differences.

5.5.3.2 Influence of reverberation devices

Sound reinforcement systems are also used for prolonging the reverberation in acoustically heavily damped rooms or enclosures (see Chapter 6). To this effect the directly picked-up sound signal is 'reverberated' and then radiated by one or several loudspeaker assemblies.

This, however, implies the risk that the reverberation equipment is not free from residual timbre changes, and that such a change will have a disturbing effect on the listener. Some of the causes of such a timbre change are:

- insufficient eigenfrequency density (see Kuhl [5.9]);

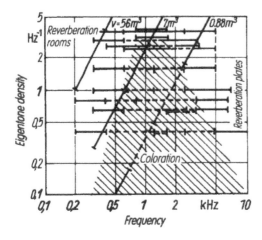

Figure 5.13 Timbre change (coloration) area of several reverberation devices as a function of frequency and number of eigenfrequencies in kHz band width [5.9].

- insufficient irregularity of the distances between the eigenfrequencies (insufficient time-related diffusion);
- unequal exciting conditions for individual eigenfrequencies (Figure 5.13).

By means of appropriate adjustment and using electronic 'room simulators' it is possible with modern reverberation equipment to generate eigenfrequency densities of sufficient diffusity as to make residual timbre changes no longer noticeable for most of the audience in halls (see also [5.2], pp. 34–35). Similar conditions are applicable to other processing equipment. In this regard it should be mentioned that timbre change effects are occasionally desirable.

5.5.3.3 *Influence of room-acoustical conditions*

If sound absorption is low, the excited energy of the reverberant sound is high. This results in a high diffuse-field level and thus in a small reverberation radius. In this case the sound pattern of the electroacoustically generated signal is superposed on that of the steady-state room, as is shown by the expression $10 \log(A\gamma_L)$ dB in equation (5.52a). This can be prevented only by enlarging the reverberation-free area by means of highly directional loudspeakers to such an extent that the reverberant sound is largely suppressed within the audience area. A reduction of the reverberation time is not in itself possible by the use of loudspeakers.

In heavily damped rooms, such as theatres and cinemas, however, the design of a sound reinforcement system is simpler, because the sound spectrum of the direct field of the loudspeakers becomes effective at the listener's location. In certain cases, as in the presence of rear-wall reflections for example, disturbing echoes or other comb-filter effects may take effect under these conditions (see section 3.2.2.3). Similar conditions apply in the open. Here the disturbing factor is mostly the absence of 'intermediate reflections', making it easier for individual loudspeaker signals or strong reflections to predominate and give rise to echoes and interference effects.

Rooms with reverberation times of around 1–1.5 s are generally favourable for the operation of sound reinforcement systems. In such halls the diffuse sound energy is still sufficient to 'blur' early-to-late sound arrival time differences between various sources, but not so great as to reduce the critical distance of the loudspeakers so much that the room sound influenced by the frequency-dependent equivalent absorption area is superimposed on the electroacoustically generated direct sound.

5.5.3.4 Influence of the directional characteristic

The influence of the directional characteristic becomes noticeable only when the listener is within the direct-sound effective range of the loudspeaker: that is, within twice the critical distance. With carefully designed sound reinforcement systems this generally applies at least to the middle and upper frequency range to be transmitted.

Since directivity increases significantly with the frequency (Figure 5.14), the loudspeakers have to be arranged in such a way that the high frequencies of the direct sound radiated by them reach the whole audience. If this is not the case, one perceives in the lateral areas only a dull sound pattern as influenced by the reverberant sound. An extensive frequency independence of the directional characteristics is for this reason a criterion for good sound reinforcement loudspeakers. But loudspeakers with frequency-independent sound-pressure and sound-power curves (which means a constant front-to-random factor) are rare in practice. Loudspeaker manufacturers are constantly trying to get nearer to this aim by means of various technical innovations (special configuration of sound columns and horns, see section 4.1). The fact that along with the directivity increasing with frequency the effect of air absorption also increases is quite helpful in this respect by counteracting too intensive a radiation of high-frequency components at greater distances from the loudspeaker. Generally, however, a minimally frequency-dependent directional characteristic can be achieved only by equalizing the performance curve: that is to say by reducing the efficiency of the loudspeaker.

5.5.3.5 Influence of coherent signal repetitions (comb-filter effect)

If loudspeakers are arranged side by side at a distance of $d = 5$ m, and the listening position is at the same distance of $d = 5$ m from the left of the two

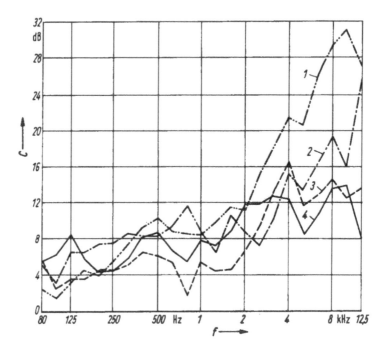

Figure 5.14 Front-to-random index of loudspeakers used in sound reinforcement systems: 1, TZ 133 sound column; 2, TZ 127 sound column; 3, TZ 128 sound column; 4, composite horn loudspeaker array (Klipsch).

loudspeakers (Figure 5.15b), there results for the direct sound signals of the two loudspeakers a path difference of $\Delta d = \sqrt{(d^2 + d^2)} - d \approx 2.07$ m. When a broadband coherent signal (e.g. a noise) radiates from both loudspeakers, there occurs at the listener's position a comb-filter-like reproduction with frequency intervals of about 164 Hz. The first 'in-phase frequency', f_1, with a corresponding pressure addition results herewith at 82 Hz. This maximum repeats itself every 164 Hz. The cancellation frequencies are also displaced from each other by 164 Hz: the smaller the path difference, the higher the first 'in-phase frequency' results. With path differences of $\Delta d < 10$ cm there results a residual tone of 3.3 kHz and higher; the comb-filter effect becomes inaudible (Figure 5.15a). The perceptibility of the comb-filter effect also disappears when using additional, though weaker, level components between the primary signals, additional coherence changes between the primary signals, or additional time shifts. These timbre changes caused by early-to-late sound arrival effects have already been treated in detail in section 3.2.2.3. Comb-filter effects are especially audible in the free field, since in rooms a masking of

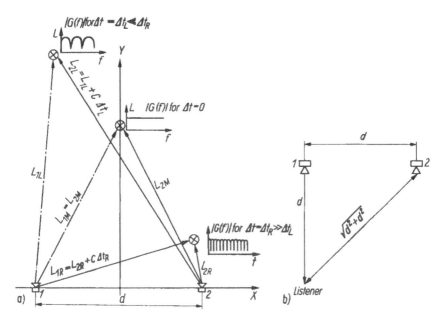

Figure 5.15 Formation of comb-filter curves: (a) dependence of the comb-filter effect on the position of the listener; (b) determination of comb-filter periodicity. Periodicity with $N(\sqrt{d^2 + d^2} - d) = d(\sqrt{2} - 1)N = \lambda N$ $(N = 1, 2, 3, ...)$. $f = c/[d(\sqrt{2} - 1)]$.

the effect occurs because of the incoherent reflections that occur beyond the critical distance.

Effects of this kind may come about not only by coherent signal repetitions, but also by natural coherent reflections. The effect makes itself noticeable upon excitation by noise or by very dense sound spectra (e.g. in speech by sibilants or fricatives) with very short repetition times of 0.5–15 ms, in the shape of newly generated sounds. This *repetition pitch* or (according to Kuttruff [5.12]) *residual tone* has in the case of phase-true repetition the frequency of

$$f = \frac{1}{\tau}$$

and is brought about by interference of the initial signal with its repetition after the time τ. With the above-mentioned repetition times of 0.5–15 ms the result is pitches of 60–2000 Hz. This distortion is very pronounced in the case of the double delay of a signal by means of **phase-linear digital delay equipment**. In the case of phase rotation of all frequency components between the initial signal and the repetition signal, the pitch of the residual tone is according to Bilsen [5.13] no longer directly proportional to $1/\tau$, but to a fraction or a multiple thereof (for instance with $180°$ to $0.88/\tau$ or $1.14/\tau$).

With music there may occur according to Müller [5.14] a melodic as well as a harmonic distortion, whereby it is possible for melody pitches to disappear as well as for residual pitches to occur. If these effects are brought about by room-acoustical reflections, they can be eliminated by damping the disturbing reflection areas by means of absorbers for frequencies above 300 Hz. In sound reinforcement engineering this occurs, for instance, if the sound is reflected by the speaker's desk and impinges together with the original sound on a microphone.

If this effect occurs as a consequence of a delay in the sound channel, the repetition signal has to be attenuated by at least -9 dB (according to [5.14] by at least -12 dB for speech and by at least -17 dB for music) (see also section 3.2.2.3).

The effect just described ought also to occur with the reproduction of coherent monosignals by any stereo system. According to Theile [5.15], however, this is not the case, since because of acquired *association patterns* the human being with binaural hearing does not perceive the timbre change that physically exists in the form of the comb-filter effect.

5.5.3.6 Influence of positive acoustic feedback

If a sound reinforcement system is operated within the range of its instability threshold, the level increase begins at peaks of the transmission curve between the determining loudspeaker and the microphone nearest to it. In the open, this transmission curve is frequently characterized by a comb-filter behaviour [5.2].

In rooms, a host of such comb-filter curves continue, forming the *transmission* or *frequency curve* (Figure 5.16). As described in [5.2], there are extensive investigations available on the statistical properties of these transmission curves. In this connection one fact is relevant for the present

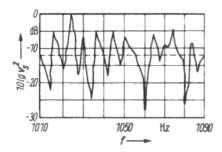

Figure 5.16 Extract from a steady-state frequency curve.

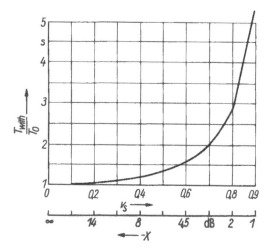

Figure 5.17 T_{with}/T_0 as a function of loop gain, v_s, or of the separation of the positive feedback loop, X, from the positive feedback threshold.

considerations: positive feedback always occurs at the respective peak of the transmission curve where the loop gain is then $v_S = 1$ (phase coincidence taken for granted) (cf. equation 5.37).

The margin of the relevant loop gain from the positive feedback threshold is decisive for the audibility of tonal changes. As a rule of thumb one can say that the average loop gain should not exceed values above 0.2. If this margin, X, is expressed in decibels, X should always be >7 dB ($v_S^2 = 10^{-X/10\ dB}$).

The nearer the loop gain of a frequency is to the positive feedback threshold, the more distinctly the feedback effects for this frequency become audible in the form of 'reverberation increase'. It has been proven that these effects can be safely avoided with a loop gain margin of $X = 15$ dB. Figure 5.17 shows the increase of the reverberation time (initial reverberation time of partly curved reverberation behaviours) as a function of the amount of actual loop gain. We can see that with a distance of $X > 10$ dB from the positive feedback threshold hardly any reverberation increase occurs. But since the peak values of a frequency curve exceed the fixed mean value (e.g. $\bar{X} = 15$ dB) by up to 10 dB, there occurs for this frequency with $\hat{X} = 5$ dB a 1.4-fold increase of the decay process, which, however, is not noticed on account of its insignificant energy.

To determine the maximum sound amplification at the respective listener's location, indication of a *feedback factor* $R(X) = v_S^2/(1 - v_S^2)$ has proved favourable (Figure 5.18). For an average loop gain margin of $X = 15$ dB one can indicate a value of $R(15\ dB) \approx 3 \times 10^{-2}$ as the maximally admissible feedback factor. This corresponds to a *feedback*

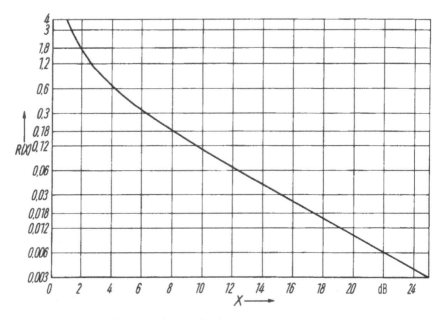

Figure 5.18 Dependence of the feedback factor, $R(X)$, on the separation of the average value of loop gain, X, from the positive feedback threshold.

index of

$$L_R = -15 \text{ dB}$$

In practice

$$L_R = -9 \text{ dB}$$

is often considered to be sufficient. For practical purposes it is possible to ascertain the average loop gain margin by increasing amplification briefly up to the positive feedback level and subsequently reducing the level until it reaches the value $\Delta L > 15 \text{ dB} - 9 \text{ dB} = 6 \text{ dB}$ (thus $X > 10 \text{ dB}$).

A feedback index of $L_R = -9 \text{ dB}$ is also recommended for the operation of *open-air* sound reinforcement systems. With this margin it is possible to avoid timbre changes and positive feedback phenomena. We refrain at this point from going into detail on procedures for positive feedback suppression, such as the use of frequency shifters, filters, and phase shifters (see also section 4.6). Detailed information on this is given in [5.2], pp. 93 ff. (see there section 2.4.4, pp. 33–34).

References

5.1 Reichardt, W., *Grundlagen der Technischen Akustik* (Foundations of technical acoustics) (Leipzig: Geest & Portig, 1968).

5.2 Ahnert, W. and Reichardt, W., *Grundlagen der Beschallungstechnik* (Foundations of sound reinforcement engineering) (Berlin: Verlag Technik, 1981).

5.3 Drejzen, I.G., *Sistemy elektronnogo upravleniya akustikoj zalov i radio – veshchatelnych studij* (Systems for electronic control of acoustics in halls and broadcasting studios) (Moskva: Svyaz', 1967).

5.4 Zwicker, E. and Feldtkeller, R., *Das Ohr als Nachrichtenempfänger* (The ear as communication receiver), 2nd edn (Stuttgart: Hirzel, 1967).

5.5 Peutz, V.M.A., 'Articulation loss of consonants as a criterion for speech transmission in a room', *Journal of the Audio Engineering Society*, 19 (1971) 11, 915–919.

5.6 Klein, W., 'Articulation loss of consonants as a basis for the design and judgement of sound reinforcement systems', *Journal of the Audio Engineering Society*, 19 (1971) 11, 920–922.

5.7 Steinke, G., 'Das Pegelprofil in der Tonstudio- und Rundfunkübertragungstechnik' (The level profile in sound studio and radio transmission engineering), *Technische Mitteilungen RFZ*, 33 (1989) 3, 49–55.

5.8 Krause, M., 'Grenzen der Aussteuerung' (Limits of modulation), *Funkschau*, 82 (1988) 4, 52–56.

5.9 Kuhl, W., 'Notwendige Eigenfrequenzdichte zur Vermeidung der Klangfärbung von Nachhall' (Eigenfrequency density required for avoiding timbre changes in reverberation), 6th International Congress on Acoustics, Tokyo (1968), paper E-2-8.

5.10 Kuhl, W. and Plantz, R., 'Die Bedeutung des von Lautsprechern abgestrahlten diffusen Schalles auf das Hörereignis' (The importance of the diffuse sound radiated by loudspeakers for the listening event), *Acustica*, 40 (1978) 2, 182–191.

5.11 Toole, T.E., 'Loudspeaker measurements and their relationship to listener preferences', *Journal of the Audio Engineering Society*, 34 (1986) Part 1: 4, 227–238; Part 2: 5, 323–348.

5.12 Kuttruff, H., Schroeder, M. and Atal, B.S., 'Perception of coloration in filtered Gaussian noise', 4th International Congress on Acoustics, Copenhagen (1962), paper H 31.

5.13 Bilsen, F.A., 'Repetition pitch: monaural interaction of a sound with the repetition of the same, but phase shifted sound', *Acustica*, 17 (1966) 3, 295–300.

5.14 Müller, L., Zur Klangfärbung durch Kurzzeitreflexionen bei Rauschen, Sprache und Musik (On the coloration by short-time reflections with noise, speech and music), 6th International Congress on Acoustics, Tokyo (1968), paper E-2-6, p. E 61.

5.15 Theile, G., 'Weshalb ist der Kammfiltereffekt bei Summenlokalisation nicht vorhanden?' (Why is the comb filter effect missing with composite localization?), *Proceedings of the 11th Tonmeistertagung*, Berlin (1978), pp. 200–214.

6 System layout

6.1 Types of system

The technical layout of a sound reinforcement system is essentially determined by the purpose envisaged and by the characteristics of the room in which it is to be installed.

According to the present state of the art we can distinguish the following types of sound reinforcement system:

- information systems;
- sound reinforcement systems with and without playback reproduction;
- sound reinforcement systems for improving room-acoustical parameters;
- sound reinforcement systems for ensuring acoustic localization of sound sources in the performance area;
- sound reinforcement systems serving as means of artistic expression.

It is also in this sequence that the technical expenditure and quality requirements increase for the systems.

An important consideration for the selection and arrangement of the sound reinforcement devices, and thus also for the selection of the technical solution to be adopted, is the spatial relation in which the addressed listener is located with regard to the original or supposed (playback) source of the transmitted sound event. This original source may be completely separated from the listener, as is the case with information systems (for instance in department stores, foyers, or stations). It may be located in the same room as the listener (in the performance area), but separated from the listener (in the reception area), or both areas may also overlap, as is the case with conference and discussion systems.

6.2 Information systems

In this context we understand by 'information systems' loudspeaker installations that serve for transmitting information to listeners distributed over a large area or many rooms. The broadcast station emitting the information is accommodated in a separate room and is thus (largely) invisible to the audience addressed. The information transmitted may be speech, music, and in special cases also a sound masking noise. Special variants are stage-manager paging systems in cultural centres as well as dispatch and loudspeaker systems in transportation terminals and factories, which are installed in enclosed spaces or in the open. A common feature of all these is that *positive acoustic feedback* is generally not a problem, if indeed possible.

The required transmission properties, such as transmission frequency range, balance of timbre and sound level as well as the maximum achievable sound level, the need for volume controls in designated areas and the availability of a priority microphone override facility, depend on the conditions for which the system is provided. Essential for the design and arrangement of the system are the room conditions. In most cases these conditions correlate with determined functional requirements, for which reason a range of typical variants of information sound reinforcement systems have been created.

6.2.1 Coverage of indoor rooms

The transmission range should be 100 Hz to 6.3 kHz or – for higher quality demands – 80 Hz to 8 kHz. If speech-only transmission systems are concerned, e.g. for an internal operating area (command/announcement/paging transmission), a transmission range of 200 Hz to 4 kHz may be sufficient. Leaving aside exceptional cases of extremely high environmental noise, the maximum sound level should be between 80 and 90 dB(A), and preferably around 85 dB(A).

6.2.1.1 Flat rooms

It is often necessary to cover one-storey rooms with a ceiling height of 2.5–6 m and considerable horizontal extension. Flat rooms of this kind are typical for

- foyers and concourses in airports, railway stations, congress and cultural centres, hotels, and theatres;
- restaurants;
- shopping centres and sales floors;
- open-plan offices;
- museums, galleries and exhibitions;
- workshops and storerooms.

For these sound reinforcement tasks various typical loudspeaker arrangements have been developed, the most important of which will now be described.

GRID-LIKE LOUDSPEAKER ARRANGEMENT IN THE CEILING AREA

The loudspeakers are uniformly distributed in the ceiling area, and radiate downwards. With this arrangement it is generally possible to obtain a uniform sound level distribution over a large area. This system is also very appropriate where very dense furnishing or partition walls exist between individual areas (e.g. exhibition stands).

To avoid flutter echoes between the room ceiling and the floor, it is necessary that either

- the ceiling is sound absorbing; or
- the floor is covered with a sound-absorbing (e.g. textile) material; or
- the floor or/and the ceiling surface is heavily structured (e.g. by furnishings or by supply lines or static construction elements standing out in relief in the ceiling area).

Calculation of the number of radiators required The spacing between the loudspeakers and thus the number of loudspeakers per surface depends on the uniformity of sound level distribution, the timbre, and the installation height of the loudspeakers. This number may also be influenced by the radiation characteristic of the loudspeakers, which in some cases is broadened by means of additional diffusers arranged in front of the loudspeakers.

Figure 6.1 shows the relevant relationships between the radiation angle and the separation of the loudspeaker from the mean head height above the floor (ear-height level). If the radiators are installed at a distance d above

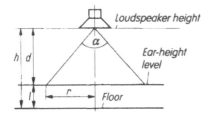

Figure 6.1 Radiation angle and loudspeaker height above ear-height level: $d = h - l$; $l = 1.2$ m for seated audience, 1.7 m for standing audience; r = radius of the coverage area; α = radiation angle.

Radiation angle	Properties
60°	Maximum balance, e.g. for lecture and conference systems
90°	Good balance, e.g. for music and speech transmission systems in restaurants, foyers, cultural centres, etc.
120°	Still acceptable balance, e.g. for information systems for music transmissions with occasional speech announcements

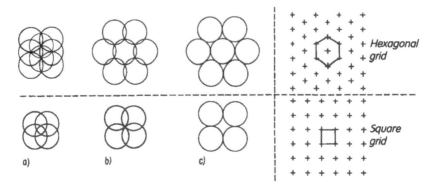

Figure 6.2 Installation grids for ceiling loudspeakers: (a) centre to centre; (b) minimum overlap; (c) rim to rim.

ear-height level and radiate downwards with an opening angle α, the number N of radiators required for covering the surface xy is (cf. also [6.1])

$$N = F\frac{xy}{r^2} \quad \text{with} \quad r = d \tan\frac{\alpha}{2} \tag{6.1}$$

The radiators may according to Figure 6.2 be arranged close together or spaced more widely. Table 6.1 indicates the irradiation area, the corresponding coefficients, F, and the radiator spacings, s, for the given overlapping degrees. The optimum solution, also economically, is given with a minimum overlapping of the radiation areas at ear-height level.

In rooms that extend mainly in one direction (e.g. corridors) it is sufficient to use radiator spacings, s, according to the indications given in Table 6.1. These spacings are about (1 to 2) $d \tan \alpha/2$ (see Figure 6.1).

Based on Table 6.1, the number of loudspeakers, N_Q (with square spacing) and N_H (with hexagonal spacing), for practical applications has been grouped in Table 6.2 for a reference area of $xy = 100 \text{ m}^2$.

We can see that with a ceiling height of 3 m above ear-height level and a radiation angle of $90°$ for the loudspeakers for full coverage one

Table 6.1 Loudspeaker coefficients as a function of ceiling height and radiation angle for square (Q) and hexagonal (H) spacing

Type of overlap	F_Q	s_Q	F_H	s_H
Centre to centre	1	r	$2/\sqrt{3}$	r
Minimum overlap	1/2	$\sqrt{2}r$	$2/(3\sqrt{3})$	$\sqrt{3}r$
Rim to rim	1/4	$2r$	$1/(2\sqrt{3})$	$2r$

Table 6.2 Number of loudspeakers required per 100 m² area

d (m)	Centre-to-centre overlap		Minimum overlap		Rim-to-rim overlap	
	N_Q	N_H	N_Q	N_H	N_Q	N_H
Radiation angle, $\alpha = 90°$						
1.5	45	52	23	18	12	13
2	26	29	13	10	7	8
2.5	17	19	9	7	5	5
3	12	13	6	5	3	4
3.5	9	10	5	4	3	3
4	7	8	4	3	2	2
4.5	5	6	3	2	2	2
5	5	5	3	2	2	2
5.5	4	4	2	2	1	1
6	3	4	2	2	1	1
6.5	3	3	2	1	1	1
7	3	3	2	1	1	1
Radiation angle, $\alpha = 60°$						
1.5	134	154	67	52	34	39
2	76	87	38	29	19	22
2.5	49	56	25	19	13	14
3	34	39	17	13	9	10
3.5	25	29	13	10	7	8
4	19	22	10	8	5	6
4.5	15	18	8	6	4	5
5	13	14	7	5	4	4
5.5	10	12	5	4	3	3
6	9	10	5	4	3	3
6.5	8	9	4	3	2	3
7	7	8	4	3	2	2
Radiation angle, $\alpha = 45°$						
1.5	260	300	130	100	65	75
2	146	169	73	57	37	43
2.5	94	108	47	36	24	27
3	65	75	33	25	17	19
3.5	48	55	24	19	12	14
4	37	43	19	15	10	11
4.5	29	34	15	12	8	9
5	24	27	12	9	6	7
5.5	20	23	10	8	5	6
6	17	19	9	7	5	5
6.5	14	16	7	6	4	4
7	12	14	6	5	3	4

needs $N_Q = 6$ radiators with square spacing and $N_H = 5$ radiators with hexagonal spacing (with minimum overlapping).

With a narrower radiation angle of $\alpha = 60°$ the number of radiators required increases to $N_Q = 17$ and $N_H = 13$. The spacing of the radiators with 90° radiation angle is $s_Q = 4.2$ m (with square spacing) and $s_H = 5.2$ m (with hexagonal spacing). With a radiation angle of 60° the spacings are 2.5 m and 3 m, respectively.

For information systems where transmission of high definition is essential, the installation height of the loudspeakers should not exceed 6 m (maximally 8 m). Greater heights are appropriate only for very heavily damped rooms. With greater installation heights several widely spaced loudspeakers will be perceived simultaneously: because of travel time differences an enhancement of spaciousness will occur that is detrimental to the definition and intelligibility of transmission.

Loudspeaker installation. It is possible to choose from a wide range of decentralized grid-like loudspeaker installations. The aesthetically most satisfactory design consists in the integration into a closed or acoustically transparent false ceiling.

For installation in a *closed false ceiling* it is possible to use open loudspeaker chassis, which are mounted directly above the acoustically transparent openings in the ceiling. The false ceiling serves in this case as a practically 'infinite' baffle. For this type of installation, special brackets are available that allow a very simple and clean mounting without requiring an additional mounting opening in the ceiling. Figure 6.3 shows a loudspeaker in a bracket of this kind, which is supplied with either a round or a square cover.

If the loudspeakers are installed above an acoustically transparent false ceiling, as used for example with a sound-absorbent covering for room damping, they must be enclosed in boxes or arranged on baffle boards so as to avoid an 'acoustic short-circuit' through the ceiling perforation. They can then be mounted without the need for a special opening directly above the ceiling board, so that the architectural appearance is not impaired. The precondition for this mode of installation is a sufficient acoustic transparency of the perforated boards or laminar false ceilings. This transparency depends on the degree of perforation of the visible surface: that is, on the ratio between the open and closed portions of the surface beneath the radiation area of the loudspeaker. However, it also depends on the individual opening, on its depth, and to a minor degree on the spacing of the effective openings. All these factors determine the size of the air mass oscillating in phase with the frequency of the sound, and are thus decisive for the upper frequency limit for which the ceiling covers are transparent. This problem is dealt with extensively in the literature [6.2, 6.3]. A thin steel plate of 1–2 mm thickness, even if it has a relatively low degree of perforation of about 15%, may be more favourable than a thicker gypsum board (plasterboard) panel with a

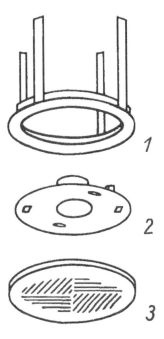

Figure 6.3 Ceiling loudspeaker in bracket: 1, bracket; 2, loudspeaker; 3, cover.

considerably higher degree of perforation. But also for thicker plates many small openings are more favourable than few bigger ones, since the latter may additionally give rise to narrow-band blocking resonances (frequency-selective attenuations or notches) in the transmission range.

The loudspeakers are generally mounted in compact boxes. If the system is to be used for music transmission, these boxes should have a volume of at least 4–6 litres (0.14 cu ft–0.21 cu ft), otherwise the sensitivity or the low-frequency transmission range decreases unacceptably.

LOUDSPEAKER ARRANGEMENT BELOW THE CEILING

If ceiling installation of the loudspeakers is not practicable, there are three alternatives available, as follows.

- The loudspeakers are installed directly under the ceiling, and radiate downwards.
- The loudspeakers are arranged at walls and supports, and radiate horizontally or at a slight angle.
- The loudspeakers are suspended from the ceiling by means of long pendants, and radiate downwards. (This solution is also adopted when the rooms to be covered are higher than 6 m, or if the ceiling area is heavily occupied by mechanical and electrical services, bridging joists, etc.)

Suspended loudspeakers Whereas for loudspeakers installed directly below the ceiling or suspended from short pendants in principle the same conditions as for loudspeakers installed into the ceiling are applicable, other installation variants require observation of additional measures.

In very reverberant, high rooms one can achieve, for instance, a significant improvement of the definition of the signal to be transmitted by suspending the loudspeakers closely above ear-height level (Figure 6.4). This holds true especially if the area irradiated is sound absorbent by its nature (which is always the case with areas occupied by audience), and if the upward radiation of the loudspeakers is deliberately reduced. Both conditions contribute to minimizing acoustic excitation of the upper part of the room.

The installation height of the loudspeakers has to be chosen in such a way that the audience is still within the critical distance of the loudspeakers. Making use of equation (5.9), and under the condition that the radiators have a radiation angle of 90°, an installation height above ear-height level of

$$d_{max} = 0.7r_R \tag{6.2}$$

should not be exceeded.

> **Example.** With a front-to-random factor of the individual loudspeaker of $\gamma_L = 6$ at 500 Hz, an equivalent sound absorption area of the room of $A = 55$ m², a total surface area of the room of $S = 600$ m², and a sound absorption coefficient of the irradiated surface of $\alpha = 0.8$, one obtains a critical distance of $r_R = 5.5$ m. The installation height should in this case not exceed 3.8 m.

The radiators may be arranged according to Table 6.1, to which end it is necessary to enter in equation (6.1) the respective distance d from loudspeaker to ear-height level.

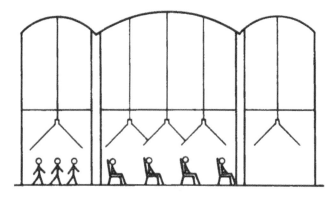

Figure 6.4 Suspended directional loudspeakers to avoid excitation of the upper reverberant space.

Horizontally radiating loudspeakers. The main advantage of this type is that, thanks to the directional characteristics, it is possible to choose a wider loudspeaker spacing. This may provide a solution in historically or architecturally listed rooms for instance, where a loudspeaker installation in the ceiling area is not possible. It should also be preferred if the risk of flutter echoes between ceiling and floor can otherwise not be eliminated.

The general installation principles are as follows.

- The loudspeakers must be installed above the head level of standing persons, in order to avoid masking by the audience or direct irradiation in the near range of the loudspeakers (<1 m). The preferred installation height is between 2 and 3 m. If a greater height cannot be avoided, the main radiation direction should be inclined accordingly.
- The spacing between neighbouring loudspeakers should not exceed 17–20 m.
- The loudspeakers should be aimed so that no initial or phase-coherent wavefront impinges on a smooth reflecting surface.

A special approach for covering extensive rooms or long corridors uses radiators with bidirectional characteristics and consists for example of two loudspeaker boxes joined back to back. As shown in Figure 6.5, these double loudspeakers are arranged in a staggered pattern so that the widest coverage area of one loudspeaker fits into the narrowest coverage area of the staggered next loudspeaker. In this way it is possible to obtain a relatively uniform sound level and timbre distribution.

Loudspeakers with bidirectional characteristics are also used for covering long and relatively reverberant corridors. Level compensation is achieved by use of the reverberant field, which with narrow sidewalls still produces acceptable results.

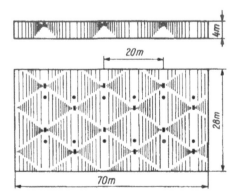

Figure 6.5 Double loudspeakers arranged in a staggered pattern for covering a large-surface room (•, loudspeaker supports).

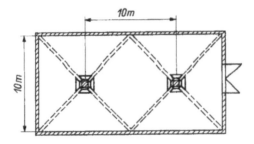

Figure 6.6 Ceiling loudspeakers concentrated in clusters radiating out and downwards in a star-like pattern.

A less favourable level distribution is achieved by means of the loudspeaker arrangement shown in Figure 6.6, which, however, is used because of its favourable mode of installation.

CALCULATION OF THE INSTALLED POWER FOR CEILING LOUDSPEAKERS IN DAMPED FLAT ROOMS

It may be taken for granted that the loudspeaker nearest to the listener essentially determines the sound level. Assuming an ear-height level of 1.5 m above the floor (average head level between sitting and standing person), the resulting sound pressure level according to equation (5.20) is

$$L = L_K + 10 \log P_{el} \text{ dB} - 20 \log(H - 1.5) \text{ dB} \quad (6.3)$$

where L_K is the characteristic sound level of the loudspeaker used, P_{el} is the installed loudspeaker power in W, and H is the room height in m.

Thus the installed power for one loudspeaker is

$$P_{el} = 10^{[L + 20 \log(H - 1.5) \text{ dB} - L_K]/10 \text{ dB}} \quad (6.3a)$$

The characteristic sound level of loudspeakers mounted on a baffle board is about

$$L_{Kf} = 90 \text{ dB}$$

of large-volume compact box loudspeakers (>6 litres) is

$$L_{KK} = 86 \text{ dB}$$

and of miniature compact boxes with a wide transmission range is

$$L_{EKK} = 84 \text{ dB}$$

These values are to be considered only as guidelines, and should be checked in actual cases.

The overall installed power results from the number of loudspeakers per area to be covered with sound multiplied by the ceiling area and the individual installed power of the loudspeakers.

6.2.1.2 *Factory and exhibition halls*

If the information is to be transmitted into large halls with a frequently high noise level, relatively powerful loudspeakers are generally required. Because of the large area and the relatively high reverberation time and spaciousness of these halls, and the impairment of sound propagation by built-in structures such as screens, craneways, signboards and publicity boards, decentralized coverage by means of suspended ceiling loudspeakers is often the only practicable solution.

In elongated and relatively narrow halls, powerful radiators such as sound columns along one longitudinal wall or – for somewhat wider halls (maximum 20 m width) – along the two longitudinal walls may also be suitable. The installation height and thus the inclination of the radiators depend on the local conditions: built-in structures etc. may have a masking effect. The installation height should, however, not be below 3 m or above 6 m.

6.2.1.3 *Complexes of individual rooms*

In large building complexes it is often necessary to supply many individual rooms from a central source. Examples are hotels with several restaurants and lounges, leisure and sports facilities with numerous training areas, administration and transportation buildings. Loudspeaker installations combined in a stage-manager paging system for individual and ensemble dressing-rooms as well as lounges and workshops for artists and technicians in cultural centres may also be included in this category. In all these cases it is necessary to provide good speech intelligibility and an adequate, often varying, loudness level for every room. In the coverage area it must be possible for the loudness level to be adjusted to adapt it to conditions that vary over time. The main control facility must, however, be able to override this adjustment for certain designated and emergency announcements. Volume controls with a power-handling capacity between 6 and 100 W are available for these purposes.

The type and arrangement of the loudspeakers are determined by the type and size of the rooms to be covered. In larger rooms and corridors one often proceeds according to section 6.2.1.1. In higher rooms one may also use centralized arrangements, for example in the form of monoclusters. Small rooms are usually provided with wall or ceiling loudspeakers. In particular cases, such as in administration rooms, table loudspeakers are also used.

Although the quality requirements of these installations are as varied as their intended uses, they all have it in common that good intelligibility is required. This is why a wide frequency transmission range is waived, particularly in the low-frequency region. With simple command or paging/announcement systems this is also permissible. For systems that are also intended to provide the listeners with music, the lower limit of the transmission range should extent to at least 150 Hz.

6.2.2 Sound coverage of outdoor and transportation areas

Typical examples for the use of information sound reinforcement outside buildings are:

- sound systems for large exhibition grounds, factory installations and other large open-air sites;
- sound and information systems for sports centres such as outdoor swimming pools, sports grounds, and stadiums;
- information systems for station platforms, bus terminals, etc.

It is common to all these systems that:

- weather-dependent propagation conditions may have to be reckoned with because of the large distances involved;
- heavily fluctuating noise levels with occasional high peak values may occur;
- interference with neighbouring areas may take place, which has to be avoided for reasons of noise protection or crosstalk to other areas of the system.

6.2.2.1 Large open spaces

Here the task is to cover large areas economically, without echo, and with an adequate signal-to-noise ratio (about 10 dB). Unlike indoor sound reinforcement, the diffuse sound caused by reflections may be almost completely neglected. For covering large distances, as is often the case with open-air sound reinforcement, one has however to reckon with an additional weather-dependent attenuation (see Figures 3.19–3.21). Since in most cases the systems concerned are used for speech transmission, a treble drop above 10 kHz is nearly always tolerable.

With sound transmissions over large distances one has to take care that listeners at close range to the loudspeakers are not exposed to excessive sound levels. In this case the definition of the transmission could be impaired by the occurrence of distortions in the ear; or temporary threshold shifts could occur (see section 3.2.3). This risk is especially relevant for a centralized arrangement of loudspeakers.

With a decentralized arrangement of the loudspeakers, however, 'double hearing' may occur when two wavefronts from two separately located loudspeakers or loudspeaker arrays impinge on the listener with a time difference of more than 50 ms, so as to be perceived separately. This may also occur if the listener hears the loudspeaker sound directly as well as via a strong reflection.

Not least one must ensure that the sound levels impinging on adjacent areas that are not to be covered are kept within the limits legally admissible by law.

CENTRALIZED ARRANGEMENT OF LOUDSPEAKERS

This is the common solution for the coverage of smaller sports facilities such as outdoor swimming pools and sports grounds. The loudspeakers are installed at a central elevated position near the paging station. In an outdoor swimming pool this may, for example, be the roof of the pool attendant's cabin around which the pools and leisure areas are grouped (Figure 6.7). The individual loudspeakers and loudspeaker arrays are dimensioned in such a way that the desired sound level, L, is achieved at the largest distance to be covered in each individual case. The installed loudspeaker power required to achieve this is calculated approximately according to equation (5.30) as

$$P_{el} = 10^{(L - L_K + D_r)/10\,dB} \cdot r^2/\Gamma^2(\vartheta) \tag{6.4}$$

where P_{el} is the installed loudspeaker power in W, L is the required sound level, L_K is the characteristic sound level of the loudspeaker concerned, r is

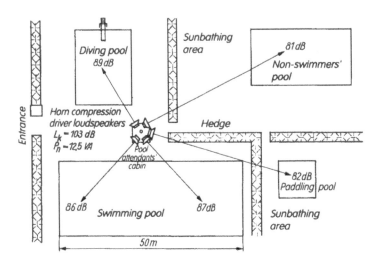

Figure 6.7 Sound reinforcement system for an outdoor swimming pool.

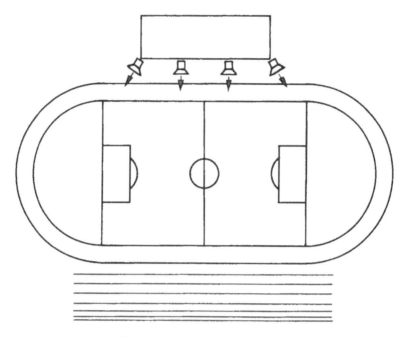

Figure 6.8 Centralized sound reinforcement system at a sports ground.

the maximum distance in m of the radiator from the most remote location to be covered, D_r is the climate-dependent propagation loss that has to be considered according to Figure 3.19 when distances to be covered exceed 40 m, and $\Gamma(\vartheta)$ is the angular directivity ratio of the radiator used.

In order to avoid inadmissibly high sound levels in the near field of the loudspeakers, they must be installed at a sufficient distance from possible listeners' locations. This requirement can in most cases be met only by an adequate height of installation.

To achieve good intelligibility, the frequency range of 250 Hz to 2.5 kHz has to be transmitted with sufficient intensity. Because of the increased directivity in the upper frequency range, in large areas this is possible only by incorporating additional treble loudspeakers to cover the far and intermediate regions.

It is not always possible to realize centralized radial coverage. On sports grounds, for instance, the loudspeakers have to be installed at the edge of the playing area, mostly in the vicinity of the grandstand (Figure 6.8), in which case the spacing of the loudspeakers should not exceed 15 m. If the system is used mainly for spoken announcements and only occasionally for music transmissions, as is usually the case with smaller sports grounds, it is possible to use compression driver horn loudspeakers. Such radiators have a characteristic sound level of $L_K \geqslant 100$ dB. With a required sound level of 85 dB, and with the loudspeaker aimed at the most distant point at $r = 34$ m, an installed loudspeaker power of $P_{el} = 36$ W is required. Since the total number of loudspeakers is four – that is, the two in the centre radiating towards the opposite stands and the two lateral ones radiating at an angle of about 30° towards the curved areas – an overall installed power of $P_{eltot} \geqslant 100$ W is required.

With larger sports grounds, higher demands for reproduction quality, and particularly for the transmission frequency range, are required. This means that radiators capable of producing a wider transmission range have to be used. Moreover, it is necessary to pay much more attention to the weather-dependent air attenuation, as well as to the availability of an adequate headroom.

When assuming the characteristic sound level for a high-quality sound column to be 96 dB, and requiring moreover an overhead of 10 dB, the calculation level has to be specified with 95 dB (compare the sports facility discussed above). Considering moreover a maximum weather-dependent propagation attenuation of 12 dB (bad sound transmission conditions due to heavy sun irradiation), there results, for the sports facility discussed above, with an identical centralized loudspeaker arrangement, a required installed loudspeaker power of $P_{eltot} \approx 60\ 000$ W. It is obvious that in this case a centralized coverage is no longer feasible with this type of loundspeaker.

DECENTRALIZED ARRANGEMENT OF THE LOUDSPEAKERS

This should always be employed when large and extensively distributed areas have to be covered uniformly. This applies to streets, large squares, sectioned exhibition grounds, industrial plants, etc. As has been shown above, however, large sports stadiums can also normally be covered with appropriate quality only by means of a decentralized loudspeaker arrangement.

An important criterion affecting the decision is in all cases the maximum spacing of the loudspeakers arranged in one line. For economy, the spacing should be chosen to be as wide as possible, whereas any double hearing or creation of echoes has to be avoided as far as possible in the interests of good intelligibility. In this respect one must remember that with spacings below 17 m echo disturbances are not usually to be expected, since then the wavefronts originating from adjacent loudspeakers arrive at the listener's location within 50 ms. With spacings exceeding 17 m a level difference of >10 dB between the near and the remote loudspeakers has to be ensured. Under these conditions the nearer (and thus louder) loudspeaker masks the more remote one, so as to eliminate the risk of echoes.

Figure 6.9 shows the sound propagation between two loudspeakers arranged at a distance a from each other. At a point between the loud-speakers, i.e. at a distance r_1 from loudspeaker S_1 and r_2 from loudspeaker

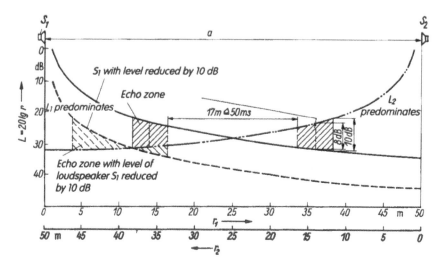

Figure 6.9 Sound level relations between two loudspeakers arranged at distance *a* from each other.

S_2, there exists a sound pressure difference of

$$L_1 - L_2 = 20\log(r_1/r_2) \text{ dB} = \Delta L$$

If the radiated sound pressure levels are equal, no echo disturbances occur in the region of

$$r_1 - r_2 \leqslant 17 \text{ m}$$

Since according to Figure 6.9 the distance between the loudspeakers, $a = r_1 + r_2$, the resulting maximum in-line spacing is

$$a = 17\left(\frac{2}{10^{\Delta L/20 \text{ dB}} - 1} + 1\right) \text{ m} \tag{6.5}$$

With $\Delta L = 10$ dB, as required according to Blauert [6.4], one obtains according to equation (6.5)

$$a = 33 \text{ m}$$

If according to Petzold [6.5] one admits $\Delta L = 8$ dB under otherwise equal conditions, there results a maximum in-line spacing of

$$a = 39 \text{ m}$$

If the level of one of the loudspeakers is reduced in comparison with that of the other, because of a lower characteristic sensitivity or different radiation characteristic for example, the in-line spacing has also to be reduced. This can be quantitatively assessed by adding the level difference caused by the different radiation to the minimum level difference of ΔL. If a level difference of 6 dB is added to the level difference of 10 dB required at the 17 m boundary (i.e. $\Delta L = 16$ dB), the maximum loudspeaker spacing is thus reduced to a value of

$$a = 23 \text{ m}$$

Hence it follows that to utilize the maximum loudspeaker spacing – without using a delay system – it is convenient to make adjacent loudspeakers radiate against each other with the same directivity characteristic and under the same radiation angle. In this way, for covering a long, narrow area, for example, it is preferable to use radiators with a symmetrical bidirectional (figure of 8) characteristic.

However, to cover larger areas, with longitudinal dimensions that are of the same order of magnitude as the transverse dimensions, it is necessary to arrange rows of loudspeakers staggered one behind the other. To this effect one normally uses directional radiators arranged so that the array following in the direction of radiation is located where the level of the first array has diminished to a level at which double hearing does not occur (Figure 6.10). Petzold [6.5] indicates for 'staggered' radiators with asymmetric bidirectional characteristics a maximum inter-line spacing in the radiating direction of 80 m, and an in-line spacing of 60 m. These values vary

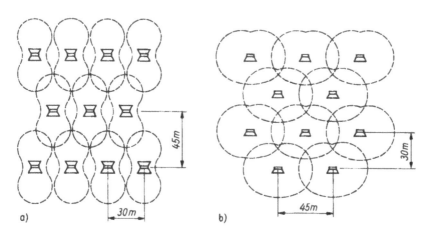

Figure 6.10 Loudspeaker arrangement for decentralized coverage of a large square: (a) loudspeakers with bidirectional characteristic; (b) loudspeakers with cardioid characteristic.

greatly with the design and level adjustment of the radiators, so it is difficult to generalize.

Figure 6.10 shows two solutions used in practice for loudspeakers with a bidirectional characteristic (Figure 6.10a) and a cardioid characteristic (Figure 6.10b). The sketches do not show that an arrangement with a bidirectional characteristic has to work *without* delay equipment, but that cardioid loudspeakers enable greater inter-line spacings, thanks to the possible delay.

A safe remedy for avoiding double hearing, and one that also allows an acoustic orientation on the action area, such as a playing area or field, consists in the use of *delay equipment* between the individual radiators or lines of radiators (Figure 6.10). Here the delay times have to be chosen in accordance with the sound travel times that are to be expected from the spacing. With a sound speed of $c = 344$ m/s at 20°C we can calculate a travel time of

$$t_L = 2.9 \text{ ms/m} \approx 3 \text{ ms/m}$$

(see equation (6.7) or Table 6.3).

Delay systems generally also enable large distances to be employed between the loudspeakers. It is important in this case, however, to ensure that the backward radiation – that is, the radiation in the direction opposite to that of the delay used – is suppressed, since otherwise the risk of echo formation caused by the travel time plus the additional delay increases significantly. It is for this reason that the radiators used under these conditions are exclusively of a unidirectional characteristic (for example, sound columns with a cardioid characteristic, or horn loudspeakers).

Table 6.3 Path–time relation at 20°C

s (m)	t (ms)	s (ft)
1	2.91	3.28
2	5.83	6.56
3	8.74	9.84
4	11.65	13.12
5	14.56	16.40
6	17.48	19.68
7	20.39	22.96
8	23.30	26.24
9	26.22	29.52
10	29.13	32.81

$s = 0.343t$, $t = s/0.343$
s in m, t in ms

If the range of transmission of the loudspeakers reaches beyond the intended coverage area, disturbance of adjoining neighbours can occur. Also – in the case of reflections at remote surfaces – echo interference may occur in the coverage area itself. Such overspill or over-coverage has to be avoided by a suitable arrangement and aiming of the loudspeakers. As a rule, the main radiating axis of the loudspeakers is aimed down onto the area to be covered. If the area is occupied by an audience, the radiated sound is absorbed; if the area is not occupied, then the sound is reflected more or less upwards according to the angle of incidence.

Important in this respect were formerly the so-called 'interference radiators', which consisted of two superposed loudspeaker systems whose phase relationship was chosen in such a way that only the near range was covered, whereas in the far range (above about 20 m) the signal was diminished by interference. This effect, however, was frequency dependent, so that the desired effect took place only for a narrow frequency range. Nowadays these radiators are therefore used only rarely.

If several rows of loudspeakers are staggered one behind the other in front of a large reflecting surface, such as a large glass façade, it is necessary to rotate laterally at least the last rows in front of the reflecting surface – in addition to the inclination in the main radiation direction – since here it is possible for the reduced attenuation decrease of a linear source to occur, which amounts to only 3 dB per distance doubling instead of 6 dB.

A still smaller propagation attenuation can be expected in the case of delay-time compensation systems employing signal delay equipment. In this case the loudspeaker rows staggered one behind the other enable the wavefronts to be constantly amplified, so that a higher energy output becomes available, which can cause overspill and over-coverage radiation. The indicated de-attenuation effects are nearly always heavily frequency dependent, since several frequency-dependent influences, such as the directional effect of the radiators, the energy summation obtained by the alignment and the attenuation decrease of the individual radiators, act together. Over large distances it is usually only the low-frequency components that are noticeable; however, they can give rise to significant disturbance.

6.2.2.2 *Transportation systems*

On platforms, under protective roofs, and to a certain extent in large enclosed spaces, the room-acoustical conditions that prevail lie between those of indoor rooms and those of open spaces. Because of the magnitude of the resulting equivalent sound absorption areas, one may generally expect free-field propagation conditions in such situations. However, it is necessary to aim at diminishing the perceived reverberation, which (especially with large halls) may often be excessive.

The *sound coverage of platforms* is one of the most frequent and also most complicated sound reinforcement tasks for transport facilities. This holds true especially if the platforms are jointly covered by a large, mainly closed dome. Under these conditions the sound reinforcement system has to meet the following requirements:

- realization of a largely uniform sound level and a consistent timbre along the total length (150–250 m) and width (7.5–15 m) of the platform;
- minimization of crosstalk to adjacent platforms (distance 7.7–20 m),
- adaptation of loudness to the constantly varying environmental noise levels (65–90 dB(A)).

These requirements can usually be realized only by means of a decentralized loudspeaker arrangement. Very directional loudspeakers with a defined beam-like characteristic and a transmission range of 200 Hz to 3 kHz are normally used. Restricting the transmission range to only that important for human speech enables better optimization of the radiation characteristic of the radiators.

Under these conditions pressure-chamber (compression driver) horn loudspeakers have proven themselves to be optimally suited. They are installed at a height of 3–5 m and at intervals of 7–15 m either in the centre of the platform or, if the platform carries many superstructures, in two rows offset from each other at the edges of the platform, in such a way that they all radiate in the same direction and that the main radiating direction is aimed at the platform floor below the following loudspeaker. With arrangement at the platform edge, the main radiating direction is swivelled by 5–10° towards the centre of the platform. It is important for such loudspeaker arrangements to avoid reflections, especially from the end of the line. Moreover, it is suitable to aim the loudspeakers at tunnel mouths, because on the one hand these areas are therefore directly covered, and on the other hand one largely avoids reflections from the platform surface, which could for instance result in reverberant sound excitation in an enclosed space.

Of late the loudspeaker arrangements described in section 6.2.1.1 have been used in station halls. These consist of cone loudspeakers, which are screened towards the top and installed with very close spacing (approx. 4–5 m) at a height of about 4 m over the central axis of the platform. This requires, of course, closely arranged mountings, which cannot always be realized. Since each loudspeaker radiates only a relatively small downward-directed acoustic power, a very direct coverage results, which helps to improve the intelligibility in a reverberant environment.

If in certain cases it is necessary to arrange the loudspeakers at intervals of more than 15 m, it is appropriate to make them operate in opposition to one another, but a separation of 25 m should not be exceeded (see equation (6.5)). The radiators must be of the same type, with equal level (power) adjustment (see section 6.2.2.1).

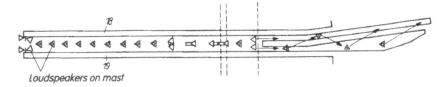

Figure 6.11 Loudspeaker chain for coverage of a platform of the main station at Leipzig.

Delay devices are also being used to an increasing degree for platform systems. For reasons of economy a symmetrical layout is normally used, in which the delay is zero in the centre of the long platform, and maximum at each end. However, it is also possible to establish a localization reference to a presumed source (such as the station staff on the platform or in the supervisory building). Also in this case one or two delay chains are formed. Figure 6.11 shows an example of the concept: the schematic structure of such loudspeaker chains using delay equipment for the coverage of the longitudinal platform 18/17 of the Central Station in Leipzig.

An important problem that is difficult to handle is the adjustment of the *optimum loudness*. Excessive loudness may produce disturbances on the neighbouring platforms, mask announcements given there, and irritate the passengers. Insufficient loudness reduces intelligibility in the area to be covered by preponderance of noise. The noise level is very variable. Measurements made in the station hall revealed the following values:

- on the empty platform, 65 dB(A);
- on the fully occupied platform (before the arrival of a train), 75 dB(A);
- during the arrival or departure of a train, 88–90 dB(A).

On a neighbouring platform and at 8 m distance from the exciting loudspeakers, a crosstalk attenuation of 4.5 dB was ascertained. This increased to 8 dB with the screening effect of a train standing on the track between the two platforms. The excitation loudspeakers were mounted at a height of 5.5 m above the platform and with an in-line spacing of 15 m.

These values indicate that *noise-dependent volume regulation* would be appropriate. This can be realized by means of an automatic volume control employing a microphone located at a noise-representative location, such as in the central area of the platform, to control the amplifier gain.

Another problem consists in obtaining appropriate *microphone signal levels*, since very differing sound pressure levels are to be expected at the microphone capsule. To avoid overmodulation of the amplification channel, as well as insufficient loudness levels, an AGC microphone amplifier is recommended for the platform paging points, all the more so because these paging points are mostly operated by poorly trained personnel and often in stressful situations. For this reason some railway administrations use prerecorded messages for routine announcements, which may be played back as required.

The audio power requirement for the loudspeakers depends essentially on the geometric conditions and on the characteristic sound level, L_K (Figure 6.12):

$$P_{el} = 10^{(L - L_K + 20\log\sqrt{a^2 + (h - 1.5)^2})\,dB/10\,dB} \qquad (6.6)$$

where L is the desired maximum sound level, P_{el} is the installed loudspeaker power in W, a is the maximum spacing of the loudspeakers in m, and h is the installation height of the loudspeakers in m.

With $a = 10$ m, $h = 4$ m, and a characteristic sound level of $L_K = 105$ dB for the compression driver loudspeakers to be used, and with a maximum sound level of 93 dB, this formula produces an installed

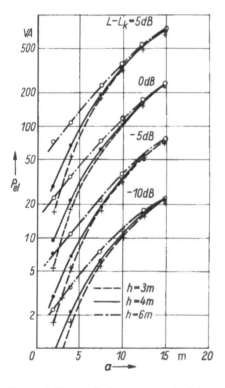

Figure 6.12 Installed power required for station loudspeakers as a function of maximum spacing, *a*, installation height, *h*, and difference between desired sound level, *L*, and characteristic sound level, L_K.

power per loudspeaker of 6.7 W. With a platform length of 200 m, 20 compression driver loudspeakers with a rating of approx. 10 W and an overall installed power of 200 W are required.

In addition to the usual platforms running along the tracks, terminal or dead-end stations also have *transverse platforms*, which are often more than 20–30 m wide. On one side these border onto the comb-like adjoining longitudinal platforms, while on the other side they are limited by the reflecting walls of the main building. Coverage of these platforms should therefore preferably be by loudspeakers with a cardioid characteristic mounted laterally on the walls of the adjoining buildings with a spacing of 20–30 m. It is important to ensure, however, that no travel time interferences occur with the loudspeakers of the adjoining platforms.

COVERAGE OF RECEPTION HALLS

Reception halls as well as corridors in station and airport buildings are covered in a similar way to foyers and, depending on the ceiling height, either by a ceiling loudspeaker installation or by laterally arranged sound columns (see section 6.2.1).

6.3 Sound amplification reinforcement systems

This category refers to systems in which sound pick-up as well as radiation of the amplified signal are effected in the same room or – in the case of open-air facilities – in the same area and immediately after amplification. Because of the inclusion of the original source (or primary source), additional problems arise:

- Any feedback of the amplified signal to the beginning of the amplifier chain consisting of microphone, amplifier and loudspeakers has to be avoided as far as possible, and positive feedback (see section 5.4) *must* be excluded.
- The listeners have to be oriented to the original source; in the ideal case it has also to be acoustically localized. This makes the listener concentrate on this source, and has a positive effect on intelligibility.
- The timbre of the radiated signal has to be adapted as far as possible to that of the original source (unless other, more important considerations, such as intelligibility and freedom from positive feedback, prevent this).

Many of the sound reinforcement systems actually in operation fall into this category. A system in this respect must consist at least of microphone, amplifier and loudspeaker.

With regard to the arrangement of the loudspeakers we can distinguish between centralized, centrally supported, and decentralized coverage.

6.3.1 *Centralized sound reinforcement systems*

The characteristic feature of centralized coverage is the concentrated arrangement of the loudspeakers near the action area where the original source, whose sound signals are to be amplified, is located. The loudspeakers are often combined in *clusters* or *monoclusters*. This arrangement has the advantages that:

- there is extensive coherence of the wavefronts of the loudspeaker sound (of the secondary sources) and generally also of the sound emitted by the original sources;
- the acoustic orientation of the listeners is largely directed towards the original source (although in the region near the action area there often occurs a mislocalization to overhead);
- no delayed sound originating from secondary sources can give rise to travel time interferences in the action area.

Because of the directional effect obtained according to equation (5.8a), this loudspeaker arrangement makes possible an increase of the critical distance:

$$r_R = \sqrt{\gamma_L}\, \Gamma_L(\vartheta) r_H$$

where γ_L is the front-to-random factor of the loudspeaker array, and $\Gamma_L(\vartheta)$ is the angular directivity ratio of this loudspeaker arrangement.

The angular directivity ratio $\Gamma_L(\vartheta)$ in equation (5.8a) may be approximately reduced to 1, since the loudspeaker is directed towards the audience area. The reverberation radius, r_H, of one loudspeaker may thus be enlarged by the square root of the front-to-random factor ($r_R \approx \sqrt{\gamma_L} r_H$). This arrangement allows a relatively large critical distance and thus a large direct-sound-dominated (reverberation-free) area to develop, which results in high intelligibility in medium-sized, acoustically difficult rooms.

If the action area is relatively small, such as a small platform of less than 17 m × 17 m area, a speaker's desk, or a boxing ring, it is possible with an appropriate arrangement of the loudspeakers to ensure correct acoustic localization without the need for delay equipment. A precondition in this regard is that the *precedence effect* must be brought to bear: that is, that the direct sound from the original source must reach the listener before the amplified signal of the central loudspeaker array, and that the level of this direct sound must not be more than 6–10 dB lower than that of the amplified signal. In general the level of the original source is not sufficient for this to occur, so that so-called *simulation loudspeakers* (or image-

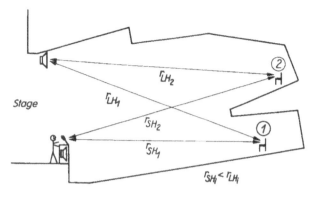

Figure 6.13 Geometric relations for centralized coverage without delay equipment.

stabilizing loudspeakers) are required for boosting the original sound in the region of the source. A typical example in this respect is the use of loudspeakers built into a speaker's desk (podium). Moreover, one must avoid a delay of more than 30–50 ms between the first wavefront arriving at the listener from the original source or its simulation, and the wavefronts of the amplified sources. Figure 3.37 shows the time and level conditions that have to be considered in this case (see section 3.2.1.5).

For succeeding without delay equipment, the following requirements are also applicable:

- The distance from loudspeaker to listener, r_{LH}, must be greater than the distance from the original source to the listener, r_{SH} (Figure 6.13). These conditions should also be complied with for the greatest possible distance between source and listener.
- Sufficient loudness of the original source has to be ensured (if necessary by means of simulation loudspeakers).

In addition to enabling an appropriate coincidence between visual and acoustical source localization, the centralized loudspeaker arrangement also offers the advantage for the listener that there are either no or only slight travel-time differences, so that no unintentional enhancement of the apparent or perceived 'spaciousness' can take place.

This condition becomes evident in Figure 6.14. We can see that, with a large time difference of the sound arriving at the lateral front seats (H_1) the effective critical distance, r_R, of the loudspeakers becomes reduced by the parallel arrangement, since any one of the loudspeakers increases the reverberant component in the area of the other one. For a listener's location H_2 in the central or rear area of the hall this is not the case, since the travel time from the two radiators L_1 and L_2 is almost equal.

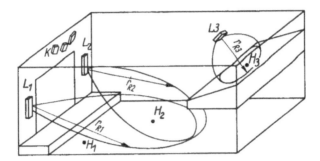

Figure 6.14 Sound-field relations with different loudspeakers arrangements: L_1 and L_2, loudspeakers at the stage proscenium (left and right) with critical distances r_{R1} and r_{R2}; L_3, supporting (infill) loudspeaker above the balcony, with critical distance r_{R3}; K, central loudspeaker cluster above the stage.

The critical distance depends not only on the room-acoustical and loudspeaker technical data, but also on the way the loudspeakers are arranged. A supporting loudspeaker L_3 in the rear area of the hall, for instance, always has a definition/intelligibility-reducing effect for the front seats. Its energy therefore has to be radiated directly onto the audience to be covered, so that it becomes largely absorbed. At a rear seat H_3, however, the front loudspeakers may also contribute an enhancement of the definition (and intelligibility) of the signal, if the signal is supplied to the supporting loudspeaker with such a delay that it arrives at the listener simultaneously with (or slightly later than) the signals from the front loudspeakers.

This problem does not arise with a monocluster (K). Here it is necessary only to observe the travel-time difference between the paths from the original source to the monocluster and from the monocluster to the listener, as well as the uniformity of sound level distribution.

SPORTS HALLS

The loudspeakers are often combined in a cluster consisting mostly of loudspeaker arrays (sound columns) with a cardioid characteristic or horn loudspeakers (Figure 6.15). This results in a radiation with a flat, circular directional characteristic. This directional effect, as for most loudspeaker arrays, is less pronounced for the low frequencies than for the middle and high-frequency ranges. Because of this, and because the radiation impedance increases in the low-frequency range, there is likely to be a significant bass emphasis in the diffuse field, especially outside the main radiating direction. This particularly affects the area below the cluster, and has to be considered when choosing and arranging the individual loudspeakers. A proven technique is to complement the horizontally and omnidirectionally radiating

a)

b)

Figure 6.15 Sound reinforcement in sports halls by means of monoclusters: (a) sound column cluster; (b) horn cluster.

loudspeaker units with special loudspeakers that radiate the middle and high-frequency range more intensively downwards. If necessary, such loudspeakers can also be aimed directly at the area below the monocluster (see also Figure 6.15b).

For halls with a platform at one long or front side, which can also be used for cultural events, monocluster systems have been constructed that can be brought into their optimum position by means of travelling trolleys (as in the old sound reinforcement system of the Gruga-Halle in Essen [6.6], for example). Among these clusters there are also variants that are divisible, and which may be located for example on the left and the right above the edge of a platform. Nearly all these clusters are adjustable in height.

For non-travelling clusters arranged in front of a platform and used as the main sound reinforcement system for cultural programmes, a subdivision into various circuits is nevertheless suitable. Thus separate level adjustment and possible filtering of individual arrays becomes feasible, enabling the loudspeakers aimed at the platform to be used for monitoring, the downward-radiating ones for optimizing the timbre and the sound in the area near the platform, and the ones radiating away from the platform for determining the sound level in the main coverage area.

In very large halls it is no longer possible to cover the entire hall by means of a single loudspeaker cluster with a sufficiently high and uniform sound level. In such cases the principle of centralized coverage has to be abandoned, although even so it is possible to use one advantageously positioned cluster to provide the main coverage [6.7].

THEATRES AND HALLS IN CULTURAL CENTRES

In Japan and the USA, but less in Europe, a sound reinforcement variant has become generally accepted that uses a loudspeaker cluster centrally arranged over the stage opening and radiating in a number of different directions into the auditorium area (Figure 6.16). Apart from aesthetic objections this implies the problem of a merely monophonic source that, although stage related, is localizable only in the region of the cluster. (In the actual case of Figure 6.16 this arrangement was soon abandoned.)

To avoid point-like localization, the centralized loudspeaker arrangement in theatres and cultural centres is often subdivided into various subarrays (Figure 6.17). These subarrays are arranged beside and above the stage opening, and thus offer the advantage of two-channel or even, in halls with stages more than 10 m wide, stereophonic coverage. Such arrangements are indispensable for theatrical utilization of the system, but with single-channel operation they have the disadvantage of producing a strong lateral position shift, especially in the front seats. This disadvantage is particularly noticeable if the lateral loudspeaker arrays are arranged at a very low level, so that the travel times to large areas of the auditorium are lower than those of the original sound from the stage. This problem can largely be solved by

Figure 6.16 Monocluster in the Theater des Westens in Berlin.

arranging the lateral loudspeaker arrays at a higher level, and by installing above the opening a powerful loudspeaker array that normally operates as a central sound reinforcement system.

This explains why 'loudspeaker stacks', as frequently used by bands, often produce the desired effect when operating from the depth of the stage, and yet when located in front of the stage opening lead to strong

Figure 6.17 Proscenium loudspeakers of the Royal Theatre, Copenhagen, with subdivided central loudspeaker arrangement.

mislocalization as well as non-uniform level distribution and an unbalanced sound pattern in the front lateral areas.

CHURCHES, AND HALLS FOR CHORAL AND SYMPHONIC MUSIC

To obtain satisfactory speech intelligibility in such very reverberant rooms, a sound reinforcement system is required that is aimed at the audience. In this case centralized radiation by one single array may be advantageous. Because of the decaying reverberant sound, a reinforcement system that consists of various loudspeakers at different locations would unnecessarily reduce the coherence and the effective critical distances, so that increased expenditure on equipment might actually produce less favourable results.

Figure 6.18 Centralized loudspeaker arrangement in a modern church building.

Figure 6.18 shows a central monocluster in a modern church. The individual horn loudspeaker systems, which are aimed into the room in different directions, are arranged behind a concealing screen.

In large gothic churches, which sometimes have extremely long reverberation times and a large number of columns dispersing the sound and screening the side-aisles, a centralized loudspeaker arrangement is generally not successful. Decentralized arrangements with highly directional lines of loudspeakers installed at close range to the audience are to be preferred here (see section 6.3.3).

6.3.2 *Centrally supported sound reinforcement systems*

Here we are referring to systems that, although operating with decentralized booster loudspeakers, nevertheless convey the sound impression of a centralized system localizable in the action area. Strictly speaking, two or more loudspeaker arrays are no longer part of a truly *centralized* system. However, if they are all arranged in one plane (for example, in the stage area to the left and right, and above the proscenium), we may still speak of a centralized system (the precondition is extensive coherence of the wavefront of the radiated sound). In practice it is often necessary to use additional supporting or infill loudspeakers to supply screened listeners' locations, such as on or under the balconies. This causes not only localization problems but also echo phenomena, if the travel time difference between the

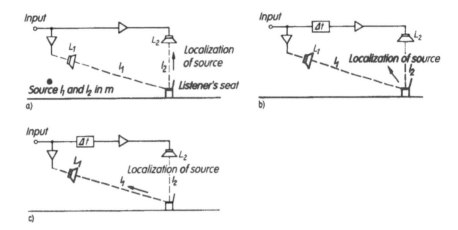

Figure 6.19 Use of infill loudspeaker for coverage of a theatre balcony or of the area
beneath it: L_1, central loudspeaker; L_2, loudspeaker near the listener's
seat. (a) Taking advantage of the increased level at the listener's
location (on the balcony). (b) Echo elimination by travel-time
compensation: $\Delta t = (l_1 - l_2)/c$. (c) Acoustical localization with a
delay slightly longer than the transmission path would require:
$\Delta t = (l_1 - l_2)/c + 0.015$ s.

signals from the various infill loudspeakers and from the remote main
loudspeakers is greater than 50 ms. In general, mislocalization and
reduction of definition may ensue. Figure 6.19 illustrates these problems.
In Figure 6.19a the level of an infill loudspeaker prevails at the listener's
seat.

This implies two problems:

1   Although the source is located near a centralized loudspeaker, it is not
the signal from the source but a sound pattern stemming from the infill
loudspeaker that is perceived as originating from its direction (of the
source) (cause: level $L_2 > L_1$ and distance $l_2 < l_1$) (precedence effect, see
section 3.2.1.5).
2   With increasing distance between the main and infill loudspeakers
definition decreases, and with distances above 17 m there is an
increased risk that the signals arriving from the loudspeakers are
perceived separately by the ear and heard as echoes.

While case (1) concerns the fidelity and incident direction of a sound
impression (mislocalization and thus distraction are the consequence of
diverging acoustical and visual sound patterns), case (2) has to be avoided,
since the echoes lead to poor or degraded speech intelligibility and floating
sound images. Without delay equipment this is possible only by means of
a specific level control: that is, one has to ensure that a disturbing

loudspeaker signal is masked by the one that corresponds to the listener's location (see section 3.2.1.2). In the resulting transition zones, however, echoes may occur that have an adverse effect on the reproduction quality. To avoid this, the infill loudspeakers have to be operated with time delay.

The delay times required for travel time compensation can be determined from the following equation:

$$\Delta t = \frac{s}{0.34} \tag{6.7}$$

where s is the sound path in m, and Δt is the travel time in ms.

If only the difference between the arrival times from source to listener's seat and from infill loudspeaker to listener's seat is compensated, a phantom source results, located somewhere between the two loudspeakers if the levels and spectra of the incident signals are approximately equal.

By introducing a further delay of 15–20 ms, localization jumps clearly over to loudspeaker L_1, provided that the level of L_2 is not more than 6–10 dB higher than that of L_1 (Figure 6.19c).

The echo elimination that has already been achieved by the travel path compensation (Figure 6.19b) continues to function, of course, with an additional delay of the signal to ensure localization (Figure 6.19c).

Also, with this loudspeaker arrangement – which hereafter will be referred to as 'centrally supported' – different applicable variations have been developed.

LARGE MULTI-PURPOSE HALLS

For economical reasons, nowadays large, multi-purpose halls are often built that are used for the performance of great symphonic works as well as for conferences and shows. These types of function all pose different acoustical requirements. A frequent requirement in this respect consists in optimizing the room-acoustical conditions for symphonic music, and adapting them for all other uses by means of an electroacoustical system. Since such halls are often very broad and relatively low, they can be covered only by means of a centrally supported loudspeaker arrangement. Because of the high divergence decrease and the directional effect of the main loudspeakers used, it is often not possible for the main loudspeakers to provide a uniform coverage of all audience areas. Nevertheless, a centralized loudspeaker array is often used as a reference source for the centrally supported sound reinforcement system. This main *loudspeaker array* already operates with a basic delay in order to ensure that, within large areas of the hall, the acoustic localization of the original source is on the stage. The delay for the subordinate (fill) loudspeakers has to take this basic delay into account, at least for the front area of the hall. At greater distances from the action area, where the angles between the original sources on the stage and the main secondary speakers arranged around the proscenium are narrow, and the primary sources are no longer perceived because of their low acoustic power, the proscenium loudspeakers can be used as reference sources.

LARGE THEATRES AND CULTURAL CENTRES

Sound reinforcement systems have been used in such halls for more than 50 years, with the objective of improving speech intelligibility in the areas on and below the balconies. To this effect there exist, apart from the main loudspeaker arrays in the proscenium region, further loudspeakers for these 'acoustically separate' areas, which nowadays are usually fed via electronic delay equipment. In this way it is possible to ensure a uniform and balanced sound coverage of all auditorium areas as well as correct acoustical localization, and to eliminate all echoes. The system can be optimally adapted to the room.

SPORTS FACILITIES

While decentralized systems have stood the test of time for the broadcasting of information signals, coverage that allows acoustical localization is desirable in certain cases. It is for instance required that:

● during victory ceremonies a directional reference to the scene be established;
● during show items of the programme, or publicity spot projections on screens, the visual presentation coincides with the acoustical perception.

This is made possible by loudspeaker arrangements that are capable not only of ensuring definition and clarity, but also of localization. Preferred solutions are, for example, centralized systems for music reproduction and decentralized systems to preserve speech intelligibility.

6.3.3 Decentralized sound reinforcement systems

Decentralized systems are sometimes characterized by the use of a large number of individual loudspeakers at close range to the listeners' seats, so as to achieve a high degree of intelligibility by eliminating the reverberant sound through direct coverage. A consistent product of this decentralization is the *loudspeaker built into the back of the seat*. Such systems, in which a loudspeaker is assigned to every listener at close range, have proven their outstanding efficiency for conferences with large numbers of delegates. Although the front seats are mostly located in the direction of the platform, so that visual contact with the speaker is not impaired, the use of delay equipment under certain circumstances should not be discounted – for example in order to avoid echo disturbances – if such a system is to be operated together with a proscenium system. With a decentralized arrangement of the loudspeakers it is generally not possible for the 'internal travelling time' of the system – that is, the travel time difference between the nearest and the remotest audible loudspeaker – to be kept below 50 ms. Therefore either the booster loudspeakers at the listeners' seats have to produce so high a level that the sound from the remote loudspeakers becomes negligibly small (so as to be drowned in the ambient sound), or the loudspeakers remote from the source and provided with a high backward attenuation have to be operated with delay, so as not to cause any disturbances in the vicinity of the loudspeakers near the source.

With special decentralized arrangements, such as seat-back loudspeakers, or coverage of flat rooms by means of ceiling loudspeakers, the loudspeakers nearest to the listeners determine the definition and intelligibility, and the remote ones are masked. With seat-back loudspeakers in particular, the inherent increase of the diffuse ambient sound component has to be taken into account. A loudspeaker radiating towards the listener is in this respect more favourable than one radiating upwards or towards the front.

In the following section we describe briefly some categories of application in which it is possible to avoid travel-time interference by decentralized systems, and also to establish a position reference to the action area without delay equipment.

SPORTS STADIUMS

The action area of a sports stadium is the field or pitch. This is the area where the attention of the spectators is concentrated, and from where the reproduction of acoustical signals should also emanate. The source to be amplified, in most cases an announcer, and in some cases also a music band, may be in the action area or at its edge. With athletics events, for instance, the stadium announcer is located at the edge of the tracks, whereas with football games he is in a booth.

Because of the amount of noise that needs to be drowned out, the required sound level is relatively high. According to Boye [6.8], an average value of 85 dB and a maximum value of 95 dB or more should be achieved.

The loudspeakers are normally installed along the edge of the field at about 15–20 m apart. The loudspeakers have a cardioid characteristic, and are aimed upwards at the stands (Figure 6.20a). Because the sound is propagated at grazing incidence over the absorbing spectators, the distribution loss for full attendance is often more than 6 dB per doubling of the distance. This is why, for high stands in particular, it is necessary to install additional loudspeakers for the upper section of the stands. If the stand is roofed, these loudspeakers are fixed below the front roof (Figure 6.20b), from where they are aimed at the rear zone of the stands.

a)

b)

Figure 6.20 Decentralized coverage of a stadium: (a) general view; (b) detail.

Thanks to the relatively long travel time of the sound from this installation point to the listener, localization shifts occur only rarely, and travel-time interferences do not occur at all.

If there is no roof, the problem is harder to solve. In this case it is necessary to erect masts in the region of the stands, at the end of which loudspeakers with high rearward attenuation are fastened and aimed at the upper third of the stands. The height of the masts depends both on the travel-time difference of the sound coming from the nearest loudspeaker, which should not exceed 50 ms, and on the need to obstruct the view as little as possible. The loudspeakers at the edge of the field – serving as reference loudspeakers – should be capable of transmitting the whole frequency range, whereas the additional loudspeakers may perhaps have a narrower bandwidth. Since music transmission is generally required, the transmission range should at least be 150 Hz to 10 kHz.

To cover the field of play itself, loudspeakers are arranged at the rear of those located at the edge of the field. Often it is sufficient to arrange these loudspeakers on one side of the field (for example, where the start and finish positions of the runners are situated). In this case it is possible that a sound level difference might occur between the two sides of the ground, which is undesirable in certain cases, but on the other hand there are no travel-time interferences to be expected on the whole field of play, which is not always the case with bilateral coverage.

Special requirements have to be met by the sound reinforcement system in stadiums for synchronized gymnastics. In some cases decentralized loudspeakers embedded in the ground have been used [6.9]. In this case any delay is not permitted, as it impedes the synchronism of the exercises.

LARGE MEETING AND SPORTS HALLS

If delay equipment is not used in rooms having a volume of more than 50 000–70 000 m^3 and a width of more than 80 m, a localization reference to the action area (platform or field of play) should nevertheless be established, although travel-time interferences are not necessarily to be expected. In the same way as with centralized systems, this is achieved by a staggered installation of sound reinforcement systems, often of single monoclusters. By varying the installation height it is possible to ensure that in the stalls, and in the first balcony or platform area, localization is preferably to the loudspeaker position nearest to the front of the platform. By means of additional loudspeakers on the platform this effect may be enhanced so as to direct the localization reference in the platform or near-stage area onto the action area. An appropriate staggering of the monoclusters (spacing <15 m) helps to prevent travel-time differences that could cause echoes.

Generally, however, such halls, which may have a volume of up to 10^6 m^3, are nowadays covered by means of loudspeakers arranged in a circular pattern and operated via delay equipment (see section 6.5.3).

CONFERENCE AND DISCUSSION SYSTEMS

Additional problems arise here from the fact that the microphone stations are within the area to be covered. Thus the risk of positive feedback increases, since the loudspeakers may be aimed at the microphone. In larger halls, moreover, it is possible for travel-time differences to occur between the podium and the loudspeakers, which significantly affect the speaker, and may even make talking impossible for the inexperienced user.

Hence decentralized loudspeaker arrangements are often chosen for systems of this kind. In smaller rooms the usual solution is a system of ceiling loudspeakers, while in larger rooms they may also be integrated in the tables or seats. These loudspeakers are wired in such a way that they are disconnected, or at least heavily attenuated, as soon as a microphone is used in their neighbourhood, so that no positive feedback can take place [6.10]. Since the surrounding loudspeakers are very near, the user is not disturbed by travel-time differences.

Another sound reinforcement variant that is applicable for discussions in very large rooms is operating in the International Congress Centre (ICC) in Berlin. Several clusters with a switchable directional characteristic (separately controlled individual loudspeakers) (Figure 6.21) are assigned to different delay channels. When a discussion microphone is used, the nearest

Figure 6.21 Conference (discussion) loudspeakers in Hall 1 of the International Congress Centre in Berlin.

cluster is switched to a condition in which it radiates uniformly and omnidirectionally without delay, thus serving as the reference loudspeaker. All other clusters of the system then radiate only in the direction away from the reference cluster and with a delay corresponding to their distance from it. Thus the talker enjoys almost undelayed reproduction while the other delegates are afforded a certain degree of acoustical localization of the person speaking [6.11]. Because of the relative distance between the talker and the loudspeaker extensive freedom from positive feedback is also ensured.

LARGE REVERBERANT ROOMS

In large, undamped rooms, such as large reverberant churches, exhibition and station halls, a decentralized sound reinforcement is often the only possible solution. The loudspeakers have to be positioned very near to the listeners so that as high as possible a proportion of the loudspeaker's direct sound becomes effective, and the radiated sound is absorbed immediately after it is heard. With high rooms in particular it is important to ensure that the sound is radiated as directly as possible onto the area provided for the audience, so as to avoid exciting the acoustically undamped upper region of the room. In churches directional sound columns are often arranged on the structural columns between the aisles, an arrangement that also satisfies aesthetic considerations (Figure 6.22).

Even closer contact with the listeners can be achieved by means of individual loudspeakers integrated into the seat backs, or tables in front of

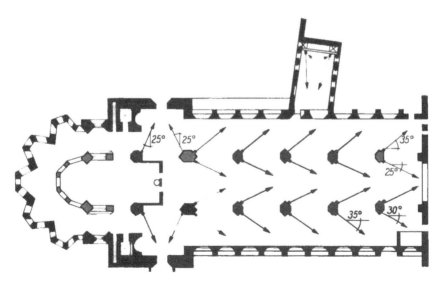

Figure 6.22 Decentralized coverage of a church nave by means of sound columns for speech.

the listeners. Here the loudspeakers are aimed directly at the listeners, who largely absorb the sound.

6.4 Systems for adjusting room-acoustical parameters

The use of loudspeaker systems implies in most cases a modification of the acoustic impression of the reproduction. It allows improvement of the definition in a reverberant room, as well as enhancing the spaciousness in a room that is too 'dry'. In many cases sound reinforcement systems are required to compensate for room-acoustical shortcomings, or to adapt the acoustic properties of a large hall when it is to be used for very different purposes. Such versatility is highly desirable for economical reasons alone. In the design phase the application basis for the room-acoustics planning has to be decided at a very early stage. Utilization for symphonic, choral or even organ music demands high spaciousness and long reverberation time, whereas utilization for conferences, shows, rock and jazz concerts, as well as sports events, requires high definition and clarity. The decision must not necessarily favour the most frequent type of use. Once it has been taken, however, there is the need to improve, by electroacoustic means, those conditions that are not optimized by room-acoustical measures.

In historic rooms that are very reverberant, such as churches, it is often desirable to make improvements to speech intelligibility by means of electroacoustical correction. It is easier to make a correction of this kind the smaller the discrepancy is between the existing room-acoustical conditions and those aimed by electroacoustical means.

6.4.1 Targets for the correction, and values to be expected

Although the basic function of a sound reinforcement system is to transmit an acoustic signal as clearly as possible to a great number of listeners, under certain circumstances it may also be desirable to increase the spaciousness and reverberation time. However, this can also mask the microstructure of the transmitted signal and the clarity of reproduction. Under certain room and temporal conditions, and thanks to the properties of the hearing system (such as its directional effect), there may be an optimal result. This also applies to the enhancement of the useful sound level against the background noise level.

In the following we are going to deal with the sound field conditions that have already been described in section 3.1.2, and whose realization is desirable.

When using electroacoustical means to influence room-acoustical conditions, the optimum subjective sound impression must be set as a target quantity. This quantity is not uniform for a given hall, but depends on the nature of the performance and on the listener's location.

One requirement for the transmission of speech and recordings is that the information should be transmitted as unaltered as possible to the listener. To enhance the emotional impact, however, increased spaciousness and a corresponding reverberation time are sometimes desired. (Modern sound film procedures, such as Dolby Surround, make good use of this, but discotheques also capitalize on it.) The dependence of reverberation time and spaciousness on the type of music has been mentioned several times already. In large halls one does not expect the same spatial impression in all areas. In balcony areas far from the stage, for instance, more spaciousness than in areas near the stage is generally considered to be optimal.

Of importance for the expected sound pattern are loudness, timbre (loss of high-frequency components at large distances from the signal source is accepted, for example, because this tallies with experience) and localizability of the source.

6.4.2 *Modification of the temporal energy distribution at the listener's location*

One of the most important possibilities for influencing definition and spaciousness is modification of both the temporal and local energy distribution of the sound pattern perceived by the listener. At the listener's location it is possible to manipulate:

- the direct sound;
- the initial energy (influencing direct sound and reverberant sound);
- the reverberation.

With most sound reinforcement systems the intention is to increase the radiation of direct sound from the loudspeakers so that the transmitted electronic signal arrives at the listener as unaltered as possible. To assess this influence in a preliminary investigation, we can use the *direct sound factor*, f_d, for an approximate calculation:

$$f_d = \frac{w_d}{w_r} = \left(\frac{r_R}{r_{LH}}\right)^2 \tag{6.8}$$

where w_d is the direct-sound energy density, w_r is the diffuse-sound energy density, r_{LH} is the distance from the loudspeaker to the listener, and r_R is the critical distance of the loudspeaker.

This factor allows statistical expectation values to be determined so as to arrive at objective quality criteria (cf. equation (3.8)). For assessing the distance r_{LH} up to which the listener may be away from the loudspeaker without the values to be expected for the measure of definition C_{50} or the measure of clarity C_{80} falling below a certain limit value, e.g. 0 dB, one may

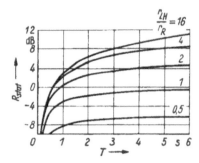

Figure 6.23 Statistical spaciousness index, R_{stat}, as a function of the reverberation time, T.

use Figure 3.7, p. 25. (If the loudspeakers are aimed at the sound-absorbing audience, the critical distance, r_R, is calculated by means of equation (5.8a): $r_R \approx \sqrt{\gamma r_H}$.)

Figure 3.7 shows that the ratio r_{LH}/r_R decreases when the directional properties of the loudspeakers are increased. This goes along with an increase in clarity and definition. Conversely, an enhancement of clarity obtained by an increase of direct sound energy brings about a decrease of spaciousness.

Figure 6.23 shows how the statistical spaciousness index, R_{stat}, depends on the reverberation time, T. As in Figure 3.7, the parameter is the distance ratio, r_{LH}/r_R. (Lehmann [3.7] indicates $R = +2$ to $+6$ dB as optimum values.) We can see that with reverberation times of $T < 1$ s the spatial impression is always impaired, independently of the distance from source (loudspeaker) to listener. With long reverberation times of $T > 2$ s (churches, etc.), however, the ratio r_{LH}/r_R should not exceed 3, because otherwise too high a spaciousness is to be expected. In the following we show briefly how these relations can be influenced by the sound reinforcement system.

INFLUENCING THE DEFINITION

Using equation (5.8a) and considering the fact that the loudspeaker is aimed at the audience area, we obtain the critical distance by multiplying the square root of the front-to-random factor by the reverberation radius.

The ratio r_{LH}/r_R given in Figure 3.7a is also *reduced* by this factor $\sqrt{\gamma}$. With the same reverberation time we thus obtain another curve from the family of curves in Figure 3.7a, which results in higher measures of definition. This can be explained by means of an example (which applies analogously to the enhancement of clarity: cf. measure of clarity C_{80} in Figure 3.7b).

> In a hall with a reverberation time of $T = 2$ s and a reverberation radius of $r_H = 5.8$ m a loudspeaker assembly is to be arranged in the proscenium area (front-to-random factor, $\gamma_L = 10$ at mid-frequencies). The resulting critical distance is $r_R = 18.5$ m.

Without a sound reinforcement system, and with a distance from source to listener of $r_{SH} = 30$ m, Figures 3.7a and b reveal a measure of definition $C_{50stat} \approx -3.5$ dB and a measure of clarity $C_{80stat} \approx -1$ dB, which may perhaps be considered as just satisfactory.

With the loudspeakers the ratio r_{LH}/r_R changes from 5.2 to 1.6, so that with otherwise equal conditions $C_{50stat} \approx 0$ dB and $C_{80stat} \approx 2$ dB, which may be considered as good.

In the literature [3.10] the quality of intelligibility is nowadays often characterized by the articulation loss of consonants, Al_{cons}, which according to Figure 3.13 can be converted into RASTI values. Figure 6.24 shows an important correlation in this respect by revealing that for $C_{50stat} > -3$ dB one can always expect ideal intelligibility values ($Al_{cons} < 2\%$). For $C_{50stat} > -0$ dB good intelligibility values still result in practice.

An analogous correlation between RASTI values and C_{50stat} is depicted in Figure 6.25. We can see that excellent RASTI values, >0.75, are always achieved for $C_{50stat} > 3$ dB, independently of the reverberation time, T (in which case, however, the disturbing noise captured with the RASTI measurements is not considered).

For more negative C_{50} values the curves approximate to individual pole zeros, since statistical definition values for a given reverberation time tend to a final value ($r_{LH}/r_R \rightarrow \infty$). The resulting values are (cf. limit curve in Figure 3.7a):

T (s)	$C_{50stat\,asym}$ (dB)
1	0
1.5	−2.3
2	−4
2.5	−5
3	−5.8
3.5	−6.4
4	−7.2
6	−9.1

The asymptotic behaviour reveals that for negative measures of definition every single additional decibel contributes essentially to improving intelligibility. This is also corroborated by experience.

Applied to the above example, with $C_{50stat} = -3.5$ dB and no additional measures, it is evident that intelligibility is no longer satisfactory, with $Al_{cons} > 20\%$. With the loudspeaker ($C_{50stat} \approx 0$ dB), however, the articulation loss according to Figure 6.24 is reduced to $Al_{cons} \approx 3.5\%$, a value that ensures good intelligibility. [With RASTI one obtains analogously, according to Figure 6.25, values of 0.2 (poor) and 0.72 (good to excellent).]

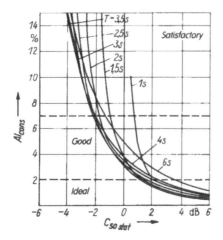

Figure 6.24 Correlation between articulation loss, Al_{cons}, and statistical measure of definition, C_{50stat}.

Since centrally supported loudspeaker arrangements (such as centre clusters and fills) require several loudspeakers, for example for enabling a uniform sound level distribution, the so-called 'internal travel time' of the sound reinforcement system must not exceed 50 ms. In this way it is guaranteed that all the radiated energy arrives effectively as 'direct sound' at the listener's position so as to enhance definition and clarity. This is

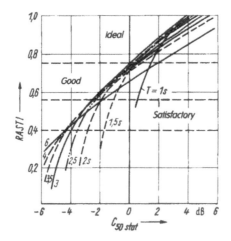

Figure 6.25 Correlation between RASTI values and statistical measure of definition, C_{50stat}.

especially important in the open, and in rooms with a short reverberation time, since because of the low reflection density it is possible here for the signals of the individual decentralized infill loudspeakers to become disturbing; if the gaps between the signal impulses (sound arrivals) are not filled, there is a risk of echoes being heard. If the loudspeakers are arranged in a cluster, and if it is possible to aim them so that the radiated sound is mainly absorbed, conditions are very favourable for achieving a high definition. If such a loudspeaker arrangement cannot be realized, for example if the achievable sound level distribution is insufficient, the same effect can also be obtained by means of appropriately arranged booster loudspeakers with a signal delay system (see section 6.5.3).

With the type of loudspeaker arrangement shown in Figure 6.14 the result may be that the front-to-random factor is not increased, but the reverberation radius (or the critical distance) is reduced. If this critical distance, r_R, can be reduced by $\sqrt{\frac{1}{2}}$, the resulting value of r_{LH}/r_R will be *increased* by this factor, as Figure 3.7a shows. The attainable definition indexes thus decrease, which also explains why after a costly but rash installation of an expanded system, the results for speech intelligibility are often worse than with the less expensive old system.

With centrally supported systems in particular it is therefore important to ensure that the loudspeakers are aimed at the audience area, and that they emit as little sound as possible into other directions of the room so as not to increase the diffuse field. This is possible, for instance, by means of well-aimed constant-directivity horns (see Figure 4.21). Provided such loud-speakers are properly used, one may proceed from the above-mentioned conditions: that is, on the basis that diffuse-field excitation is independent of the loudspeaker arrangement.

CONTROLLING THE CLARITY

In the same way as for the definition of speech it is also possible by means of a sound reinforcement system to influence the clarity of musical performances and thus also the measure of clarity C_{80}. The expected measure of clarity as a function of reverberation time has already been referred to in Figure 3.7b. If for example the critical distance increases, the ratio r_{LH}/r_R decreases with a constant distance from source (loudspeaker) to listener, so that the attainable measure of clarity becomes greater.

As is the case with the installation of definition-increasing systems, measurements for checking the results are appropriate. Such measurements are, of course, much more exact than estimates based on statistical conditions. The energy–time curves thus obtained make it possible to determine the resulting measures of clarity, C_{80}, and of definition, C_{50} (see Chapter 7). Figure 6.26 shows a reflectogram of this kind. Figure 6.26a shows the conditions without a sound reinforcement system, revealing a high articulation loss (12.15%) and a low definition (≈ -7 dB). When a

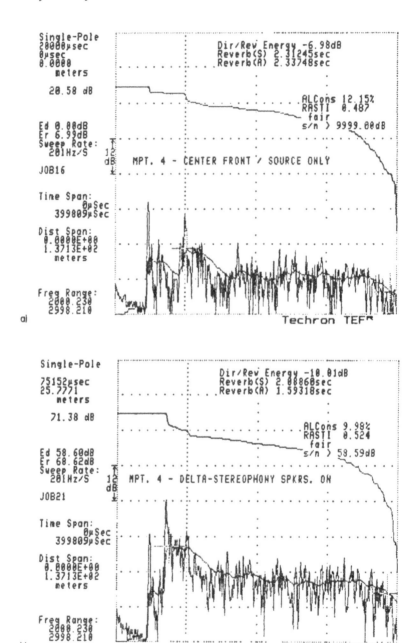

Figure 6.26 Determination of the measure of clarity from the energy–time behaviour: (a) without sound reinforcement system; (b) after installation of a centralized loudspeaker system.

centralized sound reinforcement system is added (Figure 6.26b), the proportion of effective 'direct-sound' energy increases (12 dB peak at about 30 ms after the direct sound), enabling better definition values to be obtained (not indicated in the reflectogram).

REDUCTION OF THE SPATIAL IMPRESSION

Figure 6.23 shows clearly that the ratio r_{LH}/r_R decreases with increasing directivity of the loudspeakers, but that the expected spaciousness, R, is also reduced by the measure with which definition and clarity increase. This means that a compromise has to be found between spaciousness and clarity. By using directional loudspeakers the critical distance and thus the 'reverberation-free' area are increased, which corresponds to a reduction of the definition-diminishing reverberance impression.

Changes in the reverberance impression with a given definition and clarity can be achieved, however, only by means of changes in other room-acoustical parameters, such as the reverberation level. Only in this way is it possible to manipulate optimally that part of the energy–time curve that is determinative for the spatial impression. This will be discussed in the next section.

6.4.3 *Procedures enabling expansion of spaciousness and increase of reverberation time*

Improvement of the reproduction conditions of symphonic music, chamber music and especially of choral and organ music requires a good mix of the individual voices, and thus a relatively high spatial impression. This often implies a reverberation time longer than that enabled by the room-acoustical conditions. Several of these conditions may be improved by increasing, during their radiation by the main loudspeakers, the energy components of the reverberant sound relative to those of the direct sound and of the initial reflections. With a view to enhancing the reverberance impression it is also necessary to prolong the period of audibility of these late 'reflections'. Both parameters can be influenced and controlled by electroacoustical means.

Another possibility for increasing spaciousness, which is relatively easy to put into practice by means of electroacoustical systems, consists in having late-energy sound radiated by loudspeakers distributed around the room. In this way it is possible to achieve an 'envelopment' of the listener and especially an amplification of the energy components impacting laterally on the listener.

6.4.3.1 *Use of sound delay systems for the enhancement of spaciousness*

These procedures influence mainly the energy of the reverberant-sound initial reflections.

AMBIOPHONY

This procedure makes use of time-delay equipment that simulates the discrete 'initial reflections' as well as the 'reverberation tail'. The reflection sequences have to be chosen so that no comb-filter effects, as in the form of flutter echoes for example, can occur during impulsive musical sequences. The mode of operation of a simple ambiophony system as shown in Figure 6.27 (*ambio* being derived from *ambience*) can be described as follows. To the direct sound radiated directly into the room by the original source, delayed signals are mixed via delay equipment (in the early days a magnetic tape device, for example), which are radiated into the room so as to simulate delayed reflections coming from the walls or the ceiling. This requires loudspeakers to be appropriately distributed in the room and to radiate the sound as diffusely as possible. To extend the reverberation, an additional feedback signal from the last output of the delay chain to the input is created. A system of this kind was first proposed by Kleis [6.12], and has subsequently been realized in several large halls [6.13].

ERES (ELECTRONIC REFLECTED ENERGY SYSTEM)

This technique, proposed by Jaffe, consists in a simulation of the early reflections that constitute the initial reflections that are effective for reverberant sound [6.14]. The idea is not new, but it was Jaffe who converted it into a marketable solution (Figure 6.28).

By arranging the loudspeakers in the walls near the stage of the hall, and by making use of the possible delay variation, filtering and level control of

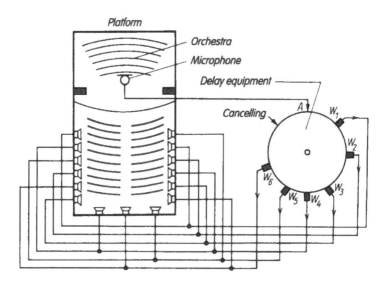

Figure 6.27 Ambiophony system.

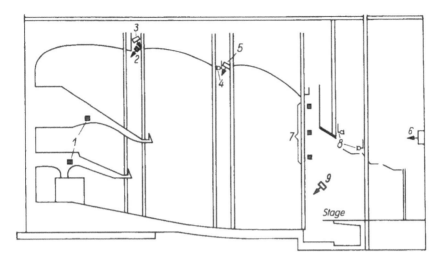

Figure 6.28 Layout of the the ERES/AR system in the Silvia Hall, Eugene Performing Arts Center, Oregon, USA: 1, one of the 14 pairs of AR (assisted resonance, see section 6.4.3.2)/ERES loudspeakers under the balcony; 2, one of the 90 AR loudspeakers; 3, one of the four ERES loudspeakers in the third ceiling offset; 4, one of the 90 AR microphones; 5, one of the four ERES loudspeakers in the second ceiling offset; 6, ERES stage-tower loudspeakers; 7, three of the six AR proscenium loudspeakers; 8, ERES microphones; 9, one of the two AR proscenium loudspeakers.

the signals supplied to them, simulated lateral reflections can be radiated. Thus the spatial impression can be adjusted so as to simulate an acoustically wider or narrower proscenium by simulating a longer or a shorter delay respectively. Because of these features of

- adaptation to acoustic requirements,
- simulation of different hall sizes,
- influencing of definition and clarity,

Jaffe and his team employ the term *electronic architecture*. It is certainly true that this intentional injection of 'reflections' is well capable of simulating the missing room-acoustical characteristics of a hall, so as to compensate for deficiencies in the acoustical–spatial structure. After installing the first system in the Eugene Performing Arts Center in Oregon [6.15], Jaffe-Acoustics has installed further systems in numerous halls in the USA, Canada, and other countries.

The use of electronic delay equipment in sound reinforcement systems has been generally accepted all over the world, and has become common practice for the introduction of delayed signals (for simulating late reflections, for example). Thus 'electronic architecture' is practised in all instances where such 'reflections' are used intentionally or unintentionally. Also, by means of the Delta Stereophony system [6.16], which will be described later on, it is possible to effect the generation and introduction of reverberant-sound reflections, so that this procedure can also be used for enhancing the spatial impression. Since the trend is towards an increasingly complex manipulation of sound fields, and since 80% of the spatial impression is determined by the initial reflections, the future task will consist more and more in simulating acoustical environments by means of delayed, electroacoustically introduced signals.

6.4.3.2 *Reverberation enhancement by using the travel time*

This procedure consists primarily in enhancement of the late reverberant sound energy, including prolongation of the reverberation time.

ASSISTED RESONANCE

For optimizing the reverberation time in the Royal Festival Hall, built in London in 1951, Parkin and Morgan [6.17, 6.18] proposed a technique that permitted extension of the reverberation time, especially at the low frequencies (Figure 6.29).

Parkin and Morgan started from the fact that, in a room, there exist a multitude of eigenfrequencies, which allow the formation of stationary waves with nodes and antinodes decaying as an exponential function according to the corresponding absorption area. This decay process is characteristic for the reverberation time of a room at a given frequency. Each stationary wave has its own specific spatial disposition. At the point where the sound pressure has its maximum (antinode) for a given frequency, a microphone is installed. The signal from the microphone is transmitted via an amplifier to a loudspeaker mounted in a remote antinode of the same stationary wave to compensate the energy loss caused by absorption. Thus it is possible to sustain the energy at this frequency for a longer period (*assisted resonance*). By increasing the amplification it is possible to prolong the reverberation time for these frequencies considerably (up to the onset of feedback). Because of the spatial distribution of the radiating loudspeakers this accordingly also holds true for the spatial impression.

These considerations apply to all eigenfrequencies of the room, but difficulties may arise with the installation of the microphones and loudspeakers at the points determined by the pressure antinodes of the

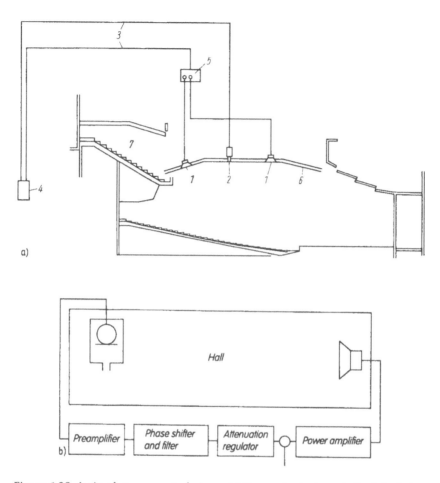

Figure 6.29 Assisted resonance technique. (a) System in the large hall of the Palace of Culture in Prague: 1, 60 loudspeaker boxes each in the ceiling and proscenium areas; 2, 120 microphones in Helmholtz resonator boxes; 3, 120 microphone or loudspeaker cables; 4, remote control, phase-shifting amplifiers for the 120 channels; 5, distributor for loudspeaker enclosures; 6, lowered ceiling for concert set-up; 7, separated balcony area. (b) Components of an assisted resonance channel (microphone in resonance chamber).

chosen eigenfrequencies. The microphones and loudspeakers are therefore installed at less critical points and instead are operated via phase shifters. Filters (Helmholtz resonators, bandwidth about 3 Hz) are also installed in the transmission path, which allow the transmission channel to respond only to the eigenfrequency in question (Figure 6.29b). The radiating loudspeakers must not be arranged farther away from the performance area than the associated microphones, otherwise mislocalizations of the source may occur because of the premature arrival of the reverberant signal.

The technique has subsequently been employed in a large number of halls (such as the Palace of Culture in Prague). In spite of the high technical expenditure and the fact that the required installation can only be used exclusively for the procedure of assisted resonance, it is one of the safest solutions for prolonging the reverberation time especially at low frequencies and without influencing the timbre.

MULTIPLE INDEPENDENT AMPLIFICATION CHANNELS

Franssen [6.19] first proposed the use of a large number of broadband transmission channels, each with amplification so low that no timbre change due to the onset of positive feedback can occur. While each individual channel operating below the positive feedback threshold provides only a slight amplification, the large number of channels results in an energy density capable of increasing noticeably the spatial impression and the reverberation time.

The reverberation time increase is determined from

$$\frac{T_M}{T_o} = 1 + \frac{n}{50} \tag{6.9}$$

For example, if the reverberation time is to be doubled (which means doubling the energy density), $n = 50$ separate amplification chains are required, according to equation (6.9). Ohsmann [6.20] has published an extensive paper on the operating mode of these loudspeaker systems, and has shown that the results envisaged by Franssen with regard to the prolongation of the reverberation time cannot be attained in practice. As a possible cause for the divergence from theory he considers the fact that 'cross couplings between the channels were not sufficiently taken into account' by Franssen (cf. [5.2]).

Under the name of 'Multi-Channel Amplification of Reverberation System' (MCR), the Philips Company offers a system that constitutes a technological realization of this procedure for reverberation and spaciousness enhancement [6.21]. The manufacturer's documentation specifies that with 90 channels a prolongation of the average reverberation time from 1.2 s to 1.7 s is obtained (Figure 6.30); still longer reverberation enhancements are said to be possible. The system has been installed in numerous medium-sized and large halls (for example, in Hall 1 of the ICC in Berlin).

ACOUSTIC CONTROL SYSTEM (ACS)

This procedure was developed by Berkhout and de Vries at the University of Delft [6.22]. Although the authors speak of a holographic attempt to

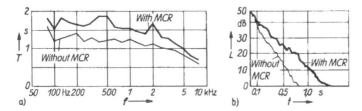

Figure 6.30 Reverberation time enhancement by means of the MCR system in
the POC Theatre in Eindhoven: (a) frequency response of the
reverberation time with and without MCR (typical adjustment);
(b) reverberation behaviour at 400 Hz. Technical data of the
system: hall 3100 m³, stage 900 m³; 90 channels (preamplifiers,
filters, power amplifiers); 90 microphones in the ceiling; 110
loudspeakers in the lateral walls, in the orchestra ceiling and under
the balcony; maximum sound level 112 dB, remote control of
reverberation in 10 steps.

prolong the reverberation in rooms, it is rather the result of a
(mathematical/physical) convolution of existing signal properties with
desirable room properties bringing about a new room characteristic with a
new reverberation time behaviour.

Figure 6.31 shows the principle of an ACS circuit for a loudspeaker–
microphone pair. The acoustician formulates the properties of a desired
room, in a computer model for example, and then imposes these properties
by means of appropriate parameters on a reflection simulator, whereafter
these reflection patterns are convolut with the real acoustical properties of
the hall.

Unlike the other systems explained above, the ACS functions without
feedback loops, so that no timbre changes caused by autoexcitation
phenomena are to be expected. It is already used in several halls in the
Netherlands, the UK and other countries, and promises with further

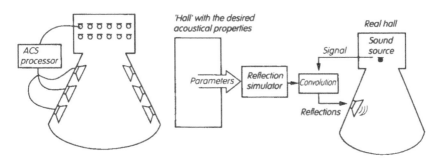

Figure 6.31 Basic block diagram of the ACS system, illustrated for a loudspeaker–
microphone pair.

technical development to become an important procedure for reverberation time enhancement in halls that require a high room-acoustical standard.

RECENT DEVELOPMENTS

The LARES system by Lexicon also allows the spaciousness and the reverberation time to be readily increased by using completely time-variant signal processing [6.23].

Another similar system, SIAP, uses processors to 'convolve' the room-acoustical parameters of a hall with the desired parameters [6.24]. It provides the reflections missing in the natural acoustics of the room. The inventor claims a sound that is so natural that it is not noticeable that a 'sound system' is involved.

REVERBERATION PROLONGATION IN THE SOUND CHANNEL

Another possibility for prolonging reverberation consists in looping a reverberation device into the sound channel. The devices applicable to this effect are described in section 4.6.3. The reverberated signals are radiated into the room via as many incoherent loudspeaker channels as possible, and are subjectively perceived as a reverberation time prolongation. Investigations have revealed that the coupling of the reverberation equipment has to be effected in the direct field of the signal to be reverberated; signals picked up by microphones in the hall can only weakly be fed back into the hall for producing the desired effect as reverberation signals, as a timbre-changing positive feedback will occur before a prolongation of the reverberation is even subjectively perceived [6.25]. The radiation of the reverberated signals through loudspeakers aimed at the audience area enhances the subjective effect of the injected reverberation signals. These limitations of the reverberation prolongation, conditioned by acoustic positive feedback, are of course not applicable in the case of playback signals. It is true, however, that playback signals are often already reverberated, a fact that has to be taken into account when subjecting them to further reverberation.

6.5 Localization of sound events

Given that sound reinforcement systems are increasingly provided with delay equipment, we are now going to expound in more detail the effects thus attainable. The technical procedure used for generating the delays will not be discussed: suffice it to say that, thanks to their reliability, digital devices are mainly used nowadays (see section 4.6.1). Moreover, all aspects of travel-time effects will be summarized, in which case reference may be made to previous sections.

6.5.1 *Travel-time phenomena*

In everyday life we are confronted with numerous travel-time (delay) phenomena of sound. Best known in this regard is the fact that the distance of the centre of a thunderstorm can be ascertained by measuring the time lapse between the lightning ($c_0 = 300\ 000$ km/s) and the sounding of the thunder ($c = 344$ m/s at 20 °C). If, for example, 3 s pass after the lighting before the thunder is heard, the thunderstorm is about 1 km away ($s = ct = 344$ m/s $\cdot$ 3 s ≈ 1000 m).

With open-air events where large projection screens are used for showing the artists, one observes at the rear seats that visual and acoustical perception drift apart. At a distance of 300 m time differences of up to 1 s will result. The delay problems cannot be compensated by technical means in this case, since the synchronization defect is dependent on the listener's position. (Or it may be that several projection screens are erected at different distances in front of the action area; see section 6.6.4.)

It has already been shown above that the definition of speech or the clarity of music reproduction can be influenced by a delayed introduction of direct sound integrating early reflections (as in the ERES procedure, for example; see section 6.4.3.1). As we also saw in that section, the spaciousness of an auditorium can be increased by ambiophony systems.

In the following we deal with the elimination of echo disturbances, the safeguarding of natural sound reproduction, and the localizability of an acoustic source.

6.5.2 *Suppression of echoes*

As is generally known, at the 'blurring' threshold of $\Delta t \approx 50$ ms the human ear begins to hear separately two successive sound events (e.g. sound impulses). If two loudspeakers are for instance installed at more than 17 m from each other, it is possible for an echo to become audible, unless the signal of the nearer loudspeaker masks that of the farther one.

Figure 6.32a shows a situation that often occurs in large halls, where two loudspeakers L_1 and L_2 are installed: one of them, L_1 (main sound reinforcement), near the stage, and the second one, L_2, under the balcony to improve the reproduction conditions in that area. If this second loudspeaker is connected directly to the same source as the first one, an echo problem will occur in the area under the balcony, if the main loudspeaker is also audible here (Figure 6.32b). The echo disappears, however, if the signal of loudspeaker L_2 is delayed by Δt (Figure 6.32b). If it is delayed still more (by Δt_2, Figure 6.32d), an echo still does not occur, but in addition the main loudspeaker assembly is acoustically localized, and not the loudspeaker mounted under the balcony. It is also possible, of course, that the signal from the main loudspeaker assembly L_1 is masked by the 'near' L_2 signal

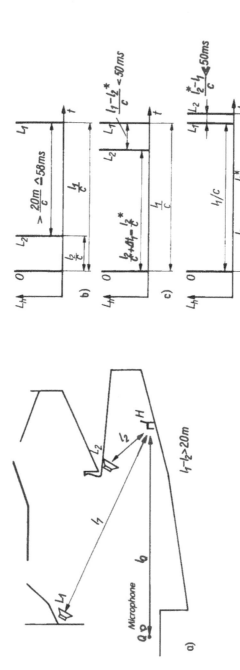

Figure 6.32 Loudspeaker system for suppressing echoes: (a) geometric conditions; (b) temporal sequence of the signals without delay; (c) temporal sequence of the signals with delay; (d) temporal sequence of the signals with delay and localization change. H, listener's location; Q, sound source; L_1, L_2 loudspeakers.

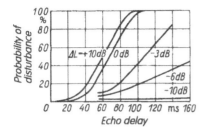

Figure 6.33 Proportion of listeners still perceiving an echo, as a function of echo delay.

(applies to large instances), so that echoes are not perceptible either without delay equipment.

Bolt and Doak [6.26] have investigated what proportion of listeners feel disturbed by echoes of varying intensity and delay (Figure 6.33). A limit curve, derived from the results for the attenuation of the level of the echo versus the direct sound, and which has to be met if echo interferences are largely to remain unperceived, is shown in Figure 6.34.

How this curve can be applied for calculating the way in which echoes can be masked, can be explained by means of an example [6.1]:

The travel path, time and level relations are shown in Figure 6.35. Without the loudspeaker L_2 there occurs at the listener's location, about 175 ms (271.13 ms − 96.21 ms) after the L_1 signal, a rear-wall signal with a level attenuated by −9 dB (−39.4 dB − (−30.4 dB)), which according to Figure 6.34 is perceived as an echo. If now the L_2 signal is introduced with a 30 ms delay after the L_1 signal (Δt = travel time along (path 1 − path 3) + 30 ms = 111.6 ms), it is not perceived, according to the precedence effect, as long as it is not louder than the L_1 signal. Since it arrives, however, about 145 ms earlier (travel time along (path 2 − path 1) − 30 ms), it should be about 16 dB louder than the proper echo signal (see Figure 6.34): that is, about −23.4 dB strong (−39.4 + 16 dB). With due consideration of the level loss of 14 dB

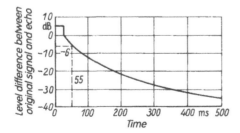

Figure 6.34 Limit curve to be complied with to minimize echo disturbances.

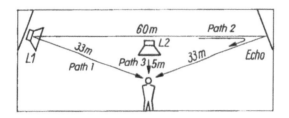

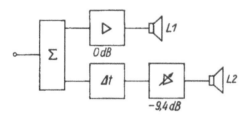

Figure 6.35 Use of time delay of loudspeaker L_2 for masking an echo.

$$\Delta t = \left(\frac{33}{0.343} + 30 - \frac{5}{0.343} \right) \text{ ms} = 111.6 \text{ ms}$$

Path 1: 33 m $\equiv$ 96.21 ms $\equiv$ −30.4 dB level attenuation in the open
Path 2: 93 m $\equiv$ 271.13 ms $\equiv$ −39.4 dB level attenuation in the open
Path 3: 5 m $\equiv$ 14.67 ms $\equiv$ −14 dB level attenuation in the open.

along the 5 m path, the level controller for L_2 should thus be adjusted to −9.4 dB, referred to 0 dB in channel 1.

This L_2 signal thus masks the echo, but is still masked itself by the L_1 signal on account of the level difference of about 7 dB (−23.4 dB − (−30.4 dB)).

So one can say that by means of sound delay it is possible to concentrate signals at audience groups in such a way that even with a distributed loudspeaker arrangement the supply of a good sound level does not necessarily imply echo disturbances. Given a suitable delay compensation, one can achieve, apart from the prevention of echoes, an extensive coherence of the sound radiated by the various loudspeaker arrays. In this way the initial energy is increased, and the definition is improved.

Sound speed is calculated from the following empirical equation:

$$c = 331.4\sqrt{1 + 0.00366\vartheta}$$

where c is in m/s, and ϑ is in °C.

For $\vartheta = 20$ °C, for instance, there results (neglecting the sound-pressure dependence of sound speed)

$$c = 343.3 \text{ m/s}$$

To ascertain the travel times as a function of the travel paths under normal air pressure and temperature conditions, Table 6.3 may be used. We can see that 58 ms are required to compensate a travel-path difference of say 20 m (cf. Figure 6.32). As travel path unit the British foot is indicated as well as the metre (1 ft = 0.3048 m). The foot scale offers the advantage that 1 ms travel time corresponds to about 1 ft travel path.

6.5.3 Improving the naturalness of sound reproduction

Unless it is intended for producing special effects, a good sound reinforcement system should be designed so that its effect does not contradict the accustomed acoustical perceptions. This requires the following conditions to the met:

- The reproduction loudness should be adapted to the environs (the noise level).
- There should be no perceptible unusual non-linear distortions.
- The system should not produce any additional disturbing noises (noise, cracking, etc.).
- There must be no occurrence of feedback phenomena.
- The linear distortions that occur should correspond to the usual room-acoustical conditions (see section 7.2.2).
- The visual and acoustical localisation of the true or imaginary source (primary sources) should coincide.

Except for the last requirement, all problems can be solved purely by technical measures (see Chapters 3, 4 and 5). Apart from technical factors, psychoacoustical questions are of primary importance for localization. The fundamentals were dealt with in Chapter 3. Here we shall be concerned with the application of the precedence effect (see section 3.2.1.5).

Figure 6.36 explains once more the fundamental effect of time delay. Without time delay one localizes a phantom source in the centre between the two loudspeakers. If the signal is delayed in one channel, the loudspeaker of the non-delayed channel is localized, and only if the level of the delayed signal is increased by more than 6–10 dB against the non-delayed signal (see section 3.2.1.5, Figure 3.38) does the localization point (the phantom source) return to the centre between the loudspeakers.

Table 6.3 (see p. 232) can be used to find the required delays. Parting from one radiator or source providing the reference level, the times and levels required by the individual loudspeakers to ensure localization of the

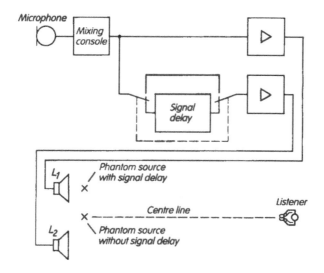

Figure 6.36 Explanation of localization and phantom source formation as a function of the time delay of one component signal [6.1].

source or the reference loudspeaker are chosen with reference to the source location and according to the travel time required in accordance with the distance from the listener.

This is illustrated by Figure 6.37. The voice of a speaker at the desk is radiated by a loudspeaker integrated in the desk. The sound level of this

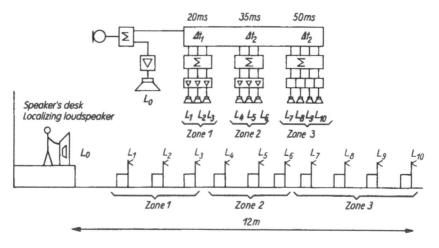

Figure 6.37 Acoustical localization of a simulation sound source (speaker's desk) with hall sound reinforcement (schematic).

loudspeaker does not have to prevail over the sound reinforcement system of the hall, but only to supply the localization reference. The loudspeakers integrated into the seats and used as the main sound reinforcement system are supplied via delay devices according to their distance from the speaker's desk, and produce a sufficient sound level thanks to their proximity to the listeners. The speaker's desk is acoustically localized, however, since its signal arrives earlier, though at a lower level, at the listener's location. In the rearmost rows it is possible for the less delayed front loudspeakers to serve as a localization reference, so that here localization is also towards the front, although the desk loudspeaker is perhaps no longer audible.

6.5.3.1 Arrangement of the loudspeakers

The acoustical localization of the reference sources (which may be not only original sources such as speakers, singers, or orchestras, but also reference or 'simulation' loudspeakers in the case of reproduction of recordings or weak-sounding sources) depends directly on the location of the loudspeakers relative to the reference source and the listener.

With small action areas a centralized arrangement of the loudspeakers near this area may be considered as a satisfactory solution. Because of the expected attenuation with increased distance, this centralized arrangement can be used only in small halls or outdoor auditoriums. Although depending largely on the shape of the room and the arrangement of the audience in front of the stage, the maximum hall volume for this kind of arrangement should be in the region of 8000 m^3.

Frequently an acoustical localization on parts of the action area is also desirable for smaller halls (such as theatres). In such cases a multi-channel intensity sound reinforcement system, which enables localization to occur in halls with a maximum width of 15 m, may function satisfactorily.

In larger rooms a sufficient sound level distribution and definition of reproduction can be achieved only by means of a centrally supported loudspeaker arrangement appropriately staggered in width and depth. This also holds true in principle for installations used for pop and rock music, although in this genre fully centralized sound reinforcement systems are for various reasons usual also for large auditoriums (see section 6.6.4). Since large and thus highly directional loudspeakers are used here, excessively high sound levels result in the near field, which are capable of causing permanent hearing damage among performers and listeners, whereas at larger distances the levels and the spectral composition of the signals no longer correspond to the desired values.

If a centrally supported or even a decentralized loudspeaker arrangement is used, and acoustical source localization is required, it is necessary for the loudspeakers, unless they are in the immediate proximity of the source, to be operated in conjunction with a delay. In the following sections we are going to compile the requirements of sound reinforcement systems ensuring acoustical localization with or without delay equipment.

6.5.3.2 *Single-channel sound delay*

If the action area containing the sound sources is not wider than 12–15 m and not deeper than 10 m, sufficient travel-time compensation can be achieved by means of a single-channel delay system to which the staggered loudspeaker arrays are connected with different delays. If the powerful main loudspeaker arrays are at a greater distance from the audience, for example above a stage proscenium, localization of the sources can be accomplished with observance of the necessary level conditions without pre-delay of the main loudspeakers.

With large action areas this is no longer possible. To enable acoustical localization under these conditions, a pre-delay of the main loudspeakers was effected for the first time in 1965 in the Jahrhunderthalle in Hoechst [6.27]. The pre-delay was adjusted in such a way that the first wavefront of the signals coming from the original source (or its simulation) located on the stage arrives before those emanating from the loudspeakers.

With wider action areas, single-channel sound reinforcement systems may give rise to problems, since the travel time of the sound from one side of the stage to the other cannot be compensated for by a basic delay. This delay time would have to be so long that a noticeable pause between the original signal and the amplified signal would occur. This would result in an echo-like double hearing.

6.5.3.3 *Multi-channel intensity procedures*

To avoid the drawbacks of single-channel sound reinforcement systems, the loudspeaker arrays near the action area are subdivided. The loud-speaker array nearest to the original source radiates the signal coming from the same with the greatest intensity (emphasis of sometimes up to 10–15 dB), whereas the remaining loudspeaker arrays radiate either not at all or with an intensity reduced according to their distance from the original source.

A disadvantage of this procedure is that source-dependent level fluctuations occur all over the auditorium. Moreover, the attainable source localization is rather unsatisfactory.

This procedure was used for many years in the Congress Palace of the Moscow Kremlin [6.28] (Figure 6.38). There the stage was subdivided into five equal-sized areas to which five loudspeaker arrays (1) were assigned to the local-stage ceiling area. Each of these 'proscenium arrays' radiated with greatest intensity the signal originating from the corresponding stage area located behind the assembly. The additional loudspeakers arranged over the depth of the room (2, 3, 4) reproduced a composite partial signal, delayed

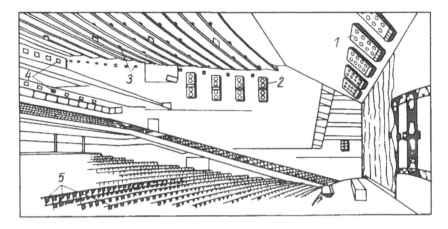

Figure 6.38 Multi-channel intensity sound reinforcement: arrangement of the loudspeakers.

according to the distance from the stage. In principle it was also possible to include loudspeakers integrated into the seats (5).

In an improved version the proscenium loudspeakers were made partially movable for better adaptation to the setting arrangement, and powerful loudspeakers were added at the front edge of the stage for better acoustical localization on the level of the action area. The procedure is now also used in Japan in a large multi-purpose hall [6.29].

6.5.3.4 *Multi-channel procedures with delay systems*

To avoid the drawback of poor sound-level balance increasing with the stage width, and to provide the possibility of acoustical source localization with central stages jutting far out into the audience area, multi-channel delay systems were developed. The action areas are subdivided, and a delay chain is assigned to each of these 'source areas'. The proscenium loudspeakers and the loudspeakers arranged in the depth of the room are supplied via these delay chains.

A delay system of this kind was installed in the International Congress Center (ICC) in Berlin. Here different loudspeaker arrays were specifically assigned to the different delay circuits. Figure 6.39 shows the schematic location of the loudspeakers coupled to channel A, which is responsible for radiation of the signals to be localized in the right-hand stage area, as well as the associated delay times (these loudspeakers are marked in black) [6.30].

More difficult than in the ICC, where there is a linear boundary between the stage and the audience area, are the conditions in the former Palast der

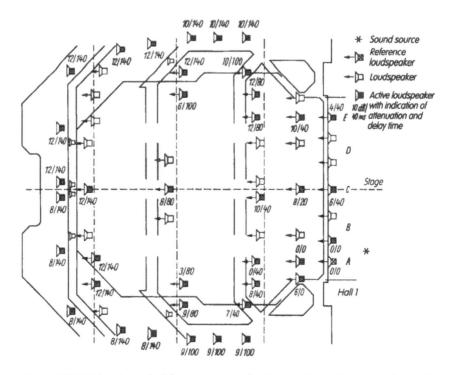

Figure 6.39 Multi-channel delay system in the International Congress Centre in Berlin.

Republik in Berlin, which was finished in 1976 [6.31]. Here it was possible to establish very different boundary formats between the various possible stages and the audience area. The width and depth of the stage could be up to 40 m, and the stage could jut far out into the encompassing audience area, so that an arena-like format could sometimes result (Figure 6.40). To solve the problems involved, the *Delta Stereophony System* (DSS) was developed [6.16].

With this procedure the action area is again subdivided into several source areas (Figure 6.41), and a delay chain is assigned to each of them. The signals originating from the source areas are fed via the corresponding delay chains to all loudspeakers in the room and some of the loudspeakers arranged on the stage. The delay times are chosen in such a way that – with due consideration of the travel time of the sound between the individual loudspeaker arrays and the listeners' seats – the first wavefront of each signal emitted by the reference source comes from the assigned source area. The calculation required to create this effect has to consider not only the critical limit areas of the source areas, but also those of the audience areas. Figs. 6.42a and b show such sequences of wavefronts arriving from the

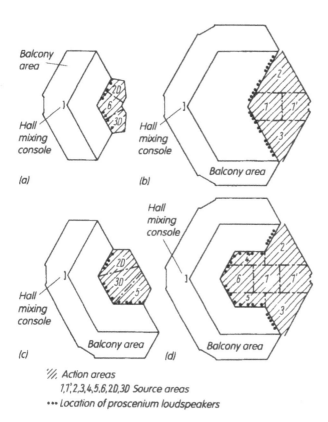

%% Action areas
1,1',2,3,4,5,6,2D,3D Source areas
••• Location of proscenium loudspeakers

Figure 6.40 Different platform configurations in the large hall of the former
Palast der Republik in Berlin: (a) small platform (1800 spectators);
(b) panorama platform (4200 spectators); (c) diagonal variant
(2800 spectators); (d) amphistage (3800 spectators).

individual loudspeaker arrays as well as the area in which the reference
sound sources may be located in the source area (sending positions 1, 2, 3
and 4). We can see that the signals originating from the source areas (OS)
always constitute the first wavefront for forming a reference sound source.
At the listener's location this reference signal always arrives prior to the
corresponding signal of the loudspeakers (arrays P, S and A).

To allow simultaneous operation of several source areas, a signal-mixing
matrix is inserted before the power amplifiers assigned to the loudspeakers.

It has to be ensured that the precedence effect, which determines the
acoustical localization, operates *throughout* the transmitted frequency
range. This means that, at the listener's location, the signal of the reference
source (that is, of the original source) or of a loudspeaker simulating the
original source does not fall over the entire transmitted frequency range by

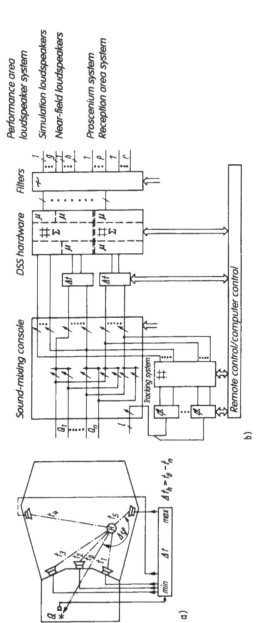

Performance area loudspeaker system

Simulation loudspeakers

Near-field loudspeakers

Proscenium system

Reception area system

Filters

DSS hardware

Sound-mixing console

Tracking system

Remote control/computer control

b)

a)

Figure 6.41 Schematic layout of a Delta Stereophony sound reinforcement system ([6.44]): (a) working principle; (b) schematic layout of the DSS. φ, angular deviation between visual and acoustical direction of perception without DSS; Q, original or simulation source; t_0, acoustic travel time of the original sound to the listener; t_h, acoustic travel time of the loudspeaker sound to the listener; Δt_n, electrical delay time for the respective loudspeakers; H, listener.

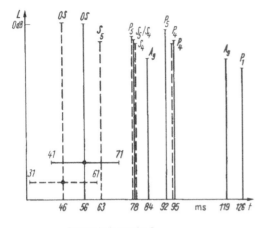

——— Transmission point 2

a) — — — Transmission point 3

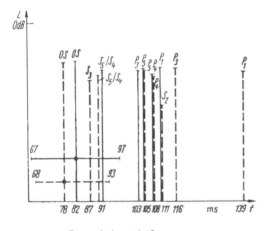

——— Transmission point 1

b) — — — Transmission point 4

Figure 6.42 Sequence of wavefronts arriving from different loudspeaker arrays.

more than 10 dB below the signal levels of the loudspeakers amplifying the sound. This condition implies that it is necessary to consider also the frequency-dependent attenuation occurring along the propagation path, as for example the absorption of high frequencies conditioned by dissipation. For this reason one should also avoid monitoring loudspeakers that are too small, as frequently required to avoid intrusion within the stage setting, since their bass radiation is insufficient. One has to keep in mind, however, that frequencies below 300 Hz do not have any bearing on safeguarding the localization. Here it may possibly occur that owing to differences in the spectra of the simulating loudspeakers and the main loudspeakers, localization becomes diverted to undesired directions (e.g. towards the main loudspeakers fully radiating the low frequencies).

The latest solution for realizing the DSS has been the development of a special processor providing controllable delay times [6.32]. Exceeding the possibilities offered by the mixing matrix (true-delay tracking of motion with true-level fade-over, e.g. by means of a panning control), it enables tracking the motion additionally with time differentiation. The mentioned 'source areas' dissolve in this case into actual 'source points'.

Figure 6.43 depicts the circuit of the processor, and shows that up to six independent sources can be combined with freely controllable delay and level values and passed on to 10 independent lines. This mode of functioning presupposes, of course, simulating loudspeakers in the stage area. Since control is effected by means of a computer, the loudspeaker positions depending on the stage setting can be stored in the computer and called up by touching a key.

A new quality in the simulation of desired sound-field structures would doubtlessly be achieved with an automatic tracking system (yet to be realized).

The DSP 610 processor has been in operation since 1992 in halls in Japan and Germany as well as in the Congress Palace in the Kremlin in Moscow after the reconstruction of its sound reinforcement system (cf. section 6.5.3.3) [6.33, 6.34].

By means of the DSS it is thus possible to influence the temporal sound-field structure. The build-up and decay of the sound field at the listener's location can be influenced in such a way that good definition values for speech and an analogous clarity for music (variable according to the genre) can be achieved (the fundamental problem of the multi-purpose hall). This effect should encompass the whole room as *uniformly* as possible.

By means of a decentralized loudspeaker arrangement, or at least one that is staggered along the depth of the audience area, the DSS enables the injection, apart from the 'primary sound', of a heavily delayed reverberation for enhancing spaciousness. This requires delay times of the reverberation of $t_{st} \approx 2\sqrt{V}$ (where V is in m^3, and t is in ms), thus being dependent on the volume of the room [6.35]. In order to avoid disintegration of the signal, with the ensuing occurrence of audible echoes (especially for long reverberation delays in heavily damped rooms), it is necessary to insert additional 'initial reflections' after the 'primary sound'. This ensures that the sound pattern at the listener's location is not determined solely by the original sound and the sound impinging from the nearest main and room loudspeaker arrays, but also that the other loudspeakers contribute to it – with varying attenuation and delay according to their distance. The

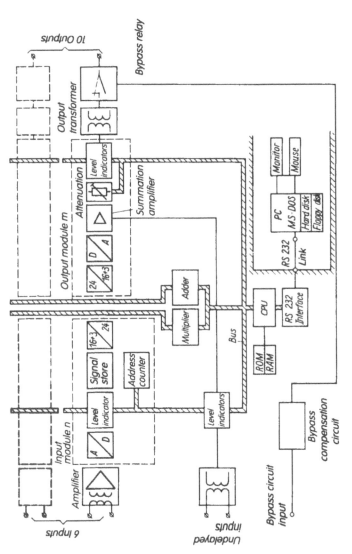

Figure 6.43 Operating principle of the DSP 610 DSS processor from AKG.

mentioned level sequences thus generate a complex manipulation of entire sound fields: that is, the sound reinforcement system simulates the direct sound with correct localization, initial reflections and reverberation, if necessary fully by electronic means. However, it must be said that the delay required can bring about an enhancement of spaciousness, which may be perceived in the near-stage areas where such spaciousness is not expected, and is thus undesirable. This may be minimized by a careful calibration of the system.

6.6 Systems for artistic purposes

Sound reinforcement systems are to an increasing degree required for realizing artistic productions in theatres, opera houses and concert halls, but also for reproducing dance and pop music.

6.6.1 Theatres and opera houses

Among other things, the electroacoustical system in theatres and opera houses is required to realize effects well known from the electronic media, but intended here to produce a localizing and sometimes also stereophonic impression. These include reverberant effects, 'spooky' voices, weather sounds, and battle noise, which are to impinge on the audience from specific directions and partly with a motion effect. Moreover, it is required for the voices of actors to be heard, either modified or as true as possible to the original, from certain directions of the stage house or the audience area, and sometimes with the imitation of an approach or departure in certain directions. Apart from that, normal sound reinforcement tasks also have to be fulfilled, and last but not least the sound reinforcement system has to replace the orchestra for ballet performances by providing the sound for the hall and coverage of the stage itself.

In a repertory theatre in particular, with frequent changes of programme, these tasks can be fulfilled only if most of the effects can be realized via permanently installed loudspeakers. As a rule, the following loudspeaker arrays are required (the example is based on a medium-sized multi-genre theatre):

- Three to four loudspeaker arrays are installed on the left and right of the stage opening; above the proscenium there is a loudspeaker assembly that is either undivided or divisible into left and right subarrays.
- In the stage house, around the opening to the backstage area, two lateral loudspeaker arrays have to be provided, plus an assembly above the proscenium.
- A powerful loudspeaker assembly is required in the region of the backstage rear wall, for instance above the rear projection booth.

- A diffusely radiating loudspeaker assembly is required in the ceiling area of the auditorium, for (among other things) emitting thunderstorm and rain noises.
- A large number of smaller individual loudspeakers are distributed as uniformly as possible over the wall and balcony balustrade areas of the auditorium. Combined in locally concentrated arrays, these loudspeakers must be effective from certain areas of the auditorium, but must be controllable by means of a panning switch to emit a signal that circulates round the hall or jumps from area to area.
- For hearing on the stage, loudspeakers have to be installed at the stage proscenium towers and at the lighting cross-bridge. In this area these must have the smallest possible travel time difference to the main loudspeaker arrays, but be effectively screened from the auditorium.

The positioning of these loudspeaker arrays is illustrated in Figure 6.44 by the example of the Semperoper, Dresden [6.36].

Broadband loudspeakers of medium directivity with a power rating of 50–300 W are most frequently used in the main arrays arranged around the stage opening. These arrays have to be capable of producing the maximum sound level desirable in the auditorium. This level is in most cases assumed to be 104 dB in the diffuse field, a value that already includes an overmodulation headroom of 10 dB. The required installed audio power of the loudspeakers can be ascertained according to equation (5.29). For high-quality loudspeaker arrays (and also in low-impedance systems) one may reckon with an average efficiency of 1%. The installed electric power required for the main loudspeaker arrays under these conditions is

$$P_{el} = 0.1 \frac{V}{T} \text{ W}$$

where V is in m^3, and T is in s.

The loudspeakers chosen for the arrays in the stage area should have a higher front-to-random factor, to ensure that as large a portion of the radiated sound energy as possible passes through the stage opening (acting like a diaphragm) into the auditorium. These stage arrays are designed so that the direct sound level of each speaker group produces 90 dB in the central stalls area of the auditorium. Equation (5.30) may be used to determine the required installed audio power.

Because of the shadowing effect of the two stage wings, and the greater the distance between the audience area and the loudspeakers, a stronger attenuation occurs in the upper and lower frequency ranges than in the mid range for the loudspeaker arrays in the backstage area (for instance, those mounted on the rear projection booth). This has to be taken into account,

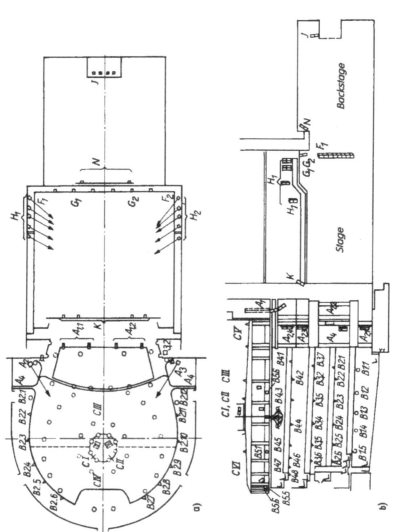

Figure 6.44 Loudspeaker arrangement in the Semperoper Dresden: (a) floor plan; (b) section. △, ○, individual loudspeaker; □□, TZ 133 sound column.

especially if the main loudspeakers are to produce a sound level enhancement in the auditorium by making use of the precedence effect through a delay system. Since the maximum level separation between the reference signal and the amplifying signal must not exceed 6–10 dB in any frequency group, a limitation of amplification takes place. An improvement could be achieved by a compensating equalising of the signals after ascertaining the acoustic frequency response of the stage loudspeakers at the listener's position.

As regards the effects loudspeakers in the auditorium, the diffusely radiating ceiling arrays should be designed so that they produce a maximum level of 90–95 dB in the diffuse field. Individual cone loudspeakers of 10 W rated power or larger arrays may be used here; the latter should be aimed towards the ceiling or reflecting surfaces in the ceiling area. The installed power can be ascertained by calculating the acoustic power according to equation (5.27) and then multiplying it by the reciprocal value of the loudspeaker efficiency.

The design of the loudspeakers in the wall area of the auditorium is especially critical. These loudspeakers are generally at close range to the audience, so that a loudness and timbre change dependent on the position of the operating loudspeakers cannot be avoided for certain groups of audience seats. To reduce this detrimental effect to the lowest value possible, the installed power and directivity of these loudspeakers have to be kept as low as possible. In a theatre with balconies it is more favourable to arrange the loudspeakers in the balustrades than behind the audience. As is the case with all loudspeakers to be installed in the auditorium, aesthetic requirements have to be solved here in addition to acoustic problems.

The installed power required for all wall loudspeakers and for individual arrays of this system has to be determined according to equation (5.31) for the sound pressure desirable in the diffuse field. For the overall system, and for a partial speaker group of the same, sound levels of at least 86 dB and 80 dB respectively should be achieved. In the near field of the loudspeakers it should, if possible, not exceed 90 dB. To calculate the installed power it is necessary to consider the free-field level caused by the three nearest loudspeakers.

The monitor loudspeakers in the stage house should be arranged so that they are audible in the auditorium either not at all or only very weakly. Their installed power should be specified for a sound level of 85 dB, and can be determined from equation (5.30).

6.6.2 Concert halls

Sound reinforcement systems are nowadays required in all concert halls, both because of the increasing use of electroacoustical techniques in modern music, and also for sound reinforcement tasks and for improving definition in halls that are often very large and reverberant. As the artistic utilization of such a system often requires the realization of spaciousness-enhancing or localizable acoustic effects, similar to the way it has always traditionally been possible before by means of instrumental groups or organs distributed over the room, the systems show certain similarities with theatre systems. The following differences have, however, to be taken into account:

- Concert halls are in general larger than auditoriums in theatres.
- Concert halls are devoid of stage houses for accommodating effects loudspeakers staggered in depth.
- The room volume per person and the reverberation time in concert halls are often considerably higher than in theatres (usually $T < 2$ s).
- The maximum sound powers are higher in concert halls: a large symphony orchestra on stage produces an acoustic power of 8–10 W, whereas an opera orchestra playing in the orchestra pit can fill the room with, at most, 6 W.
- Unlike the theatre, the use of the loudspeaker system is limited in the concert hall to a few special cases.

From this one can infer the following requirements to be met by loudspeaker systems in concert halls:

- It is necessary to use powerful loudspeaker arrays, capable of producing in the hall an acoustic power of 8–10 W (if need be with delay systems) (Figure 6.45).
- It should be possible, by means of a patch panel or matrix switching unit, to realize a single- or multi-channel sound signal of variable localization (depending on the size and configuration of the room, at 4–12 positions).
- The loudspeakers must have a largely linear frequency transmission range of at least 40 Hz to 16 kHz. This must be coupled with equalization filtering and facilities for adapting the signal to the specific requirements before radiating it.
- A hall mixing console for acoustic control and for adjusting the sound balance as well as the spatial effect is required at an acoustically representative location in the hall. Because it is relatively seldom used, the console is not permanently installed. A permanently cabled connecting point must be provided, however, to enable connection of a mobile mixing console with the necessary inputs and outputs as well as further sound processing equipment (such as a vocoder, ring-type modulator, or equalizer). Remote control of sound recording devices should also be possible from here [6.37].

Apart from the technical equipment installed, or cases where equipment is installed in concert halls for the reproduction of electronic music, there are smaller special rooms serving *exclusively* for the reproduction of electronic sounds. These include exhibition spaces serving for the reproduction of special compositions [6.38], reproduction rooms in which not only

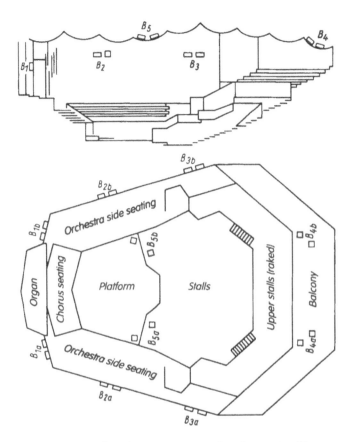

Figure 6.45 Loudspeaker arrangement for the music effects sound reinforcement system in the Neues Gewandhaus, Leipzig: □ sound cabinet (irradiation of effects).

acoustical but also visual impressions (such as laser effects) can be facilitated [6.39], and rooms for the performance of experimental music [6.40]. They mostly have a considerably shorter reverberation time than the conventional concert halls, and occasionally have altered room acoustics. Thus the sound signals can be radiated – uninfluenced by the relatively long initial and decay transients of the concert halls – with greater transient fidelity and sharpness of localization.

6.6.3 Reproduction of effects signals

In theatres and concert halls (see previous sections), but of late also in dome-shaped rooms such as planetariums, loudspeakers distributed over the room are required to produce the directional impressions demanded by the scenario editor or producer of the performance or show. Among other things, a more intensive involvement of the audience in the sound event is achieved. The sound sources to be realized may be stationary as well as travelling point or wide-area sources.

There are the following possibilities for the realization of such effects signals:

- Several relatively powerful loudspeaker arrays of diffuse radiation characteristic are arranged in or around the audience area. They are assigned to specific sound channels, or may be put successively into operation by means of a channel switchboard (group matrix) unit. With continuous switching a panoramic effect results.
- A large number of small loudspeakers are distributed over the boundary areas of the room; combined in arrays, they form a localizable area of variable width. With panoramic switching it is possible to realize an acoustical cross-fade by successively switching the individual loudspeakers on until completing the whole array, while simultaneously the first activated loudspeakers are switched off.
- Directional cross-fading is possible by means of a coupled delay–attenuation circuit, which thanks to the precedence effect enables localization of individual loudspeaker arrays with simultaneous operation of all arrays of the system.

Planetariums have quite recently been equipped with a sound reinforcement variant that consists of loudspeakers that are uniformly distributed over the surface of the dome. By computer-aided control it is possible to produce acoustical impressions whose localization corresponds to any desired point of the dome surface. This creates, apart from the click-free switching achieved also by other solutions, an acoustical motion that precludes any jumping of the directional impression [6.41] (Figure 6.46). Loudspeaker arrangements with circular, square or hexagonal spacing at an average distance of 4 m from one another have proved to be suitable. Thus the visual events shown in these dome-shaped rooms, such as the simulation of rocket launchings or an impressive acoustical background to laser and slide shows, become acoustically traceable. With such performances it is possible for the control of other demonstration media, including the very planetarium projector itself, to be undertaken by the sound reinforcement system.

6.6.4 Transmission of rock and pop music

Modern popular dance and rock music is unthinkable without amplification and sound processing equipment, as well as loudspeaker systems for sound radiation. Apart from classical instruments, synthesizers and samplers are being used to an increasing degree for generating sounds and

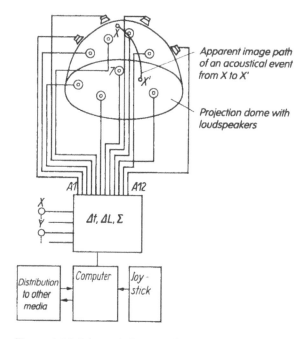

Figure 6.46 Schematic layout of a loudspeaker arrangement in a planetarium.

noises. With some instruments the tone is picked up mechanically or electroacoustically, or the sound is picked up directly, as for instance that of the human voice. The electronic signals thus obtained are amplified and radiated either via loudspeakers specifically assigned to individual instruments, so that the loudspeaker can be considered as an integral part of the instrument (inclusive of its often very strong characteristic timbre), or processed and mixed, then amplified, and finally radiated by loudspeakers that however, because of their particular timbre, also form in this case part of the 'sound' of the group. It is for this reason that such systems cannot be fully replaced by largely 'neutral'-sounding permanently installed 'house' sound reinforcement systems; they have to be used at least as reference sources on stage. Their signals should be transferred to the 'house' sound system using the outputs of the mobile stage system, which should be prepared for matching studio conditions.

The output signals are conveyed to the audience in different ways. Indoors it is quite common to install on the left and on the right of the stage large loudspeaker arrays in a tower-like fashion, but also loudspeaker walls and transportable proscenium cross-beams equipped with powerful loudspeakers are frequently found. In the open a combination of both is used.

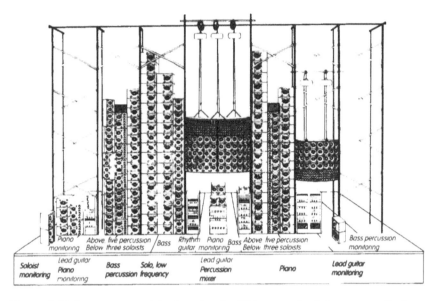

Figure 6.47 Arrangement of loudspeaker 'wall' for sound reinforcement of a large
open-air rock concert.

There are several characteristics shared by all installations of this kind
[6.1]:

- powerful bass reproduction;
- very high sound levels;
- large risk of comb-filter effects occurring;
- small microphone distances;
- powerful monitoring.

In the stalls area of indoor rooms the first two aspects can mostly still be
ensured by the tower and wall loudspeaker arrays (as in Figure 6.47). This
implies, however, that the first rows of the audience and the action area of the
musicians are covered with extremely high sound levels, which may also lead
to hearing damage, especially among the musicians (see section 3.2.3). For the
balcony areas, however, these loudspeaker arrays are often insufficient.

The assertion that the left and right tower arrays are important for the
stereo effect is not correct. Such an effect can hardly be perceived at widths
of more than 10–12 m; in the lateral direction the nearer loudspeaker
arrays mask the signal from the farther ones. Here it is certainly better to
work with multi-channel systems (such as Delta Stereophony) allowing a
'neutral' coverage of the hall: that is, a coverage that is free from eigentone
timbres. The desired eigentone timbre is then generated on the stage and fed
into the main sound reinforcement system; the musicians' system thus
constitutes the source signal in the sense of a Delta Stereophonic System.

Figure 6.48 Rock concert arrangement with delay tower in Wembley Stadium, London.

This is different from the mobile sound reinforcement systems used in the open on tours. Here the system has to cover large open spaces, sometimes with an attendance of up to 200 000 persons. Requirements of this kind cannot be met without using large loudspeaker arrays. At present the most frequently used array is that of a centrally supported sound reinforcement system. It consists of a main loudspeaker array above and at the sides of the main stage, and sub-arrays in the form of *delay towers*, the loudspeakers of which are fed with correspondingly delayed signals.

Also here a multi-channel transmission is desirable. For clarity enhancement the individual voices are distributed to two or three loudspeaker arrays, often stacked in a tower-like fashion. If a soloist is to be given special prominence, his or her signal is radiated separately via a powerful central array, whereas the accompaniment is reproduced by the lateral arrays [6.42]. In this respect it is convenient to design the central array so that on its own it can provide the whole sound reinforcement.

The objective of the sound reinforcement is to provide sufficient coverage of all areas. The towers, with their loudspeakers mounted at a height of 5 m or more, offer the advantage that the sound pressure impinging on the listener is kept within certain limits. Figure 6.48 shows as an example of open-space sound reinforcement the installation of a mobile system for a rock concert in Wembley Stadium in London. On the right of the stage we can see an additional delay tower intended to provide coverage of the large area [6.43].

It is important that each tower system is able to cover its 'part' of the auditorium with adequate clarity and definition, despite even the worst wind and weather conditions. The delays provide freedom from echoes as well as a level of sound reinforcement so as to ensure reference to the stage or the side stage.

Special circuit technologies based on the DSS concept have been used to generate sound images (so-called 'sound clouds'). Loudspeaker arrays were distributed in a coarse grid and operated with a delay from one another such that no echo formation occurred within the listening range. On open spaces of 1000 m × 1000 m and more the resulting sound impression, though intentionally not exactly localizable, truly followed visual and laser-technical patterns [6.44].

With powerful loudspeaker towers there is a risk that equal radiators interfere with each other so as to influence the directional characteristic of the whole arrangement (lobing). To avoid this, loudspeakers working according to the Bessel principle (see section 4.1.2.3) and governed by processor-controlled amplifiers have been used of late [6.42].

Problems arise when projection walls, e.g. Eidophor, are used at large events (more than 20 000 persons). At 100 m from the stage, for instance, the optical information is instantaneous, whereas the acoustical information follows about 300 ms later. This is then reminiscent of a poorly synchronized film. It cannot be remedied by acoustical measures (undelayed sound radiation near the audience area to be covered produces echo interferences with the main signal radiated near the projection wall), but only by means of an appropriate 'delay' of the video projection, a solution that requires, however, several projection facilities staggered in depth.

6.6.5 Museums and exhibitions

In museums, galleries and exhibition rooms, sound reinforcement engineering may have to cope with solving the following tasks:

- Explanation of the exhibits of a collection on demand by an *information system*. Stationary equipment (slot machines) permanently assigned to specific exhibits is still of some importance in this respect. Most frequently used, however, are portable cassette recorders enabling guidance also in various languages. New is the use of infrared systems, enabling a wireless and specific explanation of the exhibits or other details in the required languages (it is merely a question of expenditure) (Figure 6.49).
- Generation of background sound (music, noises or special acoustic effects).
- Reproduction of sound as a form of exhibit (playback recordings).

Figure 6.49 Infrared tour guide information system in a museum (by Sennheiser):
(a) wireless receiver unit; (b) high-power infrared transmitter.

- Reproduction of sound collages serving as a dramatized acoustic picture.
- Use of diatone and video techniques.

In the future, interactive operation of such systems would be desirable: the visitor to an exhibition would by his or her appearance on a certain route trigger activities providing acoustical as well as visual information. In the end there will be the development of a so-called *guidance system with cognitive orientation*, enabling man and computer-controlled information to enter into immediate and extensive interaction [6.45].

References

6.1 Davis, D. and Davis, C., *Sound System Engineering*, 2nd edn (Indianapolis: Howard W. Sams, 1987).

6.2 Cremer, L., *Die wissenschaftlichen Grundlagen der Raumakustik* (The scientific foundations of room acoustics), Vol. I (Stuttgart: Hirzel, 1978), p. 295.

6.3 Kraak, W., Fasold, W. and Schirmer, W. (eds), *Taschenbuch Akustik* (Handbook of acoustics) (Berlin: Verlag Technik, 1984).

6.4 Blauert, I., *Spatial Hearing* (Cambridge, MA: MIT, 1983).

6.5 Petzold, H., *Grundlagen der Beschallungstechnik* (Foundations of sound reinforcement engineering), Vol. IV (Leipzig: Fachbuchverlag, 1957).

6.6 Oehme, W. *et al.*, 'Elektroakustische und nachrichtentechnische Anlagen in der neuen Mehrzweckhalle Gruga in Essen' (Electroacoustical and communication systems in the new multi-purpose Hall Gruga in Essen), *Siemens-Zeitschrift*, 33 (1959) 11, 728–735.

6.7 Steffen, F., 'Beschallungsanlage der Sport- und Kongreßhalle Rostock' (Sound reinforcement system of the Sports and Congress Hall Rostock), *Technische Mitteilungen RFZ*, 25 (1981) 2, 25–29.

6.8 Boye, G., 'Freiluft-Beschallungsanlagen für Großveranstaltungen und Sportstadien' (Open-air sound reinforcement systems for mass events and sports stadiums), *Fernseh- und Kino-Technik*, 33 (1979) 1, 13–17.

6.9 'A transistorized audio amplifier system with an output power of 23 kW (Equipment Strahov State Stadium Prague)', *Tesla Electronics*, 7 (1974) 2, 58–61.

6.10 Kahmann, W.G., 'Beschallungs- und übertragungstechnische Einrichtungen im Plenarsaal des Palastes der Republik, Berlin' (Sound reinforcement and transmission equipment in the Plenary Hall of the Palast der Republik, Berlin), *Technische Mitteilungen RFZ*, 21 (1977) 3, 71–72.

6.11 Keller, W. and Widmann, M. 'Kommunikationssysteme im ICC Berlin' (Communication systems in the ICC Berlin), *Funkschau*, 51 (1979) 15, 858–864.

6.12 Kleis, D. 'Moderne Beschallungstechnik', *Philips Technische Rundschau*, 20 (1958/59) 9, 272ff. and 21 (1959/60) 3, 78ff.

6.13 Kaczorowski, W., 'Urzadzenia elektroakustyczne i naglosnia w Wojewodzkiej Hali Widowiskowo-Sportowej w Katowicach' (Electroacoustical equipment in the Voievodeship Sporting Hall Katowice), *Technika Radia i Telewizji*, 4 (1973) 1–16.

6.14 McGregor, C., 'Electronic architecture: the musical realities of its application', Sixth International Conference on Sound Reinforcement, Nashville (1988).

6.15 'Sound system design for the Eugene Performing Arts Center, Oregon', *Recording Engineer/Producer* (1982) 86–93.

6.16 Steinke, G., Fels, P., Hoeg, W. and Ahnert, W., 'Das Delta-Stereofonie-System' (The Delta Stereophony System), *Radio, Fernsehen, Elektronik*, 36 (1987) 10, 615–620 and 11, 722–727.

6.17 Parkin, P.H. and Morgan, K., ' "Assisted resonance" in the Royal Festival Hall, London', *Journal of Sound and Vibration*, 2 (1965) 1, 74–85.

6.18 Parkin, P.H. 'A special report on the experimental ("assisted resonance") system in the Royal Festival Hall, London', *Journal of Sound and Vibration*, 1 (1964) 3, 335–342.

6.19 Franssen, N.V., 'Sur l'amplification des champs acoustiques' (On the amplification of sound fields), *Acustica*, 20 (1968) 315 ff.

6.20 Ohsmann, M., 'Analyse von Mehrkanalanlagen' (Analysis of multi-channel systems), *Acustica*, 70 (1990) 4, 233–246.

6.21 Koning, de S.H., 'Konzertsäle mit variablem Nachhall' (Concert halls with variable reverberation), *Funkschau*, 57 (1985) 14, 33–38.

6.22 Berkhout, A.J., 'A holographic approach to acoustic control', *Journal of the Audio Engineering Society*, 36 (1988) 12, 977–995.

6.23 Griesinger, D., 'Improving room acoustics through time-variant synthetic reverberation', AES preprint 3041 (B-2), AES New York, USA (1991).

6.24 Prinssen, W.C.J.M. and Kok, B.H.M., 'Active and passive orchestra shells and stage acoustics. Comparison of performance characteristics and practical application possibilities, presentation of a case study', 12th Meeting of the Acoustical Society of America, Washington (May–June 1995), paper no. 4aAA5.

6.25 Ahnert, W., 'Nachhallzeit und Schallenergiedichteverlauf in über Richtmikrofon gekoppelten Räumen' (Reverberation time and sound energy density behaviour in rooms coupled by directional microphones), *Zeitschrift für elektronische Informations- und Energietechnik*, 9 (1979) 1, 67–73.

6.26 Bolt, R.H. and Doak, P.E., 'A tentative criterion for the short-term transient response of auditoriums', *Journal of the Acoustical Society of America*, 22 (1950) 507.

6.27 Meyer, E. and Kuttruff, H., 'Zur Raumakustik einer großen Festhalle' (On the room acoustics of a large festival hall), *Acustica*, 14 (1964) 3, 138–147.

6.28 Khruschev, A.A., 'Sound systems for large multipurpose halls in the Soviet Union', *Journal of the Audio Engineering Society*, 18 (1970) 1.

6.29 Yamaguchi, K., et al., 'Design of an auditorium, where electroacoustic technology is fully available: Exhibition Hall of Yamaha Sportland Tsumagoi', *Journal of the Acoustical Society of America*, 62 (1977) 5, 1213–1221.

6.30 Plenge, G.H., 'Sound reinforcement system with correct localization image in a big congress centre', 73rd AES Convention, Eindhoven (1983), preprint no. 1980 (F2).

6.31 Steffen, F., 'Akustische Probleme bei der Konzipierung der elektroakustischen Beschallungsanlagen im Großen Saal des Palastes der Republik' (Acoustical problems in the conception of the sound reinforcement systems in the great hall of the Palast der Republik), *Technische Mitteilungen RFZ*, 21 (1977) 3, 51–55.

6.32 Nadler, W., 'The DSP610 compact processor to utilize a new directional sound reinforcement system', 82nd AES Convention, London (1986), preprint 2472 (K-2).

6.33 Ahnert, W., 'The sound system in the new Culture and Congress Centre Liederhalle in Stuttgart/Germany', 92nd AES Convention, Vienna (1992), preprint no. 3300.

6.34 Ahnert, W. and Scheirmann, D., 'The Kremlin Palace of Congress', *Sound & Video Contractor*, 20 (April 1992) 36–48.

6.35 Cremer, L. and Müller, H.A., *Die wissenschaftlichen Grundlagen der Raumakustik* (The scientific foundations of room acoustics), Vol. I (Stuttgart: Hirzel Verlag, 1978).

6.36 Steffen, F., Fels, P. and Schwarzinger, W., 'Die elektroakustischen Einrichtungen der Semperoper Dresden' (The electroacoustical facilities of the Semper Opera Dresden), *Technische Mitteilungen RFZ*, 30 (1986) 3, 49–56.

6.37 Steffen, F., 'Die elektroakustischen Einrichtungen im Neuen Gewandhaus Leipzig' (The electroacoustical facilities in the Neues Gewandhaus Leipzig), *Technische Mitteilungen RFZ*, 27 (1983) 2, 25–30.

6.38 'Sphärenmusik aus der "Blauen Kuppel" des deutschen Pavillons auf der Weltausstellung Expo '70 in Osaka' (Music of the spheres out of the 'Blue Dome' at the World Exhibition Expo '70 in Osaka), *Funktechnik* (1970) 10, 378.

6.39 'Ton- und Lasershow im Großplanetarium Berlin' (Sound and Laser show in the Great Planetarium Berlin), advertising leaflet, Berlin (1989).

6.40 Polaczek, D., '"Espace de Projection" des Pariser "IRCAM"' ('Espace de Projection' of the Paris 'IRCAM'), *HiFi*, 20 (1978) 1300–1304.

6.41 DP 3941584, 'Verfahren und Anordnung für eine örtlich sowie zeitlich veränderliche Signalverteilung über eine Großbeschallungsanlage, insbesondere für audiovisuelle Veranstaltungen in Auditorien, vorzugsweise kuppelförmigen Räumen' (DP 394/584: Procedure and arrangement' for a locally as well as temporally variable signal distribution over a large sound reinforcement system, especially for audiovisual performances in auditoriums, preferably dome-shaped rooms).

6.42 Müller, M., 'PC-gesteuerte Großbeschallungsanlage von Stage Accompany' (PC-controlled large sound reinforcement system by Stage Accompany), *dB Magazin für Studiotechnik* (1988) 333–334.

6.43 Jones, S.P. and Griffiths, J.E.T., 'Wembley Stadium sound system design – a brief description of the design criteria, implementation and results of the first two project phases', *Proceedings of the Institute of Acoustics*, 12 (1990) 8, 169–174.

6.44 Hoeg, W. and Fels, P., 'Weiterentwicklung und neuere Anwendungen des DSS im mobilen Bereich der Beschallungstechnologie' (Advancement and new uses of the DSS in mobile applications of sound reinforcement engineering), *Technische Mitteilungen RFZ*, 32 (1988) 4, 75–81.

6.45 DP 3533705 'Anordnung für ein audiovisuelles Informationssystem über Ausstellungsobjekte' (Set-up for an audiovisual information system on exhibition objects).

7   Calibration and testing

Before any sound reinforcement facility is put into operation, it should be tested for electrical and acoustical functionality. This test should confirm that the system is ready for operation, and that no disturbances attributable to electrical defects, poor calibration or incorrect travel time or timbre adjustments are occurring. Moreover, the acoustical calibration has to supply the reference values for adjustments that may be needed during operation, such as maximum amplification, adaptation of the loudspeakers to the room-acoustical conditions, realization of maximum intelligibility (or spaciousness if so required by the system), and sufficient uniformity of sound level distribution.

A precondition for all acoustical measurements is that at least all those components of the system that are involved in the measurements have been checked as to their electrical functions and found to be ready for operation.

7.1 Electrical check

7.1.1 Subjective check

With smaller or mobile systems the electrical check that is required before it is put into operation may possibly be limited to a subjective check, or the electrical check may be supplemented by a subjective check.

After the system is switched on, the following electrical defects may become perceptible:

- Hum. Possible causes:
    - earth loops – that is, the system has several earthing points, between which a potential gradient exists, so that a compensating earth current flows causing the hum voltage through an intermediate resistance;
    - an open high-resistance input with subsequent amplification;
    - polarity transposition of the mains supply between different parts of the system.

- A tendency to oscillation. This can be caused by:

  - insufficient separation between the loudspeaker and microphone lines, and powerful amplification in between;
  - an open channel input with a subsequent high level of amplification.

- Intermittent drop-outs during high levels. This can be caused by:

  - overmodulation of one or several power amplifiers;
  - heavy overmodulation of a limiter;
  - excessively high-resistance termination of a power amplifier.

- Bass-emphasized, dull or intermittent reproduction ('bubbling'). This can be caused by:

  - a low-resistance mismatch of the power amplifier output (for example, a short-circuit in the loudspeaker line, or an overload due to incorrect matching transformers).

- Treble-emphasized, 'sibilant' reproduction. This can be caused by:

  - interruption of the programme line on the input side of the power amplifiers (capacitive coupling of the signal);
  - unipolar ('one-legged') coupling of the power amplifiers.

- Reduced low-frequency reproduction between two loudspeakers with reproduction of a coherent signal. This can be caused by:

  - polarity mismatch of a loudspeaker connection.

- Radio-frequency interference due to a thyristor lighting control (the maximum effect occurs at half-brightness lighting). This can be caused by:

  - insufficient separation between the power lines of the lighting system and the programme lines applied to the input of the sound reinforcement system (a minimum separation of 500 mm has to be observed);
  - missing interference suppression of the lighting control system (dimmers); this may be cleared by inductive blocking.

7.1.2 Electrical calibration

The electrical installation, commissioning and calibration of a large, permanent sound reinforcement system is essentially identical to the calibration of a studio system. The quality requirements are similarly high in both cases, and the reliability requirements to be met by sound reinforcement systems may even be higher, since they are used primarily for live performances. Quality requirements for studio sound systems have been established by various broadcasting companies [7.1, 7.2]; further information may be gathered from the relevant literature [7.3, 7.4].

Some measurements that are of particular importance for sound reinforcement systems are as follows:

- Checking of all incoming and outgoing lines with regard to:
  - continuity and line attenuation;
  - polarity (A–B transposition);
  - balance of earth-free lines (balanced lines are always preferable to unbalanced lines, especially for long transmission distances, since they make it possible for the inductive and capacitive interferences to be cancelled or rejected);
  - crosstalk;
  - S/N ratio.

- Checking of the signal levels at the inputs and outputs of components of the whole system according to a level or gain structure diagram laid down in advance. This level diagram considers the expected upper limit of the noise voltage and the maximum modulation (signal level) range

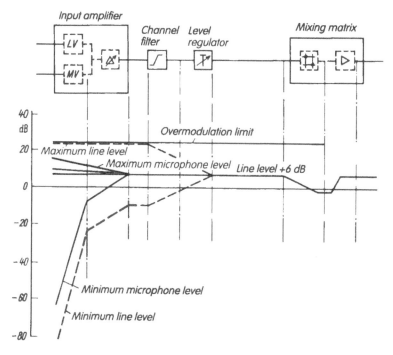

Figure 7.1 Level diagram (upper and lower limits of equalizing capacity) with assignment to the corresponding circuit elements of the sound channel.

of the system (Figure 7.1). These values must be strictly observed if the maximum dynamic range of the system is to be utilized.

- Checking of all operational functions in the mixing consoles and in the distribution and processing devices (this applies especially to all change-over switches, variable attenuators and filters).
- Checking of the system's stability and freedom from distortion by means of short-term operation with +6 dB modulation.
- Especially thorough checking of the line network of the system. Apart from serious faults that impair acoustical efficiency defects may also occur that entail drainage in the power amplifier section of the system. Because of the large power amplifiers that are frequently used these days, this may also be dangerous for people and buildings. The necessary checking includes:

  - measurement of the line impedances (with the loudspeakers both connected and disconnected): the impedance of the supply lines of low-resistance and high-resistance loudspeakers should not deviate by more than ±10% and ±5% respectively from the transformed loudspeaker impedance (see section 4.1.1.1);
  - polarity reversal, especially within the individual loudspeaker arrays (often wrongly called phase reversal).

- Checking of the control systems in the course of electronic calibration. These systems are frequently of great importance for effects equipment, and their influence will increase considerably along with the advancing introduction of automated systems.

7.2 Acoustical calibration

This involves, first of all, checking the criteria that were laid down during the planning and design of the sound reinforcement system, such as:

- reverberation time;
- intelligibility at different locations that are representative for the room;
- sound level distribution without the electroacoustical system;
- noise level in the hall;
- feedback immunity of the system.

Reverberation time measurements make it possible to determine the equivalent sound absorption area required for calculating the diffuse sound level, as well as the critical distance (see section 5.1.2). So reverberation time measurements have to be carried out for all room conditions with which the sound reinforcement system is intended to function. It is advisable also to ascertain the frequency response of the reverberation time, for the empty as well as the full room if possible, at various measuring points. In theatres, room variations of interest also include differing conditions of the stage house, such as open and closed stage, as well as extremely variable stage settings. They may also include differing hall and stage configurations, as may be found nowadays particularly in multi-purpose halls.

Intelligibility and sound level distribution without the electroacoustical system are frequently ascertained by means of the RASTI or impulse test (see sections 3.1.2.3 and 7.4). Room-acoustical data may often be taken from the measurement data of the acoustics consultant.

Acoustical check measurements include ascertaining the noise level and its frequency response in the listening area, and determining the feedback threshold. Again, differing room variations may need to be considered.

Several further measurements, which exceed the scope of the basic measurements, concern the functioning of the sound reinforcement system itself. These include:

- measuring the sound level distribution in the auditorium and in the performance area;
- measuring the acoustical frequency response in the listening and performance areas;
- checking the 'coherence' – that is, the staggered arrivals of the incident first wavefronts from the individual loudspeakers, as compared with the source of the original sound, at representative locations in the reception area;
- measuring the noise produced by the system itself.

The procedures used for measuring these quantities are dealt with below.

7.2.1 Measurement of sound level distribution

In larger rooms, or in auditoriums with complicated boundaries and in larger open-air spaces, the sound level distribution needs to be measured. Sound levels are measured at a number of measuring points arranged in a grid over the performance and listening areas. The size of the grid depends on how difficult it is to cover the areas to be checked. Critical locations, such as marginal seats and seats under a balcony, must also be considered (Figure 7.2 and 7.3).

The measurement is generally intended to determine the sound pressure level that results from the direct sound level, L_d, and the diffuse sound level, L_r:

$$L = L_W + 10 \log \left(\frac{\gamma_L \Gamma_L^2(\vartheta)}{4 \pi r^2} + \frac{4}{A} \right) \text{ dB}$$

where $L_W = 10 \log(P_{ak}/P_0)$ dB; P_{ak} is the acoustical power emitted by the loudspeakers; $P_0 = 10^{-12}$ W is the reference sound power; γ_L is the

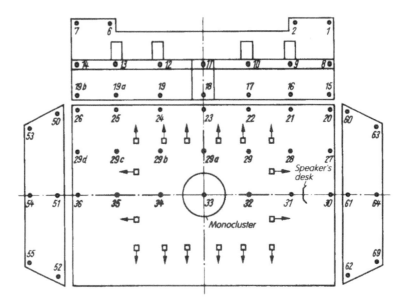

Figure 7.2 Grid of measurement positions in the Stadthalle, Rostock.

front-to-random factor of the radiator; and Γ_L is the angular directivity ratio of the radiator, referred to the measuring point.

The measurement must therefore be made so that both components – the direct and diffuse sound levels – are recorded by the measuring device, just as the human ear does. This is achieved with sufficient accuracy by devices whose sound level meter is capable of providing an indication with a dynamic range according to the IEC [7.5].

The signal used for measurement should be broadband pink noise. Such a signal is largely insensitive to the acoustical interferences that usually occur in closed rooms or in front of large reflecting surfaces. Also, it can cover the whole transmission range of the loudspeaker(s) in a representative fashion, since it displays an energy distribution that is constant with equal relative bandwidth, so as to correspond largely to the human sense of hearing. If measurements in the low-frequency range are likely to be affected significantly by disturbing noises (such as ventilation or traffic noises), the received measurement should be filtered with the 'A' weighting frequency curve, to suppress the interference in a way similar to that employed by the human ear (see section 3.2.1.2).

The levels should be measured both individually, for the main sound reinforcement components, and for all loudspeaker arrays cooperating in the system. When measuring the components, only the areas located within their coverage need be included, or in which it is possible for these components to

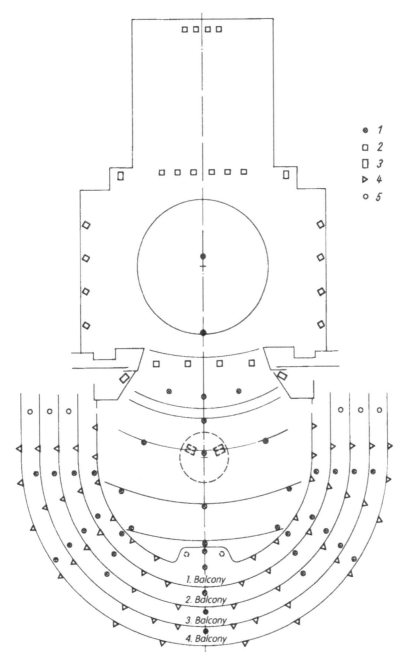

Figure 7.3 Grid of measurement positions in the Semperoper, Dresden: 1, measuring point; 2, TZ 133 sound column; 3 TZ 133 sound columns arranged one on top of the other: 4, 5, K 20 compact loudspeakers in different arrangements.

give rise to interference. With multi-channel sound reinforcement systems, neighbouring areas should also be considered, to register mutual influences (such as possible mislocalizations and interferences).

The difference between the highest and the lowest sound level measured at the listeners' locations should preferably be less than 6 dB, but certainly less than 10 dB.

7.2.2 Measurement of the reproduction frequency response

7.2.2.1 Measurement procedure

Because of the different frequency-dependence

- of the sound pressure level (direct sound level) and sound power level (diffuse sound level with normalization of room properties),
- of the equivalent sound absorption area of the room to be covered,
- of the radiating conditions varying according to the arrangement (installation conditions) of the loudspeakers,
- of the receiving conditions at the listener's locations as a consequence of interferences of the incident direct sound with reflected sound or sound emanating from other loudspeakers,
- of the directional characteristic of the loudspeakers,

the reproduction level in the reception area can sometimes be strongly affected. These linear distortions in the frequency response of the loudspeakers arranged in the reproduction room may generally be equalized by means of so-called *room adaptation filters* (equalizers). It is therefore necessary to measure the reproduction frequency response of the individual loudspeaker arrays within their main coverage areas. Random interferences at single locations can lead to misinterpretation: for example in the rear marginal zones, because of interference with the loudspeaker sound reflected by the rear wall. Several measurements should therefore be made in every coverage area in a grid pattern; the average values of these measurements are then taken as a basis for the frequency response correction. The grid of measuring points should be arranged similarly to that used for determining the sound pressure distribution, but may be somewhat wider meshed.

In smaller rooms with a high proportion of diffuse sound, measurements may be made at a few representative positions; but here also it is important to make sure that the values obtained are not just chance results. The measurement must cover both the direct sound pressure components from the loudspeaker and the diffuse sound components.

The recommended measuring signal is, again, broadband pink noise. This allows a direct indication of the reproduction curve with most of the filters used for ascertaining the reproduction frequency: these are generally third-octave filters, but for rough estimates one-octave filters are also used. As well as third-octave-filter real-time analysers, portable octave-filter real-time analysers and sound pressure level meters with attached third-octave filters are also used.

A sweep sine signal may also be used. This signal is evaluated by means of a time windowing technique and time delay spectrometry so that, given an adequate width of the window, frequency-dependent measurement of direct sound, surround sound and overall sound is possible (see section 7.3.4).

MLS sequences are also being increasingly used as measurement signals (cf. 7.3.5). So that linear distortions originating in the transmission channel can also be measured, the test signal is sometimes inserted into the input of the mixing console.

7.2.2.2 Equalization of the acoustical frequency response

If the frequency response thus ascertained fails to show the desired behaviour, it is corrected by means of room equalizers, by compensating the emphasis and de-emphasis of the spectrum through adjustment of an inverse filter curve (Figure 7.4). The so-called *graphic equalizers* are best suited to this task (see section 4.6.5.3). The reproduction curve obtained should be smooth, but need not always be linear or equal at all locations. If a linear frequency response is required for the intended use of the system [7.6], this should apply only to the area close to the stage. In large halls a treble drop at larger distances from the original sources corresponds to the natural listening experience, so that for a linear (flat) frequency response the reproduction would be perceived as too 'sharp' or 'sibilant'.

Note, however, that certain influences in the frequency response caused by acoustical interference can not be compensated for by attempting to boost the fall-off region.

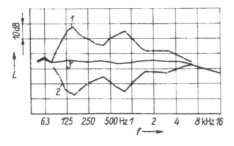

Figure 7.4 Adjustment of an equalizing curve by means of third-octave filter analysis: 1, average-value curve of the initial analysis; 2, reproduction curve of the third-octave equalizer; 3, resulting reproduction curve.

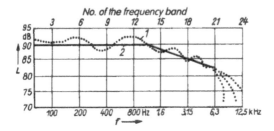

Figure 7.5 Target equalization curve at 33 m from the performance area [4.18]: ··········, uncorrected 'house curve'; ————, corrected 'house curve' at 33 m distance. Variations of the frequency response above 6.3 kHz are possible.

Schulein [7.7] gives a subjectively produced optimum adjustment of a reproduction curve in an auditorium of 800 m³. Up to about 2 kHz this curve is fairly frequency-independent but in the frequency range up to 10 kHz the sound level falls off by about 6 dB/octave, and above this frequency up to 20 kHz by 10 dB/octave.

According to Davis and Davis [4.18], at roughly 100 ft (33 m) the frequency response should be linear up to 1 kHz, and above 1 kHz should drop so that at 10 kHz a 10 dB lower value is achieved. In this case it is admissible for a greater fall-off to occur above 6 kHz, without this being registered as a shortcoming (Figure 7.5). This frequency response is similar to that given by Schulein.

In the region beyond the critical distance in a room, a treble drop of this kind occurs with nearly all sound sources because of the frequency behaviour of the sound power level prevailing in this area. The sound power drop that occurs with increasing frequency is due to the fact that directivity increases as a function of decreasing wavelength of the radiated sound (see section 4.1.1.2). This is characteristic of almost any sound source.

A fall-off in level that increases with frequency (that is, a high-frequency fall-off) therefore corresponds to natural conditions. This characteristic must, however, be considered in conjunction with the precedence effect. If a loudspeaker located far from the original source exhibits the same frequency response as a loudspeaker located near to the source, the level difference between the near and the far loudspeakers will be greater in the higher-frequency range than in the lower. This is why decentralized loudspeaker systems operating with delay equipment often tend, in the upper frequency range, to show signs of perceptible echo formation, or 'chattering'.

Tolerance ranges for reproduction curves for different applications are given by Mapp [7.6] (Figure 7.6). Different limit frequencies have been established for the roll-off in the upper frequency range, and loud rock music, generally familiar from conventional recordings (CD or cassette), requires the most balanced level behaviour.

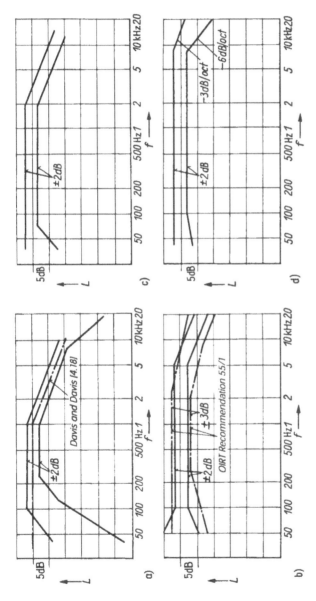

Figure 7.6 Equalization curves for different applications [7.6]: (a) recommended reproduction curve for speech reinforcement systems according to Davis and Davis [4.18] at 33 m distance from the action area; (b) recommended reproduction curve for studio monitoring equipment; (c) international reproduction curve for film projection equipment; (d) recommended reproduction curve for high-level rock music.

The described verification and correction of the reproduction curve, or of the acoustical frequency response, are nowadays part of the usual routine that is followed when commissioning sound reinforcement systems. For systems operating with numerous stage, effects and main loudspeakers this may require such a sound check to be carried out prior to every performance at a few typical locations, in order to check the effectiveness of the loudspeaker and filter adjustments. To this end the switchable multi-channel display units of the central control console are sometimes used together with filter banks as real-time analysers.

Still more extensive are those methods for checking the frequency response in which an automatic frequency response correction is effected by means of controllable filters.

7.2.2.3 Equalization for suppression of positive feedback

Given that positive feedback sets in at the frequency at which the transmission curve (at a live microphone) shows the strongest maximum, the positive feedback limit may be shifted to higher values by levelling this peak by means of a filter. This was first indicated by Boner and Boner [7.8]. With the help of appropriate equalizers it is thus possible to eliminate consecutively several feedback modes so as effectively to increase the positive feedback limit.

An equalization curve for positive feedback suppression is determined, for instance according to Davis and Davis [4.18], by means of a *regenerative equalization test*. In Figure 7.7 the real-time analyser (4) serving as display unit is fed via the microphone (1). This analyser indicates immediately any amplitude peak that appears in the reproduction curve as a consequence of incipient positive feedback. The required spectrum is produced by the generator (5), sufficiently broad and linear for the sound reinforcement system, and fed into the feedback circuit via the mixing

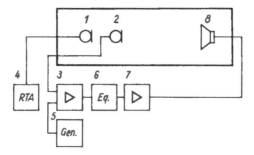

Figure 7.7 Measurement arrangement for regenerative equalization test [7.9]:
1, measuring microphone; 2, system microphone; 3, summation amplifier; 4, real-time analyser; 5, generator; 6, equalizer; 7, power amplifier; 8, loudspeaker.

amplifier (3). By means of the equalizing filter (6) it is possible to attenuate sufficiently the frequency-dependent amplitude peaks indicated by the analyser. The analyser and the generator must harmonize, so that a third-octave analyzer for instance has to be coupled with a pink-noise generator, whereas various other modern analysers have to be combined with a linear frequency sweep sine generator.

7.2.3 Checking the coherence of the arriving wavefronts

With spatially distributed loudspeakers, whether or not delay equipment is used, there is a risk that the wavefronts of a radiated signal may arrive at different times at the listener's location. If the time difference between two wavefronts exceeds 30 ms the definition of the transmission may be reduced, and with time differences of more than 50 ms echo disturbances may occur.

The time differences to be expected have usually been calculated when the system is in the planning stage (see section 6.5). However, in the completed system these time differences can deviate from those calculated, and thus become disturbing, if

- the installation positions of the loudspeakers have had to be changed (perhaps for constructional reasons);
- strong reflections are registered;
- the sound pressure level conditions between the individual loudspeakers or loudspeaker arrays have changed;
- the times of delay equipment were incorrectly adjusted.

For checking the time differences it is for instance possible to analyse energy–time curves. (Other modern measurement procedures are described in section 7.3.) To achieve this, one loudspeaker (for source simulation) and the nearest loudspeaker array are connected consecutively. In this way it is possible to determine separately the influence of the individual arrays staggered in depth. Figure 7.8 shows computer-simulated energy–time curves. They show the sequence of the individual impulses, which here are very distinct with frontal incidence. If the measurement is carried out across the main radiating direction of a widely spaced loudspeaker array, essentially broader, more diffuse energy–time curves are obtained. Ideally a Dirac impulse is used for excitation, or – for practical reasons – a Gaussian impulse. These impulses may for instance be available on a magnetic tape to trigger the measuring equipment (such as a storage oscilloscope), when they are replayed (see also section 7.3).

Evaluated accordingly, these *reflectograms* may also be used to determine the expected measures of definition or clarity (see Chapter 3).

The measurement procedure is sufficiently exact to permit the determination or exclusion of travel-time interferences (echoes, comb-filter effects

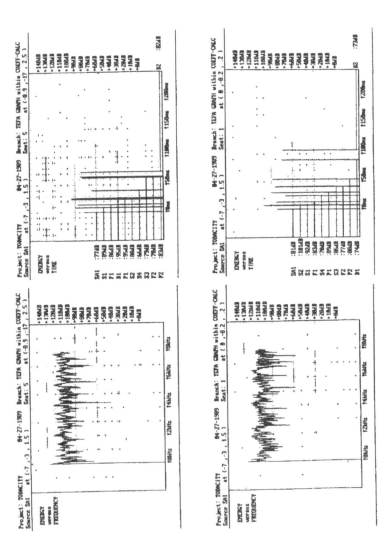

Figure 7.8 Impulse charts for checking the coherence of the arriving wavefronts.

in general, insufficient definition), but not the detection of the sources responsible for the mislocalizations. In addition to ascertaining the energy–time behaviour, in this case a frequency analysis has to be done. The intervals used should if possible correspond to one frequency group. Since such frequency group filters are not included in the usual range of measuring equipment, an octave analysis of the energy–time behaviour should be performed. There is no risk of mislocalization if, within 30 ms after the triggering impulse, a repetition impulse or intermediate reflection occurs with a level that does not exceed that of the triggering impulse by more than 10 dB, nor by more than 6 dB within the range 30–60 ms.

7.2.4 Subjective assessment

Apart from the objective measurements (see section 7.3) used when a sound reinforcement system is commissioned, an overall assessment of the system may also require recourse to subjective procedures: for example if no suitable objective procedures are available, or cannot yet be considered sufficiently reliable; or if very subjective, complex qualities such as operating features of the system or the employment of certain effects devices have also to be included.

Well known in this regard is the *determination of syllable intelligibility* in a room by means of test words (see section 3.1.2.3).

Far more complex is the *subjective overall assessment* of a sound system by a listening team. Assessments of this kind are especially important if the system is also used for influencing room-acoustical parameters. For obvious reasons the assessment in this case is similar to the subjective verification of the acoustical parameters of reproduction rooms, as described in [7.10] and [7.11] (see also [7.12]). Figure 7.9 shows a questionnaire for such subjective tests, based on the recommendations of the OIRT (International Organization of Radio and Television) [7.13]; the terms used are defined in [7.10].

The listening team generally consists of 5–20 persons who test various seat groups in the auditorium, if possible simultaneously. Various different programme examples should be used, and the assessors should change places after every test section.

7.3 Measurement procedures and systems

A host of measuring devices and procedures are available for the acoustical calibration of sound reinforcement systems. Some of these are quite conventional, while others are new types that have come into use only in recent years. Because of their diverse possibilities, and because their results deviate from one another and also from those determined by classical procedures, we treat them below in some detail.

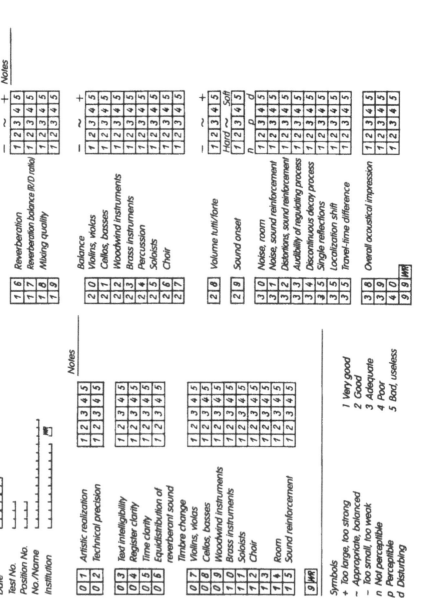

Figure 7.9 Example of a test sheet for subjective evaluation of reproduced sound events.

7.3.1 *Sound pressure measurements*

The following measurement procedures serve for determining the scalar quantity, sound pressure. The transducer used is a pressure microphone with a mainly omnidirectional characteristic (see section 4.2.2). This directional independence, and thus the proportionality between sound pressure and voltage produced, are no longer valid when the wavelength of the received signal approaches the region of the diaphragm size. With a one-inch microphone the deviation begins at about 2 kHz, and at 6 kHz there may be a difference of 3 dB between the rating of a direct sound wave incident from the front and that of a diffuse sound field with equal effective sound pressure. For this reason half-inch or quarter-inch microphones are normally used for measurement purposes, but with these microphones there is a deterioration of pressure proportionality in the frequency range above 10 kHz. Measurement microphones are usually free-field equalized: that is, at high frequencies beyond the omnidirectional characteristic the diffuse and laterally incident sound energy is underestimated.

Analysers suitable for producing comparable measurements have an effective-value indication, in the same way as sound level meters standardized according to the IEC [7.5]. For time weighting the inertia grades of the sound level meter S (slow), F (fast) and I (impulse) are used. These have also been specified by the IEC and allow a comparable indication that coincides widely with the subjective impression. Simple devices for rough indication may have other time constants. They should, however, not be called measurement instruments.

The pressure-proportional voltage is usually analysed by means of third-octave filters, or less frequently by means of octave or half-octave filters. The mid-band and critical frequencies of these filters are specified according to ISO in overlapping series (Table 7.1). Real-time analysers are being used to an increasing extent for measurements in sound reinforcement systems. These devices contain filter banks that are constantly in steady state and comprise the whole audio frequency range (often from 20 Hz up to 20 kHz). All filters have a level indication, either on picture tubes or by means of LEDs or LCD screens. The LED indication offers only a quantized resolution, and is generally switchable in steps of 1 dB, 3 dB and 6 dB. The displayable dynamic range varies according to the step size.

Pink noise is used as a measurement signal for systems with high frequency irregularity in rooms, or in open spaces with reflecting boundaries. The accuracy of reading of the measuring device depends on the analysis bandwidth: the higher the frequency resolution, f_R, the lower the time resolution, T_R.

$$T_R f_R = 1 \tag{7.1}$$

Table 7.1 Preferred series of third-octave and octave mid-band frequencies according to ISO

Mid-band frequency (Hz)	Third octave	Octave	Mid-band frequency (kHz)	Third octave	Octave
31.5	x	x	1 000	x	x
40	x		1 250	x	
50	x		1 600	x	
63	x	x	2 000	x	x
80	x		2 500	x	
100	x		3 150	x	
125	x	x	4 000	x	x
160	x		5 000	x	
200	x		6 300	x	
250	x	x	8 000	x	x
315	x		10 000	x	
400	x		12 500	x	
500	x	x	16 000	x	x
630	x		20 000	x	
800	x				

Modern real-time analysers can store several spectra, and recall and compare them. They are often also equipped with a connection for a computer, and a printer interface for the print-out of measurement data.

7.3.2 Fast Fourier analysis

The Fourier transform makes use of the possibility that any (complex) waveform in the time domain can be transformed into a number of individual sine-wave oscillations in the frequency domain [7.14]. The Fourier transform of a short impulse (Dirac pulse) theoretically exhibits a frequency-independent spectrum (i.e. wide-bandwidth flat spectrum), and so when applied to a system to be tested, it enables the frequency response (of the system) to be directly determined. In practice, however, such impulses prove to be too low in energy to produce, in the lower frequency range, an amplitude sufficient for measurement. (Increasing the amplitude in this range in loudspeaker transmission, for example, may exceed the linear range of the loudspeaker and lead to distortion.) This is why random noise is often used as the signal for the measurement of sound reinforcement systems, since it can be considered as a sequence of such impulses.

It is therefore possible to subdivide the overall spectrum to be measured into discrete sections, which are then separately calculated and subsequently reassembled in the transformed plane. This procedure is called *discrete Fourier transform (DFT)*. The inverse function is accordingly designated *IDFT*. The discrete Fourier transform makes it possible to use digital computer techniques for realizing the transform. By employing certain algorithms a very fast execution of the transform (*fast Fourier transform, FFT*) becomes possible, so that it can be used in general measurement practice.

The accuracy of the discrete transform depends on the number of samples per time unit. This is a function of the processing capability of the computer used. A 10-bit computer with a screen resolution of 400 lines, for instance, is capable of realizing a time resolution (or time window) of

$$T_R = \frac{2^{10}}{2.56 \times 20\ 000} = \frac{400}{20\ 000} = 0.02 \text{ s}$$

If a smaller (i.e. shorter) time window is required, it is necessary to [7.9]:

- accept a poorer frequency resolution; or
- increase the bandwidth that is looked at; or
- find an analyser that is capable of more than 400 lines of resolution (which is quite possible nowadays).

Given that, according to equation (7.1), the product of frequency resolution, f_R, and time resolution, T_R, must always be 1, we obtain in this case a frequency resolution of

$$f_R = \frac{1}{0.02 \text{ s}} = 50 \text{ Hz}$$

This means that the overall frequency range is sampled with

$$N = \frac{2f_{max}}{f_R} = \frac{2 \times 20\ 000}{50} = 800 \text{ samples}$$

This shows that an exact analysis of the lower frequency range, which is important for sound reinforcement systems, cannot be realized in *one* analysis operation. A subdivision of the overall frequency range into the partial ranges of 20 Hz to 200 Hz, 200 Hz to 2 kHz and 2 kHz to 20 kHz, however, produces a sufficient precision for many practical cases.

According to the sampling theorem by Shannon [7.15], a full description of a function is possible only if two samples per Hz of the spectrum to be tested are carried out: that is, the highest registered frequency, f_{hrf}, is described by the condition

$$f_{hrf} \leqslant \frac{1}{2\Delta t} \tag{7.2}$$

where Δt is the sampling interval.

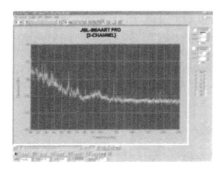
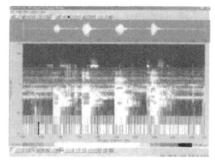
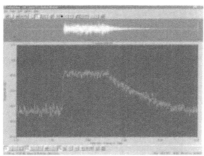

Figure 7.10 Typical results produced by the JBL Smaart Pro.

Real-time analysis, as required for various postprocessing tasks (as when correlating a two-channel analysis, for example), thus demands a considerably higher sampling rate than is possible in the given example, or else requires restriction to the lower part of the spectrum.

Some fundamental drawbacks of FFT analysers in the past (see also [7.9]) have been as follows:

- The transform is applicable only to the analysis of linear systems.
- A desirable small time window produces only an undesirably small dynamic range.
- The postprocessing of the data, though easily feasible, requires a very expensive analyser.

Since 1997 a revolutionary FFT analyser for acoustic measuring purposes has been in use. The Smaart Pro by Sam Berkow [7.16] is a software solution capable of fully utilizing the speed of modern computers, and requiring only a good sound card to produce real-time Fourier transforms. Thus two-channel FFT spectral transforms, spectral signal evaluation and reverberation time measurements are possible with any modern notebook computer (Figure 7.10).

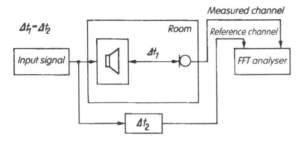

Figure 7.11 Measurement arrangement for sound analysis by means of the SIM technique [7.17].

7.3.3 Source-independent measurement procedure

Although requiring a high technical expenditure, a *two-channel real-time FFT analyser* with corresponding postprocessing offers expanded measurement possibilities. A natural signal, such as speech or music, is applied to the input of the system as well as to one channel of the FTT analyser (Figure 7.11). The other channel receives the signal influenced by the room.

The travel time of the signal between the exciting loudspeaker and the measuring microphone is compensated for exactly by means of a delay line inserted between the source and the second input of the analyser, so that extensive coherence between the channels is established. The correlation of the two spectra then produces the transmission function of the system under test. Since one cannot count on having a full test spectrum available at any moment of measuring, a storage circuit is required that continues to supply the missing spectral components until the whole transmission function becomes available in a readable fashion, such as on a display screen.

Because the measurement signal is independent of the source, this measurement technique is known as *SIM* (source-independent measurement) [7.17]. It enables the reproduction spectrum to be constantly checked, even when there are spectators in the auditorium, without disturbance to them. This procedure for ad hoc correction in occupied rooms will in the future acquire special importance for those places in which the acoustical conditions with and without spectators tend to differ greatly.

The so-called CORREQT system, which also makes use of a two-channel FFT analyser, functions in a similar way [7.18].

7.3.4 Time delay spectrometry (TDS analysis)

Another method of signal analysis is used in *time delay spectrometry, TDS*, as suggested by Heyser [7.19]. Suitable equipment was in the past supplied by Brüel & Kjaer, and is now mainly offered by Goldline.

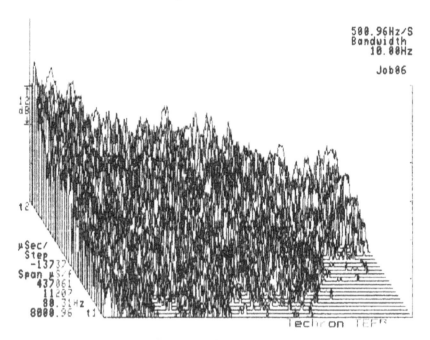

Figure 7.12 Time–energy–frequency plot.

For conventional measurements of system response by FFT or real-time analysers, impulse sound and noise sources are mostly used. The existing ambient noise, however, is also registered, giving rise to potential measuring errors on account of insufficient signal-to-noise ratio.

The test signal used with the TDS procedure, however, is a continuously varying sine wave (*sweep*), which is supplied on the reception side via an exactly synchronous filter to an evaluating and indicating device. Thanks to this tracking filter it is possible to suppress noise interference and room reflections.

If the signal sweep is repeated several times, with the time shift between the sweep and the start of the receiving tracking filter altered each time, it is possible to measure various time-displaced frequency responses. For a clearer description, these curves are shown in a 3D perspective representation, the axes of which are time, frequency, and amplitude (or energy) (Figure 7.12). (In accordance with the quantities time, energy, frequency, Crown International designated its TDS analyser as a TEF analyser [7.20].)

By means of appropriate postprocessing it is possible to display certain planes (projections) of such a 3D 'mountain'. These are the energy–time plane (impulse pattern), the energy–frequency plane (frequency response), and a section of a determined level of the frequency–time plane. Based on the phase relationship between the radiated sine signal and the response signal received with temporal displacement, it is possible to show a representation of the frequency-dependent real or imaginary components of the transmitted signal, mostly in the shape of a locus diagram (Nyquist diagram).

The time resolution, T_R, and the frequency resolution, f_R, depend on the sweep rate, S, in Hz/s and on the bandwidth, B, of the tracking filter in Hz:

$$T_R = \frac{B}{S} \quad \text{and} \quad f_R = \frac{S}{B} \tag{7.3}$$

If the frequency window accompanying the sine sweep has, for example, a bandwidth of $B = 50$ Hz (Hamming filter or the like) at a sweep rate of $S = 1000$ Hz/s, the frequency resolution is 200 Hz, which corresponds to a virtual time window width of 5 ms (± 2.5 ms). With such an adjustment it is possible to capture frequencies down to 100 Hz. If, however, the sweep rate is increased to, say, 5000 Hz/s, which corresponds to a reduction of the measuring time to 1/5, the frequency resolution is reduced to 1000 Hz. Thus it is possible to display time windows of 1 ms, while correct measurement is given only for frequencies above 500 Hz. However, this is insignificant for the testing of loudspeaker combinations in the high- and medium-frequency ranges.

This also means that with a frequency resolution (spacing of the samples on the amplitude–frequency curve) of 20 Hz, as required for the analysis of sound reinforcement systems in the lower frequency range, a sweep rate of $S \leqslant f_R^2 \leqslant 400$ Hz is necessary. This implies a bandwidth of the tracking filter of $B = S/f_R = 20$ Hz. The resulting time resolution is

$$T_R = \frac{20 \text{ Hz}}{400 \text{ Hz/s}} = 0.05 \text{ s}$$

This means that the width of the virtual measurement window amounts to 50 ms and that all time functions within this measurement window are taken into account. The 50 ms width corresponds to a path length of approximately 17 m, which means that the measuring object has to be suspended at a distance of about 9 m from reflecting walls so that possible reflections cannot interfere with the low-frequency measurement results. (With a lower limit frequency of 500 Hz – that is, with $f_R = 1000$ Hz – $T_R = 1$ ms, which corresponds to a path length of 34 cm; here room reflections no longer disturb, since they can easily be gated out.)

The signal-to-noise ratio is considerable. With a bandwidth of the measurement range of, say, $B_w = 15$ kHz ($f_l = 40$ Hz, $f_u = 15$ kHz), the resulting time resolution is $T_R = 1/B_w = 66.6$ μs.

On the screen there are displayed $400 \times T_R = 26.6$ ms. The signal-to-noise ratio is accordingly $S/N = 20 \log$ (measurement interval/time resolution) dB. With $S = 5000$ Hz/s the resulting $S/N \approx 93$ dB. This is a ratio that cannot be achieved with normal impulse or noise measurements.

This is why for example the usual change-of-set and repair works in both the stage and auditorium areas can be carried on without disturbing the measurements.

The signal recorded in the frequency window (bandwidth, B), and thus in the virtual time window, is subjected to an FFT analysis in which, thanks to application of the Hilbert transform and contrary to the impulse sound test, phase information is also available. The energy–time curve is thus the complete complex time function, and not just the real component of this function.

Moreover, the TDS analysis allows:

- phase measurements as normal and Nyquist plots;
- group travel-time measurements,
- 3D measurements (energy, time, frequency);
- plotting of the real (impulse) and the imaginary components of time function measurements;
- measurement of the direct-sound frequency response, and also of built-in electroacoustical transducers;
- Schroeder integration reverberation time measurements as well as reverberation time measurements ascertained from energy–frequency plots;
- measurement of directional characteristics.

7.3.5 Maximum length sequence measurement (MLS)

A measurement procedure that is increasingly being used is similar to the well-known impulse sound test, which been used for a long time in room acoustics. With the latter test an impulse of known amplitude and phase is radiated via, for instance, the loudspeaker to be measured, and then the impulse response is measured to assess the transducer data. The properties of the measuring room, and environmental noises that cause disturbances, have to be eliminated. This can be done by gating out with the help of an appropriate window technique. Useful and disturbing signal components are separated by means of an iterative procedure, in which the measuring object is repeatedly excited by the same impulse. By using triggered impulse response averaging it is then possible to separate the always equal useful signal from the always differing disturbing signal; however, the required measuring time increases considerably.

This problem can be circumvented by exciting the measuring object by means of so-called *maximum length sequences* (MLS). Binary MLS signals are periodic, two-level, pseudo-random sequences of length $L = 2^N - 1$, in which the number N is an integer (Figure 7.13). An analogue MLS sequence is fed into the linear system to be measured; the resulting transmission response is digitized, and then this is cross-correlated with the original sequence in the computer. The result of the cross-correlation is essentially the Fourier transform of the transmission function of the system measured.

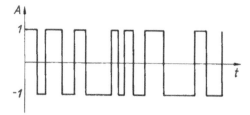

Figure 7.13 Binary MLS sequence.

From this system response it is possible to derive all other measurement quantities [7.21].

The advantage of the MLS procedure is that the signal-to-noise ratio of TDS measurements is maintained without the broadband frequency response being narrowed in the evaluation phase, independently of the chosen time window. This ultimately results in a substantial reduction of the measuring time. As compared with the older TEF procedure, the drawback of the MLSSA or SYSID procedures that have already been introduced is the shortage of suitable software. This drawback, however, is gradually being remedied [7.22]. The latest TEF20, and the two-channel TEF-PAD package, also contain an MLS option [7.20].

7.3.6 *Assessment of measurement procedures*

Measurements with real-time analysers and excitation by pink noise are currently the most viable procedures for ascertaining acoustical frequency response as required for characterizing timbre change and thus for adjusting equalizing filters.

More modern procedures using the fast Fourier transform may be employed as a complementary method: for checking the effect on reproduction of changes in the equalization setting, or of a pre-set configuration. There is a special role here for two-channel FFT analysis, which is used, among other things, with the SIM procedure. The Smaart Pro software solution is the latest step in the development of two-channel analysers. Thanks to its ease of use and excellent price/performance ratio it is going to become an outstanding acoustical measuring tool.

Nevertheless, MLS and TDS measurements using the MLSSA or TEF 20 analysers continue to be of great importance, because they are suitable for operation in environments with a high level of background noise. With a chosen measuring bandwidth suitably adapted by careful selection of time and frequency resolution it is easily possible to measure the different sound-field components, including the diffuse field.

7.4 Objective determination of intelligibility

To ascertain the intelligibility of speech, or the clarity of music, achieved with or without a sound reinforcement system, in addition to subjective tests (see section 7.2.4) requiring a relatively large number of listeners it is also possible to use objective measurement procedures.

These include procedures for determining the definition and clarity measures described in principle in section 3.1.2: for example, by means of the impulse sound test or the TEF as well as the MLSSA or the SYSID techniques.

In addition to these indirect procedures for evaluating the effect of the ratio between reverberant energy and direct energy, and thus permitting conclusions regarding intelligibility, there are also measurement techniques that make it possible to assess intelligibility impairment by other disturbances. These include the determination of the modulation transmission function [3.12], and the RASTI procedure developed from it.

7.4.1 Measurement of definition and clarity measures

With the *impulse sound test* (IST), the measures obtained from the energy–time function (the impulse response) are determined by introducing impulses via the system or parts of it (partial loudspeaker arrays) and evaluating these as described in section 3.1.2.

The impulses should have a width of 0.5–2 ms for adequate resolution, and should correspond approximately to a Gaussian curve so as to suppress sidebands and overshooting as far as possible. Thus a high level of energy can be made available for measurement, in which a Gaussian curve serving as an envelope can include a sinusoidal modulation, offering the possibility of studying the frequency dependence of the reflectograms.

The TEF analyser (see section 7.3.4) and the MLSSA board (see section 7.3.5) may also be used for determining the impulse response.

The impulse response has to be measured at a large number of test positions to take account of the varying conditions in the different parts of the listening area. The grid of the measurement positions has to be chosen according to the expected variability of the areas. There should be at least one measurement position on and below each balcony. There should also be measurement positions near the performance area and to the sides. To avoid obtaining a negative overall result by choosing exclusively critical measurement positions, the aim should be to establish a more or less uniform grid of positions in the hall.

The results are often evaluated nowadays by means of a computer connected on-line via a suitable interface.

7.4.2 Intelligibility measurement by means of the RASTI technique

Section 3.1.2.3 summarizes the fundamentals of the RASTI technique, as devised by Houtgast and Steeneken [3.12]. Brüel & Kjaer manufactures suitable measuring equipment, which can be interconnected with a computer to produce copious data for characterizing the speech transmission behaviour of a system [7.23] (Figure 7.14). The software package of the TEF analyser [7.20] as well as the MLSSA procedure permit a RASTI investigation. From the RASTI values thus obtained it is also possible to derive the corresponding articulation losses according to Peutz [7.22] (cf. Figure 3.13).

Thanks to a paper by Schroeder [7.24] it became clear that the complex *modulation transmission function* (MTF) can also be obtained from the impulse response of a passive linear system:

$$m(\omega) = \frac{\int_0^\infty h^2(t)\exp(-j\omega t)\, dt}{\int_0^\infty h^2(t)\, dt} \tag{7.4}$$

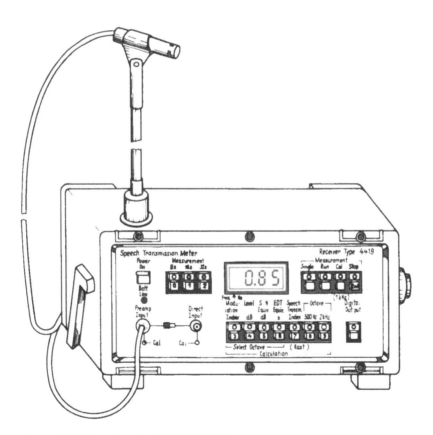

Figure 7.14 Drawing of a B & K RASTI meter.

The numerator in equation (7.4) represents the Fourier transform or the frequency spectrum of the square of the impulse response, while the denominator is the overall energy of the impulse response. The usual MTF is the amount of the above quotient, i.e. $|m(\omega)|$, and lies between 0 and 1.

The most recent implementations of the TEF analyser and the MLSSA procedure make use of this interrelation to ascertain the MTF from the energy–time curves. From this, RASTI and STI values are derived by means of weighting the frequency bands involved according to their contribution to intelligibility. See also [7.25–7.27].

Summing up, we can say that RASTI or STI values for objective evaluation of the speech intelligibility of a transmitted signal can easily be measured by means of the Brüel & Kjaer measurement equipment type 3361, the MLSSA procedure, or the TEF System 20 analyser. The MLSSA or the TEF devices offer the advantage that though the noise level is registered as a quantity influencing the results, it does not impair the measurement as such; in this respect the RASTI measurement equipment 3361 may be problematic. The latter, however, is much simpler, and can also be operated by laymen who just want to obtain an overall impression of the achieved sound quality.

References

7.1 *Meßanleitung Tonanlagen Nr.* 3 (Measuring instructions sound equipment No. 3) (Rundfunk- und Fernsehtechnisches Zentralamt der Deutschen Post, April 1973).

7.2 *Pflichtenheft 5/3 der ARD* (Specifications 5/3 of the German Broadcast Organisation ARD) (Munich: Institut für Rundfunktech, 1980).

7.3 Webers, J., *Tonstudiotechnik* (Sound studio equipment) (München: Franzis, 1985).

7.4 Dickreiter, M., *Handbuch der Tonstudiotechnik* (Handbook of sound studio equipment), 3rd edn (München: Saur, 1987).

7.5 IEC-Publ. 651 Ed. 1979 *Sound level meters.*

7.6 Mapp, P., *Audio Systems Design and Audio Systems Engineering* (Klark Teknik, 1986).

7.7 Schulein, R.B., 'In situ measurement and equalization of sound reproduction systems', *Journal of the Audio Engineering Society*, 23 (1975) 3, 178–186.

7.8 Boner, C.P. and Boner, C.R., 'Behaviour of sound system response immediately below feedback', *Journal of the Audio Engineering Society*, 14 (1966) 3, 200 ff.

7.9 Davis, D., 'Audio measurement'. In: Ballou, G., *Handbook for Sound Engineers: The New Audio Cyclopedia*, 2nd edn (Indianapolis: Sams, 1991), Ch. 36, pp. 1365–1411.

7.10 *Begriffe und Definitionen zur Beurteilung von Schallereignissen mittels der subjektiven Bewertung der Qualität von Hörereignissen* (Concepts and definitions used for evaluating sound events by subjective evaluation of hearing events), OIRT Monograph no. 31 (Prague: Organisation International de Radio et Television).

7.11 Fasold, W., Sonntag, E. and Winkler, H., *Bauphysikalische Entwurfslehre Bau- und Raumakustik* (Theory of structural–physical design, architectural and room acoustics) (Berlin: Verlag für Bauwesen, 1987).

7.12 OIRT Empf. 68/1 *Methodik zur subjektiven Bewertung der akustischen Eigenschaften von Rundfunk- und Fernsehstudios mit Publikum, Konzertsälen und Mehrzwecksälen für Musikdarbietungen* (Methodology for the subjective evaluation of the acoustical properties of radio and television studios with audience, and of concert halls and multi-purpose halls for music performances) (Miskolc-Tapolca, 1981).

7.13 IEC-Publ. 225 Ed. 1966 *Octave, half-octave and third-octave bandfilter intended for the analysis of sound and vibration*.

7.14 Küpfmüller, K., *Die Systemtheorie der elektrischen Nachrichtenübertragung* (The system theory of electric communication) (Stuttgart: Hirzel, 1952).

7.15 Shannon, C.E., *The Mathematical Theory of Communication* (Urbana: University of Illinois Press, 1949).

7.16 Leaflet on JBL Smaart Pro, JBL Professional, 1997 AES Convention, New York.

7.17 Meyer, J., 'Equalization using voice and music as the source', 76th AES Convention, New York (1984), preprint 2150 (I–8).

7.18 Deloria, K., Computerized EQ systems, REP (June 1990), pp. 50–54

7.19 Heyser, R.C., 'Acoustical measurements by time delay spectrometry', *Journal of the Audio Engineering Society*, 15 (1967) 4, 370–382.

7.20 TEF System 20 Analyser, by Crown International, PO Box 1000, Elkhart, IN 46575-1000, USA.

7.21 Rife, D.D. and Vanderkooy, J., 'Transfer-function measurement with maximum-length sequences', *Journal of the Audio Engineering Society*, 37 (1989) 6, 419–444.

7.22 Davis, D. and Davis, C., 'More on intelligibility', *Synergetic Audio Concepts Tech. Topics*, 14 (1987) 8.

7.23 *RASTI-Sprachübertragungsmesser Typ 3361* (The RASTI speech transmission meter), Data sheet, Brüel & Kjaer, DK-2850 Naerum, Denmark.

7.24 Schroeder, M.R., 'Modulation transfer functions: definition and measurement', *Acustica*, 49 (1981), 179–182.

7.25 Mapp, P. 'The effects of spectators, audiences and buildings on sound system performance', 98th AES Convention, Paris, 25–28 February 1995, preprint 3964 (F4).

7.26 Mapp, P. 'A comparison between STI and RASTI speech intelligibility measurement systems', 100th AES Convention, Copenhagen (11–14 May 1996), preprint 4279 (T-3).

7.27 Mapp, P. 'Practical limitations of objective speech intelligibility measurements of sound reinforcement systems', 102nd AES Convention, 22–25 March 1997, preprint 4410 (B2).

8 Case histories

Because of the rate of technical advances in this field, sound reinforcement equipment, like many electronic systems, rapidly gets out of date. It therefore has to be upgraded and renewed more frequently than the building or system in which it is installed. This chapter provides case histories of some typical sound reinforcement installations, demonstrating certain fundamental solutions. Some of the examples show why and how an installation was modified from its original layout.

8.1 Outdoor speech and music reinforcement

8.1.1 Football (soccer) and athletics stadiums

In this field of application, the sound reinforcement system is mainly required for transmitting commentary on the course of the game, plus information and announcements for the spectators. It is also used for speech and music transmissions both to the spectators and to the participants: during presentation ceremonies and addresses, and also for gymnastic shows. The system thus serves not only to safeguard the event itself, but also to ensure the safety of spectators and participants. This is why such systems are extremely important, especially for large stadiums, and their functionality is closely related to that of the stadium.

The noise produced by the spectators is crucial to the layout of the system. In those parts of football stadiums where fans are noisily supporting their respective teams, the background sound level that has to be overcome may well reach values of 108 dB, and for about 70% of the time is between 80 and 98 dB. For the event as a whole, an equivalent continuous sound level of 86 dB has been measured. In an emergency this sound level has to be overcome or at least equalled for announcements to be intelligible.

However, the system must not become a disturbing factor for the environment; in other words the sound level should be fully effective only within the stadium. Another special feature is the weather-resistant design. Since the stadium has generally to be operated in winter as well as in summer, the relevant requirements are often higher than for other outdoor systems, which may be put out of operation during the winter months.

8.1.1.1 Medium-sized athletics and football stadium of a larger city

As an example we are going to consider a stadium that is typical for a larger city.

DIMENSIONS AND OTHER FEATURES

The multi-purposed stadium has an almost uninterrupted 5 m high earth embankment carrying the stands and rising at an angle of 30° beyond the edge of a continuous synthetic-surface track and a 2 m wide rim. The top of the embankment is wide enough to serve as access and exit area and, at least near the inner edge, as a spectator standing area (Figure 8.1).

One of the long sides of the embankment is finished off by a roofed stand with a depth of 9 m at the sides and of 11 m in the central part where the VIP box is situated. The length of the roofed area is 85 m. The main field has a width of 70 m and a length (in the curved area) of 160 m, and is surrounded by a track of 30 m width.

SOUND REINFORCEMENT SYSTEM

Sound reinforcement was to be provided for:

- the sloping embankment (terrace) and a 10 m wide flat, top area;
- the roofed stand, including the strip in front of it;
- the main field.

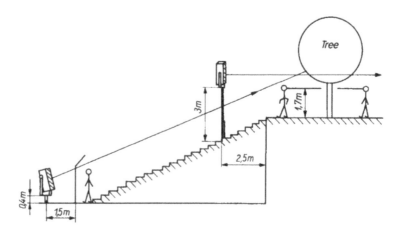

Figure 8.1 Loudspeaker arrangement on the normal stand profile of a medium-size stadium (sectional view).

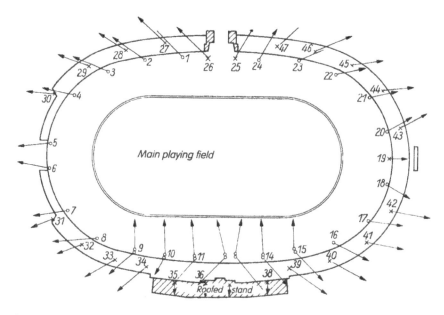

Figure 8.2 Loudspeaker distribution in a medium-size stadium: o, MTS 100 sound column; ×, TZ 228 sound column.

Throughout the whole audience area a level of at least 98 dB was to be achieved. To this end there are 24 special outdoor sound columns with a power-handling capacity of 100 W and a characteristic sound level of 106 dB (MTS 100), arranged at 2 m distance from the foot of the embankment in such a way that they radiate obliquely to the transverse axis of the stand from below towards the embankment, as can be seen in Figures 8.1 and 8.2. The oblique radiation permits uniform coverage, especially of the front stand areas. Thanks to the low level of installation, these sound columns neither impair vision nor disturb the environment noticeably. They are, however, unsuitable for covering the top area of the embankment. For this reason they were complemented by smaller sound columns (TZ 228 according to [8.1]) mounted on masts in the upper third of the embankment and radiating horizontally towards the top of the same, as shown in Figure 8.1. These sound columns have a power-handling capacity of only 50 W. Also, they impair vision only slightly, thanks to their reduced dimensions.

In the curved areas the loudspeaker arrangement differs from the normal profile. While on the right-hand side it was necessary to reduce overspill onto the announcement board so as to avoid echoes from its surface, the embankment in the left-hand curved area is lower, and contains the main access to the stands as well as to the playing area. Here it was necessary to aim the loudspeakers towards the access area.

For coverage of the roofed stand, loudspeakers were fastened in the usual way to the front edge of the roof. They are inclined at 10° downwards towards the spectators' area so as to avoid echoes from the rear wall of the stand.

The playing area is covered from the side of the main stand by means of sound columns attached to the rear of the sound columns arranged at the foot of the stand.

For coverage of the approaches to the stadium at the lateral exits, two 12.5 W compression driver horns were installed.

The overall amplifier power available for sound reinforcement in the stadium is 2 kW. The loudspeakers are subdivided into four groups: the first group forms the lower ring at the foot of the stands, with 1.125 kW; the second group is the upper ring, with 425 W; the third group covers the roofed stand, with 100 W; and the fourth group covers the playing area, with 350 W.

Control of the system is from a booth at the edge of the playing area. Commentary points are located in front of the main stand and within this booth. Delay lines may be used to ensure localization of an announcer's or commentator's position.

8.1.1.2 *Large athletics and football stadium in Stuttgart, Germany*

Built in 1933, the Neckarstadion was remodelled and enlarged in 1973–74. Up to its further extension in 1992–93 it offered capacity for 70 643 spectators: 18 028 roofed seats, 17 579 unroofed seats, and standing room for 35 018 persons.

As Stuttgart was chosen to be the venue of the 4th IAAF World Championships, it was decided in 1991 to completely roof the seating and standing areas of the stadium. In addition to a new lighting installation, it was also necessary to install a new sound reinforcement system, since the old one was out of date, and no longer corresponded to modern requirements. Currently the re-roofed *Gottlieb Daimler Stadium*, as it is now called, offers 35 000 seats, and standing room for an equal number of persons (Figure 8.3). There is an option for a further enlargement by 15 000 to 20 000 seats.

ROOM-ACOUSTICAL CONDITIONS

Normally it is not at all problematic to install sound reinforcement equipment in open-air facilities, since no reverberation component is involved therefore only the amount of the direct sound level, which should exceed the noise level, has to be considered. This was not the case in the Gottlieb Daimler Stadium: because of the size of the curved stadium roof, reverberation times of about 4.7 s and 2.5 s were expected for the empty stadium and the full stadium respectively. This was corroborated after

Figure 8.3 Gottlieb Daimler Stadium, Stuttgart: view into the stadium with loudspeakers visible.

completion by measurements. Syllable intelligibility of announcements was impaired accordingly. Thus the audience area could not be considered as open space.

SOUND REINFORCEMENT SYSTEM

The new system was intended to fulfil the following objectives:

- a high standard of music and voice reproduction:
    - frequency response of the loudspeaker systems 100 Hz to 12 000 Hz;
    - syllable intelligibility > 80%;
    - maximum sound level at the spectators' locations 105 dB;
- ample coincidence of visual and acoustical information (directional hearing), especially when the projection screen is used, but also in the stand areas;
- high degree of uniformity of sound distribution in the stadium (<±2 dB);
- high-quality electrical transmission of the sound signals (no crosstalk via lighting control cables running in parallel);

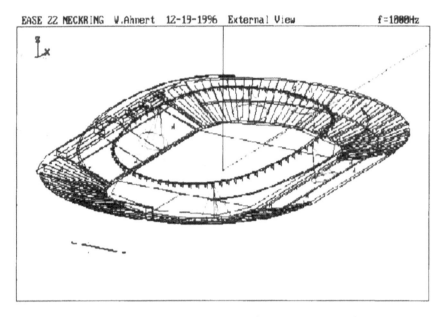

Figure 8.4 EASE presentation: coverage by Renkus-Heinz SR4 loudspeakers.

- remote monitoring and remote control of the distributed power amplifiers.

To comply with these requirements, 80 Renkus-Heinz SR4 loudspeaker systems were installed with two different inclinations, aimed so that the stand area is uniformly covered (Figure 8.4). The signals are passed through delay lines, to enable localization towards the speaker's booth or other areas of the stadium oval.

On either side of the projection screen four EAW loudspeaker systems are installed (Figure 8.5), each consisting of one MH242 and one MH662 with bass horns BH 662 and AS-790. These eight loudspeaker systems provide acoustical orientation towards the screen, and cover the curved section in front of the screen. When the screen is in use, the loudspeaker systems of the adjacent stand area are disconnected, and the other loudspeaker systems are supplied via a corresponding delay system.

All loudspeakers are supplied via distributed Crest power amplifiers, which are remotely monitored and remotely controlled by means of the Crest-NexSys-System. In the central sound control booth, where the stadium announcer's desk is situated, there is a DDA Q series mixing console, allowing injection of speech and music signals. The signals are supplied to the individual sound reinforcement systems, with assignment of the corresponding delay times and level attenuation, via a Nexus system from Stage Tec, Germany. These times and level values were simulated by

Figure 8.5 View to the video wall, with EAW loudspeakers for localization.

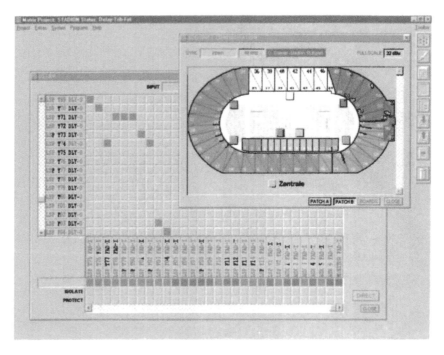

Figure 8.6 Operating surface of NEXUS system in Gottlieb Daimler Stadium, Stuttgart.

ADA Acoustic Design Ahnert in the CAD acoustics program EASE, and transferred to the Nexus. Figure 8.6 shows the operating surface of this control unit. The complete system was installed by Siemens AG.

8.1.2 Open-air stage, Bregenz, Austria

Theatre performance in the open has its origins in antiquity. In those times the aim was to build theatres to an acoustically optimum design, as is shown by the works of the Roman architect Vitruvius, and to use auxiliary devices (such as actors' masks in the shape of acoustic funnels) to improve speech transmission. Nowadays, in larger open-air stages, it is well-nigh impossible to do without electroacoustical methods, if only because of the large distance from actor to listener, and the environmental noise. Moreover, in this way it is possible to introduce sound effects and musical signals. The natural stage at Bregenz on Lake Constance has a long tradition. Open-air theatre has played in Bregenz since 1945, and for more than 15 years on the new stage (Figure 8.7). This stage is on an artificial island in the lake, whereas the audience (4000 persons!) are seated on the lakeshore [8.2].

Figure 8.7 General view of the open-air stage, Bregenz (aerial photo).

ROOM-ACOUSTICAL CONDITIONS

The maximum distance between a rear listener's location and the back of the stage area is about 100 m. Depending on the setting, the stage can be between 40 and 70 m wide; the stage set structures rise 18–20 m up above lake level. Action takes place everywhere, and often simultaneously on various scenes, so that visual turns of up to 60° or 80° are required. Because of the ambient noise (noise from the town, lapping of the waves, noise from the ships cruising on the lake, etc.), word intelligibility cannot be ensured without an electroacoustical system. Attempts were made, however, to have the orchestra that is always present in the summer (the Vienna Symphony Orchestra) play without electroacoustical support. The permanently installed orchestra pit was therefore subject to some complex modifications, but these did not achieve the desired results. The orchestra therefore continues to rely on electroacoustical support.

SOUND REINFORCEMENT SYSTEM

The Delta Stereophony System is used (see section 6.5.3.4). Unusually, there are no main reinforcement loudspeakers 'above' the stage; all the loudspeaker systems have to serve not only for sound-level transmission,

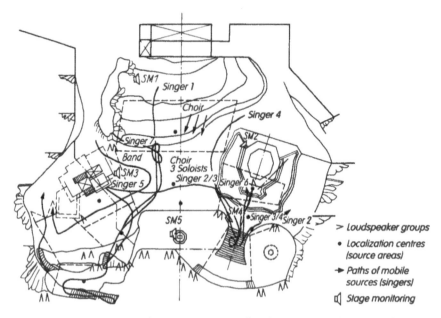

> *Loudspeaker groups*
> • *Localization centres (source areas)*
> → *Paths of mobile sources (singers)*
> ᶑ *Stage monitoring*

Figure 8.8 Ground plan of the stage setting for the 1984 production ('The Bird Seller').

but also for localizing purposes. Depending on the production there are thus up to 20 powerful mobile loudspeaker groups arranged within the stage set so that they can cover the auditorium area as a whole, as well providing a localizing source for particular sound events (Figure 8.8). Up to 12 wireless microphones and sometimes up to 50 stationary microphones can be connected. These are assigned to the source areas, and their signals are distributed via a matrix (60 inputs and 18 outputs) to the stage-edge and main sound reinforcement systems.

In this way it is possible to achieve uniform coverage of the whole auditorium. The sound engineer, who is located in the control booth behind the auditorium, supervises the whole acoustical performance. By means of a special computer program he can adapt the necessary level and delay distribution of the loudspeaker systems to the requirements of the stage setting, which generally changes every year. An additional mixing console for directional effects ensures the automatic cross-fading of the movable sources, so that the individual sources can simultaneously be controlled or sequenced by push-button.

8.1.3 Open space in front of the Congress Centre Suhl (CCS)

In connection with the World Championships in Sports Shooting in Suhl 1987 it was necessary to provide the space in front of the former town

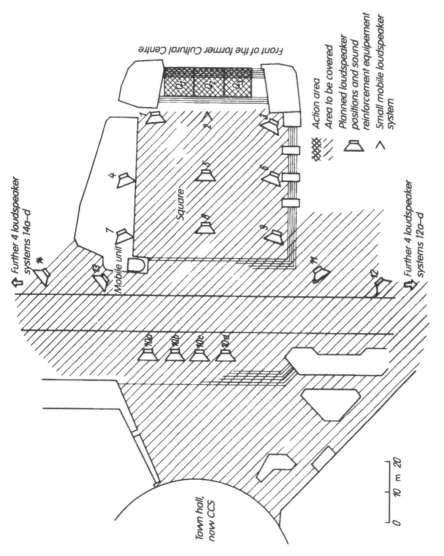

Figure 8.9 Ground plan of the space available in Suhl, indicating the loudspeaker locations.

hall, now the CCS, with a sound reinforcement system to provide adequate coverage of the opening ceremony, the presentation ceremonies, and a procession.

ROOM-ACOUSTICAL SITUATION

The dimensions of the open space (Figure 8.9) implied that one had to reckon with free-field propagation and all those special conditions and circumstances as described in section 3.1.3.

TECHNICAL SOLUTION

For the mobile stage, in the area opposite the former town hall in front of the Cultural Centre, a mobile sound reinforcement system was installed, which functioned as 'source simulation' for the Delta Stereophony System. Loudspeaker systems of type Z 229 were arranged on masts spaced in a grid of about 15 m so that they covered the area in question uniformly, as shown in Figure 6.10b. Their connection points were located in weatherproof underground boxes, and the cables terminated in a sound control booth in the Cultural Centre opposite the town hall.

For processions and parades on the street included in the project, the sound reinforcement system was extended by several outlet connection boxes and loudspeaker systems. A delay circuit (Figure 8.10) was provided

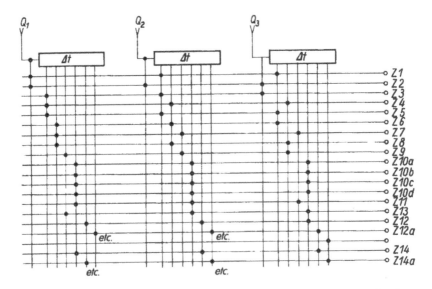

Figure 8.10 Switching pattern for coverage of open-air spaces to ensure localization and avoid echo phenomena.

to ensure the acoustical localization not only of the stage in front of the Cultural Centre (with and without speaker's desk), but also of any other action area.

All other equipment was accommodated in the sound control booth of the Cultural Centre or (for events requiring more sophisticated sound reinforcement equipment) in a mobile transmission unit.

8.2 Rooms intended mainly for speech reproduction

8.2.1 *Conference rooms*

Conference systems may either be permanently installed in a room, or consist of mobile systems installed in rooms that are used only temporarily for conference purposes. The following example describes a permanently installed system.

CONDITIONS OF THE ROOM

With a height of 3–6 m, the rooms concerned are relatively flat, and have a volume of 1000–2000 m³. Since they are used exclusively for speech transmission, they are quite heavily damped, so that their reverberation time is between 0.4 and 1.2 s. The furniture consists mainly of an oval or U-shaped table around which the conference members are grouped (Figure 8.11). For the president or chairman of the conference there is usually a prominent location which is notable for its technical equipment; sometimes it is complemented by a location for a technical operator.

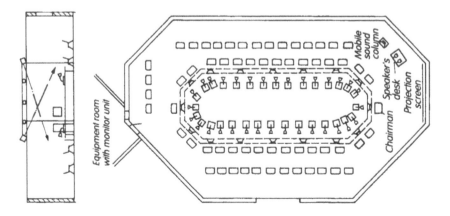

Figure 8.11 Ground plan and sectional view of a conference hall.

SOUND REINFORCEMENT SYSTEM

The basic element is the *delegate's unit* (Figure 8.12), which contains a microphone, a signalling button and often also a small built-in loudspeaker. In many installations the microphone of the delegate's unit is connected by a *central unit* only after the request for participation has been signalled by pressing the signalling button. Simultaneously with the connection of the microphone, the small loudspeaker of the delegate's unit is disconnected. The *chairman's unit* enables the chairman to interrupt the contributor and intervene in the discourse. There are also *monitoring* or *passive units*, which do not allow intervention in the discussion but enable monitoring and, if need be, recording of the discussion. The system is supplied by a 24 V power unit.

Such systems are increasingly being replaced by *fully automatic microphone systems* (e.g. system AMS from Shure, or the Voice-Matic Microphone Mixing System from Industrial Research Products), with which the active microphone is connected while inactive microphones are attenuated to reduce background noise and reverberation. The operating threshold of the connected microphone may be dynamically adapted to the original voice volume of the speaker concerned.

Delegations frequently consist of one negotiating member and several consulting members who are not authorized to participate. This is why all participants are not provided with a delegate's unit, but all are enabled to monitor the discussion. To achieve this there is often a loudspeaker system installed in the ceiling above the conference table. This is complemented by additional peripheral ceiling loudspeakers arranged in an outer ring.

For multilingual conferences a link to the interpreters' installation is required. In addition, conference rooms may offer facilities for visual reproduction, such as boards, reproduction of maps, projection of films, videos and slides, which require reproduction facilities for their accompanying sound in the immediate neighbourhood of the optical reproduction surfaces. Similar criteria apply to speech reinforcement for explanatory lectures. For this purpose, full-range loudspeaker systems are installed at both sides or behind the projection screen.

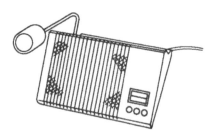

Figure 8.12 Delegate's unit of a conference system.

The maximum sound level for such systems in the diffuse field is about 85–90 dB.

8.2.2 Lecture halls

Lecture halls are a typical category of larger room in which an electroacoustical system serves essentially for the amplification of speech emanating from a central area. Generally there are also demonstration sound examples or the accompanying sound of a movie or a slide presentation to be reproduced. In lecture halls of classical design with heavy sloping of tiers, an electroacoustical system for voice reinforcement is required only for volumes exceeding 1000 m³. With a typical specific volume of 5–6 m³/seat this results in a minimum number of 200–300 seats.

8.2.2.1 Classical lecture hall

This has a heavily sloped auditorium rising, amphitheatre-like, around a small platform area equipped with boards and projection screens as well as a lecture table. The slope of the auditorium and the width of the lecture table will depend on the architectural conditions and on the number of seats. The advantage of such room configurations lies in the relatively small distance between the lecturer or the demonstration material and the listeners. This enables good visual and aural communication.

A typical example of this type of installation is a medium-sized lecture hall for a teacher training institute (Figure 8.13). It has an almost trapezoidal ground plan with cut-off edges and an auditorium rake (incline) of 3 m with a depth of 11 m. Its volume is 1530 m³, and the number of seats is 320. This results in a specific volume of 4.8 m³/seat, which corresponds to the medium range for such rooms. Since this hall is the main auditorium of the institute, it serves not only for teaching purposes, but also for other functions such as ceremonies and speeches.

ROOM-ACOUSTICAL CONDITIONS

The lecture hall was designed so that its reverberation time in the medium frequency range is about 1.1 s. For damping purposes and for decorating the main wall surface it is possible to cover the boards and projection screens behind the platform by means of sound-absorbing curtains. The ceiling accommodates large reflectors to provide useful sound reflection. The main absorber is the occupied auditorium.

SOUND REINFORCEMENT SYSTEM

The sound reinforcement system serves for speech reinforcement and for reproduction of recorded and amplified music. It consists of two

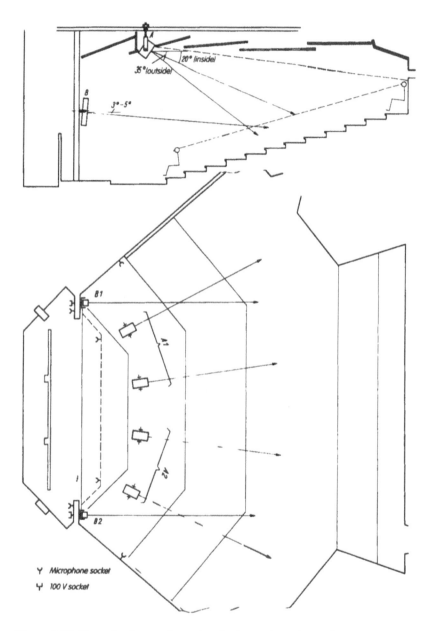

Figure 8.13 Ground plan and sectional view of a classical lecture hall with loudspeaker systems.

loudspeaker groups, each of 50 W load capacity and containing two broad-band sound columns (TZ 133). These are amply able to cover the trapezoidal room. As can be seen in the sectional view, the loudspeaker systems of these groups are partly flush mounted into the front ceiling area, which, thanks to its baffle-like function, helps to improve sound radiation in the lower frequency range.

The ceiling groups are complemented by two sound columns primarily designed for the speech range, and arranged at both sides of the boards and projection screens. These loudspeaker systems serve mainly for improving the timbre in the front area of the auditorium, which is 'masked' to a large extent from the ceiling groups.

In addition it is possible to use mobile loudspeakers, the connection sockets for which are provided in the platform area. This facility is required mainly for cultural purposes, such as for stage monitoring or for special demonstrations. The installation is designed to be capable, without these mobile loudspeaker systems, of realizing a maximum sound level in the diffuse field of 94 dB plus 10 dB headroom. The symmetrical arrangement enables two-channel operation for input of corresponding demonstration examples and for cultural functions (such as ceremonies). Microphone connection sockets are arranged at the left and right of the platform, as well as at the lecturer's table and in the front ceiling area.

8.2.2.2 *Lecture hall in a flat room*

For postgraduate education in particular, lecture halls have of late also been installed in flat rooms with a level floor. These offer the advantage of enabling the seats to be arranged not only in tiers, but also in U or circular fashion, to facilitate seminar-like discussions. Such rooms are thus a cross between a conference room according to section 8.2.1 and a classical lecture hall.

Figure 8.14 shows a typical example of a lecture hall of this kind. It has a length of 28.6 m, a width of 12.7 m, and a height of about 4 m, which results in a volume of about 1400 m^3. As the maximum number of working places is planned to be 204, the resulting specific volume of 6.8 m^3/place is relatively large.

The ceiling is transversely subdivided by 'lighting beams' 300 mm deep and 600 mm wide. As well as the lighting, these beams also contain the ventilating elements and the loudspeakers for the distributed sound reinforcement system, a central cluster arrangement being inappropriate under these conditions. At the front wall there is a large lecture table arranged on a small raised platform. This table can also seat a chairman. Behind the table there is a screen for film and video projection and beside the table a speaker's desk.

By means of a folding wall it is possible to partition off a fully functional section of 12 m length and about 600 m^3 volume. Simultaneous use of both

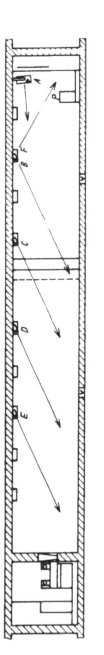
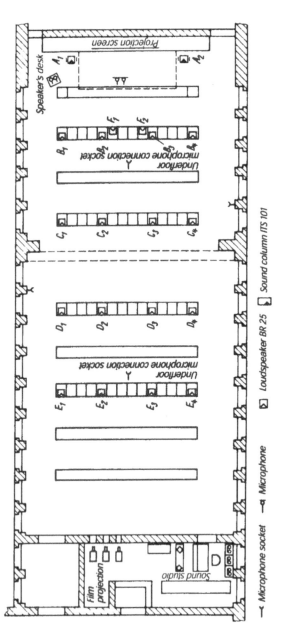

Figure 8.14 Lecture hall in a flat room with loudspeaker systems.

↳ Microphone socket   ⊸ Microphone   ▣ Loudspeaker BR 25   ▢ Sound column ITS 101 parts of the room is, however, out of the question because the folding wall provides insufficient sound attenuation.

The hall is designed to have a reverberation time of about 1 s. Sound absorption is provided by a 1 m wide strip of mineral wool behind perforated wooden panelling at the rear wall of the room; this is specifically intended to prevent echoes, which could be quite disturbing considering the length of the room. Additional absorption is provided by the window curtains at both longitudinal walls of the room. Low frequencies are absorbed by the absorbent-backed wooden panelling of the ceiling and the front and rear walls of the hall.

SOUND REINFORCEMENT SYSTEM

As can be seen from Figure 8.14, four loudspeaker cabinets are installed as main systems in about every second architectural beam so that their main reference axis is directed about 25° downwards towards the rear wall. This pattern is interrupted in the area of the folding wall, both because the spacing between the beams is slightly greater, and because in this way the loudspeakers can be placed more favourably for utilizing the rear part of the hall. The last beam before the rear wall does not contain any loudspeakers, because at this location they would merely give rise to echo disturbances.

The loudspeakers for the main sound reinforcement system are small-volume bass reflex cabinets in combination with a dome tweeter. They are equipped for this purpose with transformers for connection to the 100 V system, and their nominal load capacity is 25 W. To improve coherence they can be operated via a delay system, but testing revealed a high definition of speech transmission and astonishingly good acoustical forward localization without the delay system.

For monitoring at the lecture table there are two small bass reflex cabinets in the second beam in front of the wall carrying the projection screen and the board, which radiate towards the lecture table.

To provide accompanying sound for film, slide or video presentations there are two 100 W sound columns arranged at either side of the projection screen. Microphone connection sockets are provided at the speaker's desk, at the lecturer's table, at the walls and in the central floor area of the two partial rooms; these latter sockets being mainly used for recording round-table talks.

The system is supplied and controlled from a central sound control booth at the rear of the hall, from where there is visual contact with the hall.

8.2.3 *Plenary halls*

8.2.3.1 *Plenary hall in the former Palast der Republik in Berlin*

The plenary hall was completed in April 1976. Apart from its function as plenary hall of the Parliament of the former GDR, it was also used as a venue for important conferences.

The hall has a maximum width of 36 m with a relatively small depth of 24 m and a maximum height of 11 m. Its volume is 9700 m^3. On the floor of the hall there are 541 working desks, and at the rear, 6 m above the floor, there is a balcony seating 245 visitors (Figure 8.15). The distance between the working desks of the deputies in the hall and the presidency and the speaker at the front is relatively small. The configuration of this hall thus corresponds to that of modern, medium-sized conference and plenary halls [8.3].

ROOM-ACOUSTICAL CONDITIONS

The front wall behind the presidency and the rear wall of the hall are covered with a layer of porous, absorbent material behind perforated chipboard faced with decorative glass-fibre fabric. As bass absorbers, Helmholtz resonators are arranged in the lateral ceiling elements. Thanks to these absorbers, the furniture and the heavily upholstered seats, the occupied hall shows a largely frequency-independent reverberation time of 1 s (up to 2 kHz) [8.4].

SOUND REINFORCEMENT SYSTEM

The main function of the sound reinforcement system is to provide speech amplification of high intelligibility in the main seating area, in the presidency and on the balcony. Speaking stations are at the speaker's desk, at the desk of the conference chairman, and – during discussions – at seats in the presidency and in the main area. In addition, there was a requirement for high-quality music transmission. Finally, because of the high-profile nature of the room, the sound reinforcement system was as far as possible to be visually unobtrusive.

To meet these requirements two sound reinforcement systems were installed. One of these corresponds to a central system consisting of a central ceiling loudspeaker array combined with a near-field loudspeaker group. The ceiling loudspeakers are arranged in suitable recesses above the ceiling of the hall. Each recess is equipped with one 50 W broadband sound column (Z 128) and one 25 W sound column for speech transmission (Z 127). The recesses are visually separated from the hall by an acoustically transparent wire mesh, which, thanks to its white colouring, harmonizes with the plaster relief elements of the ceiling. The near-field group had to be

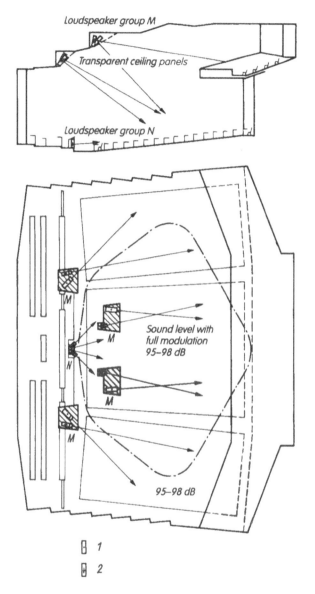

Figure 8.15 Arrangement of the loudspeaker groups in the plenary hall of the former Palast der Republik and the sound levels attainable: 1, Z 128 sound column, 50 W; 2, Z 127 sound column, 25 W.

concentrated in the speaker's desk, a solution that, though by no means optimal on account of the width of the hall and the relatively reduced size of the speaker's desk, has beneficial results for the acoustic localization of the speaker's desk. Figure 8.15 shows the arrangement and the coverage of these loudspeaker groups, as well as the sound level distribution achievable with full modulation.

Each of the partial groups arranged in the recesses as well as the near-range group is assigned to one 100 W amplifier. The ceiling array (M) and the near-field group (N) can be adjusted separately. They may also be filtered separately in order to optimize the timbre or the positive feedback threshold of the group in question.

The second sound reinforcement system is decentralized. Each working table in the main hall and in the presidency contains one loudspeaker (frame diameter 100 mm, power-handling capacity 0.5 W), which is matched to the 100 V network. Similar loudspeakers with a power-handling capacity of 3 W are installed in front of each seat on the balcony, in the backrests of the chairs or in the balustrade. This loudspeaker system serves especially for congress coverage with discussion from the hall. For this purpose a discussion system was installed, with 50 mobile speaking stations. It functions similarly to a conference system. When a discussion microphone is activated, the central discussion management system switches off the loudspeaker incorporated in the same table. The contributor can hear himself or herself undisturbed through the other loudspeakers in the neighbourhood, which continue to receive the signals without electrical delay. Syllable intelligibility measurements have revealed for the centralized and the decentralized systems a proportion of correctly understood syllables of 84–92% and of 88–93% respectively, which in both cases means very high quality.

8.2.3.2 *Plenary hall of the Academy of Arts in Berlin*

The building at the Robert-Koch-Platz was handed over to the Academy of Arts (East) on the occasion of the festivities commemorating the 750th anniversary of the foundation of Berlin. In the course of the restoration works the culturally and historically interesting plenary hall was reconstructed as well. Although it cannot be considered as a typical plenary hall serving mainly for discussion, it is worth considering here as a special case.

ROOM-ACOUSTICAL CONDITIONS

The room seats 250 persons, and is almost square in plan (Figure 8.16). The high ceiling provides a large reverberation space, which implies a long reverberation time, especially in the low-frequency range. In Figure 8.17 we can see the undesirable increase of the reverberation time towards the

Figure 8.16 View of the plenary hall of the Academy of Arts in Berlin.

low frequencies. This increase could be corrected by the installation of additional bass absorbers.

SOUND REINFORCEMENT SYSTEM

To allow the reproduction of two- and four-channel music, the room was equipped with loudspeaker groups in all four corners. In addition, two further loudspeaker systems were installed above the platform as 'proscenium arrays', which may be complemented by subwoofers (Figure 8.16).

The associated power amplifiers are installed in an adjoining room of the stage, and are remotely controlled. The control room, which is arranged below the hall level, and from which the hall can be seen only via a closed-circuit TV system, contains a control console (N 20 by Neumann, Berlin) and all the other technical and amplifier equipment: one AEG eight-track tape machine, two M15A AEG two-track tape machines, several Akustika 100-V racks including Gielmetti distributors for sound transmission into the house, one TDU 8000 delay unit, one Quantec room simulator, and other effects devices.

There is a separate recording room with visual contact to the control room.

(a)   Reverberation time analysis of RAUMMESSUNG
      by H. GROTHE /INT
      on 22.09.1998
      At AKADEMIE DER KUENSTE DER DDR, PLENARSAAL
      Remarks:
      Averaged values

      Jobs: 00,01,02,03

Band 0 Freq.	125Hz	RT	2.38sec	A
Band 1 Freq.	250Hz	RT	1.89sec	A
Band 2 Freq.	500Hz	RT	1.53sec	A
Band 3 Freq.	1000Hz	RT	1.39sec	A
Band 4 Freq.	2000Hz	RT	1.25sec	A
Band 5 Freq.	4000Hz	RT	1.03sec	A
Band 6 Freq.	8000Hz	RT	0.73sec	A
Band 7 Freq.	16000Hz	RT	0.57sec	A

(b)   Reverberation time analysis of RAUMMESSUNG
      By H.GROTHE /INT
      On 22.09.1988
      At AKADEMIE DER KUENSTE DER DDR, PLENARSAAL
      Jobs: 00,01,02,03

REVERBERATION TIME
vs. FREQUENCY

BANDWIDTH 1/1 OCTAVE

vertically: TIME
Range from 0sec
          to 3sec
horizontally: BANDS
Range from 125Hz
          to 16000Hz

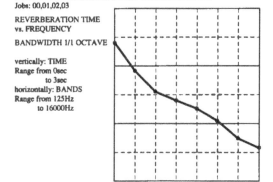

(c)   Reverberation time analysis of RAUMMESSUNG
      By H.GROTHE /INT
      On 22.09.1988
      At AKADEMIE DER KUENSTE DER DDR, PLENARSAAL
      Jobs: 00,01,02,03

REVERBERATION TIME
vs. FREQUENCY

BANDWIDTH: 1/1 OCTAVE
RECOMM. RT:1.3sec

vertically: TIME
Range from 0.0sec
          to 2.6sec
horizontally: BANDS
Range from 31Hz
          to 8000Hz

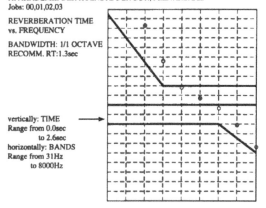

Figure 8.17 Plenary hall of the Academy of Arts. Reverberation time as a function of frequency: (a) table; (b) function chart; (c) curve within the tolerance zone.

8.3 Rooms for speech and music performances

8.3.1 *Great hall in the former Palast der Republik in Berlin*

The hall was completed in April 1976. In the former GDR it was used for congresses, for show and concert performances of various types, and for banquet and dance functions. The maximum volume of the hall is 36 000 m^3. It can be arranged into different configurations (see Figure 6.40). Its maximum capacity is 5000, of whom 1462 are seated on the balcony. A configuration that is often used for shows and cultural functions employs a platform reaching far into the audience area (Figure 6.40d). This configuration, which seats 3800 spectators when the balcony is used, poses an especially difficult problem for sound reinforcement, since sources at both ends of the platform may give rise to echoes at large distance.

ROOM-ACOUSTICAL CONDITIONS

The hall is of amphitheatre-like design, and in its maximum-capacity configuration is almost devoid of wall surfaces, since all the audience areas, with their heavily upholstered seats in the stalls and on the balcony, rise up to the balcony and hall ceilings. In its maximum-capacity form, the reverberation time of the hall without electroacoustical support is approximately 1.5 s.

ORIGINAL SOUND REINFORCEMENT SYSTEM [8.5]

This consisted of:

- main loudspeaker groups A_1–A_4 and B_1–B_4, equipped with double sound columns of type Z 128 [8.1], which were arranged above the edges of the platform and could be retracted into the variable ceiling (Figures 8.18, 8.19);
- stage system comprising groups B_5 and C, equipped with ceiling-mounted drop-down sound columns types Z 133 and Z 128;
- near-range group D, equipped with 100 V sockets in the platform area for mobile loudspeakers Z 128, Z 133 and Z 127, as well as compact loudspeakers arranged below the rail;
- decentralized ambiophony and effects system comprising the groups E with compact loudspeakers of type K 20 (10 W, cone loudspeaker of 215 mm frame diameter in 20 litre compact casing) arranged in the hall ceiling and in the ceilings under the balconies;
- centralized effects system group F, equipped with sound columns of type Z 128;
- seat system: small loudspeakers of 125 mm frame diameter incorporated in the seat backrests.

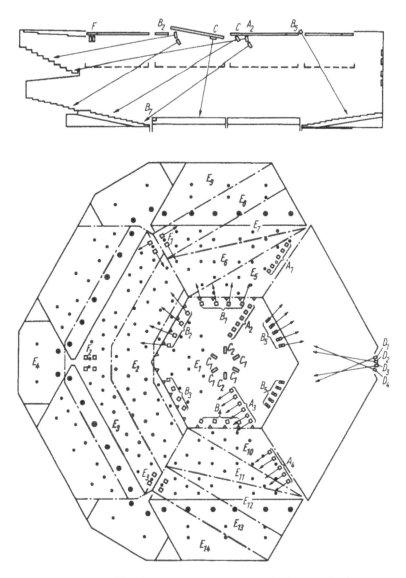

Figure 8.18 Original loudspeaker arrangement in the large hall of the former Palast der Republik in Berlin [8.5].

For the main groups, part of the stage system and the ambiophony system, the Delta Stereophony System was employed for the first time. Only in this way was it possible to solve the problems posed by the large platform reaching far out into the audience area. In the conference configuration

Figure 8.19 View of the hall with original loudspeaker arrangement.

the speaker's desk was equipped with loudspeakers serving as a localizing source for the stage.

The effects system was equipped with a panning control, which made it possible to pan a signal in the area of the balcony balustrade and above that in the ceiling area. To achieve this, groups of 10 ceiling loudspeakers were successively switched out of the ambiophony system and reconnected pairwise (one in the ceiling and one under the balcony) after passing of the signal. In total, 60 loudspeakers were involved in the system.

The seat backrest system was 'intermeshed' in such a way that every second loudspeaker was assigned to another amplifier so that listening could be continued if one of the amplifiers failed.

The hall was equipped with a central sound control room containing a large mixing console, the distribution, sound processing and recording equipment, the radio microphone receivers, and part of the power amplifier installation. The sound reinforcement systems were supervised and adjusted from a sound mixing and control console situated in the hall on its main axis in the area of the permanently installed upper rising stalls. This location is situated below the balcony and thus is not quite optimal, but had to be chosen because all other locations are variable and thus liable to be masked (cf. Figure 2.4).

The main loudspeaker groups and the mobile stage system of the installation were reconstructed in 1986, in order to enhance the maximum achievable sound level, extend the frequency transmission range, and reduce the size of the loudspeakers, which continued to be lowerable from the ceiling.

EQUIPMENT AFTER RECONSTRUCTION (SEE ALSO [8.6])

The new installation consisted of:

- main loudspeaker groups P_1-P_8, A_1-A_8 and R_1-R_8, consisting either of one loudspeaker UPA 1A (Meyer Sound Lab.) or of two stacked loudspeakers of this type (Figure 8.20) (the original design was modified so that the new high-power loudspeakers are also swivelling and retractable into the ceiling (Figure 8.21));

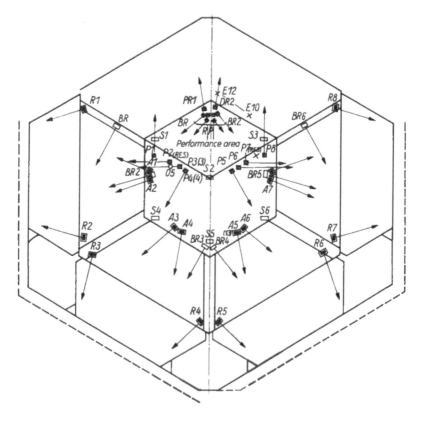

Figure 8.20 New loudspeaker arrangement in the large hall of the former Palast der Republik in Berlin.

Figure 8.21 Insertion of a UPA 1A high-power loudspeaker system in the new lifting devices at the original installation points.

- presidency, stage and near-range system, also consisting partly of stacked loudspeakers UPA 1A;
- the decentralized ambiophony and effects group E and the backrest system remained unchanged, whereas the effects group F was abandoned in favour of the new balcony system R.

Figure 8.22 View of the hall with the new sound reinforcement system and simulation loudspeaker systems at stage level.

The Delta Stereophony System was operable in conjunction with the main groups, with part of the mobile stage system, and with the loudspeaker group E connected. The position of the mobile simulation (image shift) loudspeakers is shown in Figure 8.22. Thanks to their power and compactness it was now easier than before to form centres of localization.

Since the new loudspeaker systems are of low-impedance design, the amplifiers had to be arranged close to the loudspeakers to avoid high power losses. Thus the amplifier locations in the sound control room were superceded, and the new – remotely controlled – amplifiers were installed in the ceiling area.

Further modifications of the electroacoustical systems resulted from the exchange of the original EMT 440 delay units (shift register generation) for computer-controlled units of type TDU 8000. To this end a new matrix for summation and distribution of all possible source signals with more than 40 loudspeaker positions was developed and manufactured. This installation thus counted among the most modern large systems, providing high-quality sound reinforcement by means of the Delta Stereophony System. The advantages of the DSS were proven by several demonstrations.

The excellent new system was in use for four years only. Since 1990 the hall has been closed because of asbestos contamination of the building.

8.3.2 *Kulturpalast, Dresden*

The Kulturpalast was inaugurated in 1969. It has served until now as a concert hall for the two Dresden symphonic orchestras, as well as for important performances of light entertainment of all kinds, congresses, balls, and festive banquets. With the exception of symphonic concerts, for which the use of electroacoustical techniques is an exception, all other types of performance require support by a sound reinforcement system.

ROOM-ACOUSTICAL CONDITIONS

With a maximum length of 40 m, a width of 50 m, and a height of 11–15 m, the hall has a volume of 19 000 m³. The maximum seating capacity is around 2300. The seats are arranged in the stalls and on a very steeply rising balcony, the sides of which extend far down towards the stage (Figure 8.23).

The reverberation time varies between 2.3 s without audience and 2.0 s with audience. Thanks to a relatively low diffuse sound level, this reverberation time is not subjectively too noticeable, but with symphonic music there seems to be a lack of spaciousness and reverberation time. (This property is also found in other rooms of more than 18 000 m³ volume.)

SOUND REINFORCEMENT SYSTEM

At the time of its inauguration the hall was equipped with a sound reinforcement system that relied mainly on radiation by two loudspeaker groups each arranged over a width of 4.50 m at the lateral corner points of the hall ceiling. Each group consisted of eight columns of 100 V design with a nominal load capacity of 50 W each. Six of these were aimed frontally at the rear stalls and balcony areas, and two were arranged behind the corner points of the ceiling for radiating obliquely into the side hall areas.

This arrangement soon required an installation of complementary loudspeakers, to cover the near-stage and other 'masked' areas, such as those beneath the low lateral extensions of the balcony, and the stage itself. The pronounced directivity of the closely arranged loudspeaker groups also entailed timbre alterations in the reproduction within the hall.

Moreover, it soon became evident that with the very short rehearsals and the sometimes very expensive performances, operation of the system only from the control booth was possible either not at all or only with great difficulty. This is why, after a relatively short period, a hall control console was installed under the central balcony. This control console was gradually enlarged, so that soon it was serving – and continues to be used – as the sole sound control unit for many performances.

After 20 years of operation it became necessary to renovate the sound reinforcement system of the hall in order to prepare it for the multi-channel

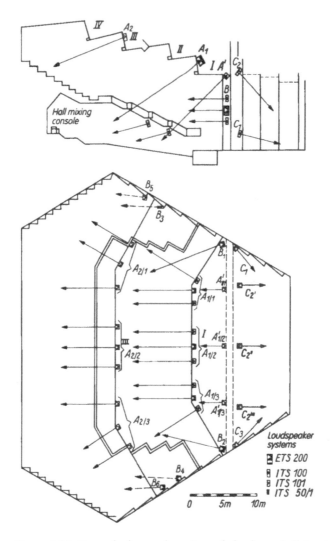

Figure 8.23 Ground plan and section of the large hall in the Kulturpalast, Dresden, with redesigned sound reinforcement system.

delay Delta Stereophony System. The loudspeaker arrangement conceived to this end is briefly described below (Figure 8.23).

The main reinforcement system in the area of the stage opening consists of five partial groups complemented by three further partial groups to cover the balcony. The main loudspeaker groups have a frequency range that is sufficient for nearly all purposes. The stage opening groups (B_1 and B_2) are each equipped with one special broadband loudspeaker. Powerful central groups ($A_{1/1}$, $A_{1/2}$ and $A_{1/3}$) ensure that the signal does not break down into left and right parts. The less powerful groups ($A_{2/1}$, $A_{2/2}$ and $A_{2/3}$) are used for covering the balcony, whereas additional support loudspeakers ($B_3 - B_6$) are arranged to supply the masked lateral areas below the balcony.

Covering the front stalls area and stage monitoring pose particular problems. For both areas mobile loudspeakers are arranged on the stage and at the edge of the same, which are not shown in Figure 8.23. Since such an arrangement is sometimes utterly impracticable, such as for ballet performances, it is planned to install additional loudspeaker groups (A_1'). Stage monitoring corresponds to loudspeakers $C_1 - C_3$, which may also be used for transmitting stage directions during rehearsals.

This more decentralized arrangement has improved the hall by providing an adequately uniform timbre distribution over all the various audience areas. It requires an electronic delay system to avoid echo-like travel-time interferences. Because of the ample stage width of 30 m the electronic delay system should be of multi-channel design to preclude any travel-time interferences attributable to cross-propagation in the hall, and to ensure adequate coincidence of acoustical and visual localization, particularly for the front audience areas. All this can be appropriately solved by using the Delta Stereophony System.

8.3.3 Town hall, Rostock

This hall was finished in 1979 [8.7], and is well provided with stage and lighting equipment. In addition to its main functions – sports and conference events – it is thus also suitable for numerous other cultural performances, as well as for banquets and balls.

The volume of the hall is 35 000 m^3. By using various performance areas at different locations, and by dividing the stage with sound-deadening curtains, it is possible to realize diverse hall configurations. The maximum capacity of the hall is 5000 spectators. The central area is suitable for most indoor games (maximum length 66 m). The two 18 m deep lateral stands can be lengthened into the stalls area by means of additional retractable seating.

ROOM-ACOUSTICAL CONDITIONS

Because of its main use, the hall is heavily damped by means of broadband absorbers in the ceiling areas and at the front and rear walls. Stage and concealing curtains allow further damping. Depending on configuration and attendance, the reverberation time is between 1.2 and 1.5 s.

SOUND REINFORCEMENT SYSTEM

The hall is equipped with sound columns of types Z 128 (50 W) and Z 127 (25 W). In all, the installed loudspeaker power is 2.2 kW (not including the

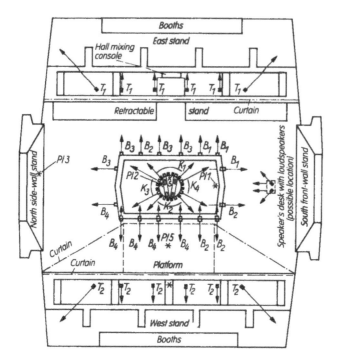

Figure 8.24 Ground plan of the Town Hall, Rostock: □, Z 128 sound column, 50 W; ■, Z 127 sound column, 25 W; K_1–K_5, loudspeaker groups in monocluster; B_1–B_4, loudspeaker groups on lighting bridge; T_1–T_2, loudspeakers to cover the stands; *P11–P15, reference locations for calculating the travel times.

connections provided for mobile stage loudspeakers). The sound reinforcement system is arranged around a central monocluster, which may be varied in height according to requirement (up to a maximum of 12 m) above the playing area. The monocluster can be subdivided into the radially aimed groups K_1 (250 W), K_2 (250 W), K_3 (150 W) and K_4 (150 W), and the mainly vertically aimed group K_5 (100 W) (Figure 8.24, see also Figure 6.15a). Thus it is possible to cover the different reception or action areas, as required, in the stalls and in the front stand areas of the hall.

Around the monocluster there is a central lighting bridge. This carries a further 16 sound columns, subdivided into groups B_1 (150 W), B_2 (150 W), B_3 (250 W) and B_4 (250 W), and aimed radially at the main as well as the front and rear stalls. Some of these sound columns, which are aimed at possible action areas at the southern and western stand sides, may be swivelled by 180° so as to cover the stalls area in front of these stands.

To cover the upper stand areas, which are masked when the monocluster is pulled up, a group of six sound columns T_1, T_2 (150 W each) was

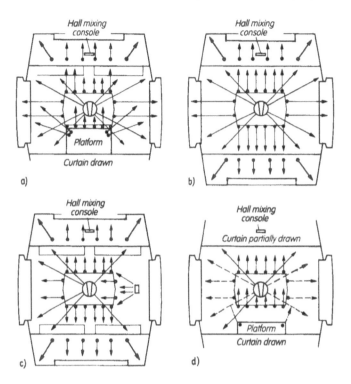

Figure 8.25 Working positions of the loudspeakers in the Town Hall, Rostock, for different hall uses: (a) platform configuration; (b) sports configuration; (c) mass event; (d) banquet configuration (cinema).

arranged in the central area of the two lateral stands. (For certain ball games, such as volleyball, the monocluster must be 12 m high to comply with international rules.)

The system was designed so that it can, if necessary, be used without delay equipment. When using a three-stage delay system, however, acoustical localization on the action area becomes possible, and intelligibility is enhanced thanks to the higher coherence of the wavefronts arriving at the audience.

To provide the reference sound levels necessary for localization, mobile stage loudspeakers are provided, along with the necessary power outlets. Furthermore, a speaker's desk with built-in loudspeakers is available. Figure 8.25 shows schematic representations of the hall in various configurations.

Control of the system is from a mixing console located above the lateral stand opposite to the platform used for most of the functions (Figure 8.26). This station is also equipped with a studio tape machine, so that one-man operation with playback facility is possible from here.

Figure 8.26 View across the hall mixing console into the Town Hall, Rostock.

A sound reinforcement control booth with visual contact to the hall is available to accommodate the power amplifiers, the sound recording equipment, the receiving equipment for the stage radio microphone transmitters, the distribution and supply equipment, and an additional mixing console for premixing and recording tasks.

8.3.4 Congress Hall of the Kremlin Palace in Moscow

This hall was completed in 1961, and was used in the former Soviet Union mainly for mass meetings, such as party conventions, and for conferences, opera performances, and music shows. Its dimensions are: maximum height 18 m, length without stage 60 m, maximum width 55 m, volume approximately 54 000 m³. Its maximum capacity is 6000 seats. The hall was at that time one of the largest in the world (Figure 8.27).

ROOM-ACOUSTICAL CONDITIONS

The room was designed by the Soviet acousticians Furduyev and Kacherovich mainly for speech transmission. Thus the reverberation time

Figure 8.27 Congress Palace in the Moscow Kremlin: partial view into the hall.

of about 1.5 s is relatively constant in the frequency range between 125 and 4000 Hz. Towards 63 Hz it increases slightly, whereas at frequencies above 4000 Hz it decreases due to the increasing air absorption. This behaviour was achieved by using perforated metal sheets in the ceiling area and perforated wooden panels in the wall areas, each backed with mineral wool. Moreover, the chairs are fully absorbent, so that the reverberation time is widely independent of the degree of attendance.

Sound transmission in the hall was already then exclusively committed to the use of electroacoustical equipment, which objective is assisted by relatively low reverberation times.

SOUND REINFORCEMENT SYSTEM

The objectives to be fulfilled by a sound reinforcement system for this hall were defined in 1960 by Furduyev [8.8] as follows:

- high-quality sound reinforcement system, ensuring utmost reliability for all usage configurations;
- direct-sound amplification for stage and cinema operation;
- installation of a five-channel system for complete coverage and for realizing an intensity-based stereophonic sound reinforcement;

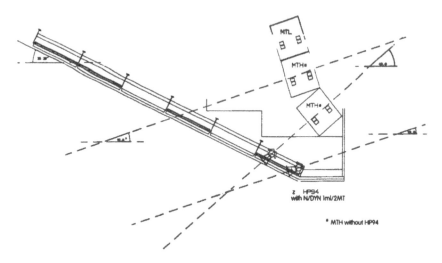

Figure 8.28 Arrangement of proscenium loudspeakers.

● input facility for spaciousness-enhancing signals, especially from the ceiling, side wall and rear wall areas of the hall;
● use of ambiophony and other reverberation equipment to modify the reverberation behaviour.

This is the project that was put into practice (see sections 6.4.3.1 and 6.5.3.3, and Figure 6.38). Save for the replacement of loudspeaker systems in 1985, no modifications were made for 30 years. In the course of the opening-up of the country in the second half of the 1980s it became necessary to adapt the sound reinforcement system to the standard that had come to be accepted in Western countries. This led to a redesign of the system, in which one of the authors of this book and, after 1990, also the engineering office ADA (Acoustic Design Ahnert) were involved.

Based on the principle of a five-channel Delta Stereophony System (section 6.5.3.4), the main loudspeakers were replaced by modern Electro-Voice units. Measurements had revealed that the perforated visual ceiling in the proscenium area, behind which the proscenium-opening loudspeaker systems were to be arranged, was fully transparent to frequencies up to 2 kHz only, which accounted for the low-frequency bias of the old system. Consequently, the MTH4 and MTL4 systems used were adapted by modifying the crossover in such a way that the high-frequency treble horn could be arranged separately from the medium-frequency and bass sections of the MT4 in front of the grid, and thus was aimed directly in the hall (Figure 8.28). The distance of the HF horn in front of the mid-range and bass speakers (about 2 m) could largely be compensated for by an appropriate electronic alignment. For the first time it was possible to supply the room with a broadband sound spectrum that included the high frequencies.

Because of the great depth of the hall, however, there was a marked attenuation of the high-frequency components on the balcony, so that five individual horns of type HP940, appropriately arranged under the ceiling and in front of the balcony edge, were required to compensate for that loss. These horns are, of course, driven via corresponding delay circuitry.

New loudspeaker systems (DML-1122, TL-606DX and TS8-2, all from Electro-Voice) were installed in the wall and proscenium areas, and mobile Deltamax systems and SSM loudspeaker systems from Apogee Sound Systems are used as localization equipment in the proscenium area.

These loudspeaker systems are required by the Delta Stereophony System to ensure that localization of the sound event remains at stage level, and does not tend to migrate to the ceiling. The electronic control system, including the DSP610 DSS processor from AKG Vienna, is accommodated in the control booth behind the rear wall of the hall (Figure 8.29), from where contact to the huge, 9 m wide analogue sound mixing console is provided. This console had to be refurbished, with new input and output cabling. It is planned to replace it with a modern digital mixing console that will allow two-man operation. The power amplifiers with their associated monitoring units and equalizers are arranged on a newly constructed platform in the roof area of the building, directly above the hall (Figure 8.30).

8.3.5 *Culture and Congress Centre, Stade*

The Culture and Congress Centre, Stade, was inaugurated in 1989 [8.9], and is a typical small-town cultural centre. Its scope of utilization is very wide ranging, from symphonic concerts via opera and theatre performances, the organization of pop and rock concerts, conferences and congresses, to fashion shows and banquets. As is often the case with multipurpose facilities, the house does not have any artistic ensemble of its own but depends on guest performances, so that its technical installation has to comply with a great variety of requirements.

ROOM-ACOUSTICAL CONDITIONS

The room has no balcony, and the stalls are essentially flat. The ground plan is fan shaped, and can be extended by opening the variable rear wall and thus integrating the foyer. This arrangement increases the volume from 4000 m^3 to 6000 m^3, and is especially suitable for balls (Figure 8.31). Reverberation time in the lower frequency range is about 1.8 s, and decreases to 0.6 s by 10 kHz. The low-frequency reverberation-time characteristic is about 1.5 s in the occupied hall. Because of the size of the stage house the reverberation time can be above 2 s for frequencies below 500 Hz. With heavy stage monitoring this

Figure 8.29 Moscow Kremlin: arrangement of delay and switching devices.

Figure 8.30 Moscow Kremlin: amplifier platform above the ceiling of the hall.

may have an adverse effect and impair the clarity of the sound balance. This effect may be avoided by means of absorbent decorations. The height of the hall is relatively low: 15 m in the proscenium area and 7.5 m in the rear area below the 'technical zone'. The shape of the room has a relatively unfavourable effect on the uniformity of sound distribution, a problem that is aggravated when the hall is enlarged by the foyer. This implies that balancing electroacoustical sound reinforcement is required in nearly all cases.

SOUND REINFORCEMENT SYSTEM

To meet the many different requirements, the hall was equipped with a comprehensive sound reinforcement system.

The loudspeaker systems consist of various loudspeaker combinations by Electro-Voice. The main system comprises the proscenium group with three pairs of stacked loudspeakers P_1, P_2 and P_3, the support loudspeaker group R_1-R_6 to cover the low hall area below the 'technical zone', and loudspeakers F_1-F_6 to cover the foyer area when it is opened up. The loudspeakers used were the (then) newly developed type DML 1122 systems. Complementary loudspeakers $PS_{1/1}-PS_{2/2}$ in the lateral proscenium areas and mobile loudspeakers connectable via outlets NS_1-NS_{12} in the stage area may be used to amplify the reference signal. The same applies to the near-range loudspeaker groups BR_1-BR_3.

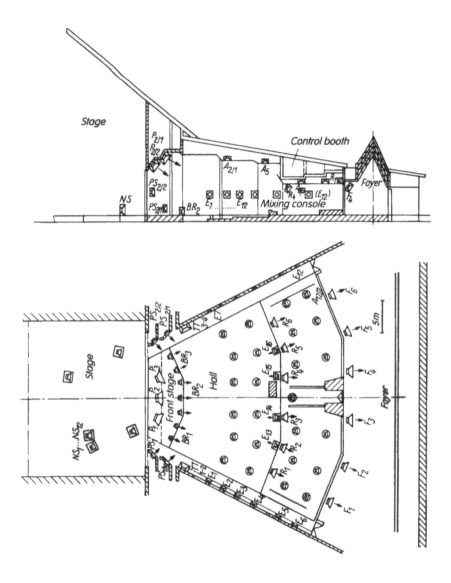

Figure 8.31 Ground plan and longitudinal section of the Culture and Congress Centre, Stade [8.9].

To boost the bass range, special woofers (EV SH 1800) were installed in the proscenium wall. For special stereo sound effects and ambiophonic sound amplification, a series of smaller loudspeaker combinations (EV S 80), $A_{1/1}$–$A_{12/2}$ and E_1–E_{16}, were arranged in the ceiling and in the side walls of the hall. Stage monitoring is effected by special monitors (EV FM 1202).

The delay system necessitated by the large number of loudspeakers is designed according to the Delta Stereophony System.

The volume of the sound reinforcement system can be controlled from a location in the centre of the hall. This uses a Harrison mixing console, which can be remotely controlled and automated, and caused some problems for the operating staff during the training period. Nevertheless, this technical solution offers the advantage that only one mobile control console is needed in the hall, from where the programming and control instructions are transmitted to the main processing equipment in the control booth (arranged in the 'technical area'). For less technically complex performances the system is operated from the main console in the control booth.

8.4 Rooms intended mainly for reproduction of classical music

8.4.1 Great Hall in the New Gewandhaus in Leipzig

The New Gewandhaus, Leipzig, was inaugurated in 1981 and serves, like its predecessor buildings, primarily for the performance of works of classical music. The system in the great hall was also designed both to meet the requirements for the performance of amplified music and electronic sounds, and to provide speech amplification and intelligibility improvement when the hall is used for conferences.

ROOM-ACOUSTICAL CONDITIONS

The Great Hall has a volume of 21 500m^3, and reaches its maximum height more or less above the concert platform. Its maximum capacity is 1900 seats, arranged – as in other modern concert halls – on steeply rising stalls, on 'lofts' located laterally near the orchestra platform, and on a rearward organ 'loft'. In the rear and lateral hall area there are further stepped 'lofts' for the audience (see Figure 6.45).[1]

In view of its main use, the hall has a reverberation time of 2 s, which is optimal for concerts, and a spaciousness that, with an index of spatial impression of 3.1–5.5 dB [8.5], is also favourable for this genre. Because of these room-acoustical conditions, the likely articulation loss according to Peutz [8.10] is 17% in the diffuse field: this value implies unsatisfactory intelligibility, which tends to deteriorate further if noise levels are high.

---

[1] 'Loft' in German means an elevated area adjacent to the performing area or stage.

SOUND REINFORCEMENT SYSTEM [8.11]

The sound reinforcement system comprises the *music effects system* for realizing electronic music, and the *congress system* required for intelligibility-enhancing voice reinforcement.

The congress system

This fulfils the following functions:

- good speech intelligibility at all seats;
- reproduction that is orientated towards the stage platform.

To compensate for the high reverberation and spaciousness with speech transmission, the loudspeaker systems used consisted of sound columns of relatively high directivity, and the aim was to find an arrangement that ensured a short distance from loudspeaker to listener. Originally it was intended to use the loudspeaker arrangement depicted in Figure 8.32, with the loudspeakers having the following functions:

- Loudspeaker groups A_4 and A_5 were intended to cover the gallery and the balcony respectively. Both loudspeaker groups were to incorporate three sound columns, individually installed in casings above the ceiling, which was acoustically transparent at these spots.
- A removable monocluster was to be installed above the front platform. This was to cover the stalls and the first rows of the lateral lofts by means of three sound columns (A_1), the two orchestra lofts by means of two sound columns each (A_2), and the organ loft by means of two sound columns (A_3). This concept was, unfortunately, not realized.
- Sound columns installed in the speaker's desk were to provide a better acoustical orientation on the speaker.

The monocluster above the platform could not be realized for several reasons, mostly attributable to design and aesthetic considerations. Thus it was necessary to find another sound reinforcement approach to improve coverage in the vicinity of the platform. Available for this purpose were the loudspeakers of the music effect system (groups B_{1a}, B_{1b}, B_{5a} and B_{5b}), and the loudspeakers in the speaker's desk and mobile loudspeaker systems.

In the final installation the following loudspeaker groups were employed: B_{1a} and B_{1b} (left and right of the organ), no delay; B_{5a} and B_{5b} in the ceiling, 50 ms delay; A_4, 100 ms delay; and A_5, 140 ms delay.

Sufficient speech intelligibility (syllable intelligibility >80%) is thus also achieved on the relatively large orchestra lofts; however, in this case acoustical localization corresponds to the organ front. Supplying the organ loft with sufficient definition is not possible with this arrangement; if the remaining loudspeaker groups of the music effect system are activated,

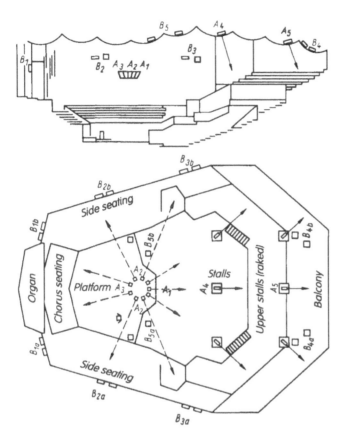

Figure 8.32 Loudspeaker groups in the New Gewandhaus, Leipzig: ground plan and section.

intelligibility is reduced because of the increased spaciousness thus obtained. Thanks to the relatively wide coverage angle of the loudspeakers used for the music effect system, their level must not be too high if spaciousness under these conditions is to be kept within acceptable limits in the platform area.

Music effects system This system was designed to meet the following requirements:

- sound reinforcement of single instruments or groups with and without electronic coloration/sound modification;
- mono or multichannel reproduction, e.g. of prerecorded music;
- migration of one- or two-channel signals above or around the room, for example with room-related live, electronically amplified performances.

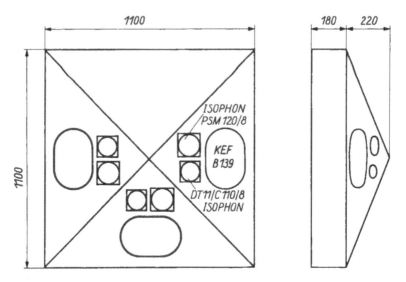

Figure 8.33 Structure of a loudspeaker system for music effects supply.

The following requirements had to be met:

- The loudspeakers were to have a sound characteristic that was compatible with the musical requirements.
- Except for some mobile units on the platform, the loudspeaker systems were for aesthetic reasons to be concealed in the boundary surfaces of the room.
- The openings in the walls and ceilings should be of minimum size so as to avoid noticeable enlargement of the sound absorption of the room.

To meet these conditions it was necessary to find special solutions for the loudspeaker system design and for the switching technique.

The correct reproduction of prerecorded moving (panned) signals required eight loudspeaker groups (corresponding to the number of tracks of the planned multitrack machine) to be installed with uniform spacing in the walls of the room. Since installation in the rear wall was out of the question because of its low height, the loudspeaker systems covering this area were installed in the ceiling. Moreover it was necessary to install two loudspeaker groups to the left and right in the ceiling above the platform and the front stalls area, so that two-channel signal panning across the ceiling was possible. Figure 8.32 shows the distribution of these loudspeaker groups associated with system B.

Figure 8.33 shows the design of this system as installed at the front wall and in the ceiling. In the side walls loudspeaker systems of modified design were installed, since the structure was heavily subdivided for room-acoustical

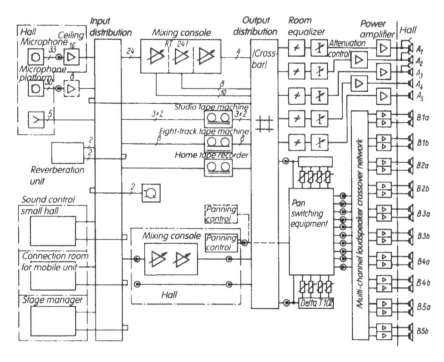

Figure 8.34 General layout of the sound reinforcement system in the New Gewandhaus, Leipzig.

reasons, and therefore allowed only a relatively low built-in depth. On the platform special mobile loudspeaker units of type M4 Mk III by Lambda Akustik, Vienna, can be connected via 6 dB lines.

Special measures were necessary to enable the sound level of the music effect system to be raised to the magnitude of the orchestra fortissimo level. For various reasons (installation conditions, transmission quality) the sensitivity and efficiency of the loudspeaker systems are relatively reduced (sound pressure level 110 dB and largely frequency dependent), and so a procedure was adopted that uses one-channel or at most two-channel switching of two delay lines with outputs that can be attenuated according to the selected delay time. Figure 8.34 shows the schematic layout of the installation. In this way it is possible to determine and modify one or two acoustically localizable sources in the wall and ceiling areas of the hall. The overall level in the diffuse field of the hall is generated by all the loudspeakers employed in the system. This arrangement differs from the Delta Stereophony System in that it does not create an action area surrounding the stage, but rather enables all-round acoustical localization over the receiving area. It enables the loudspeaker systems to be added in such a way that, thanks to the precedence effect, only the one or two loudspeakers radiating without delay are apparently effective. The switch-over is controlled by means of panels on which the control buttons are arranged schematically in accordance with the distribution of the loudspeaker groups to be localized in the room. A graphical distribution of the delay is thus possible. This 'panning' switch may be located either in the sound control booth or in a mobile mixing console, which can be installed in the hall.

8.4.2 Suntory Hall, Tokyo

In 1986, five years after inauguration of the New Gewandhaus in Leipzig, another concert hall was inaugurated: the Suntory Hall in Tokyo. In much the same way as in the Gewandhaus, Leipzig, the sound reinforcement system was to allow the input of amplified music as well as providing adequate intelligibility for speeches held before the concerts and at conferences and gala events [8.12].

ROOM-ACOUSTICAL CONDITIONS

The Suntory Hall is the first amphitheatre-like concert hall in Japan (Figure 8.35). It has a capacity of 2002 seats, arranged – as in comparable modern halls – on 'vineyard terraces'. Behind the platform there is the choir loft, and behind that is one of the largest organs in Japan (74 registers, 5898 pipes). The side walls are of oak wood, and the ceiling is of plaster stucco.

The characteristic volume/seat value is 10.5 m^3/seat. The platform is 22 m wide and 12.3 m deep (including the piano lift). It is subdivided into 21 platform lifts, enabling any desired orchestra formation to be optimally adapted to the room-acoustical conditions. These formations can be stored in the computer and recalled when required. The design conception of the hall is similar to that of the Gewandhaus Leipzig.

SOUND REINFORCEMENT SYSTEM

The loudspeaker systems used are exclusively of Altec Lansing manufacture. Above the front platform lift there is a monocluster (Figure 8.36), which can be retracted into the ceiling when not used. It contains seven 515-8G units with cabinets of special design, five Mantary MRII594 horns, five Mantary MRII564 horns and ten 288-16K HF driver systems. The monocluster cannot adequately supply the rear part of the platform, the rear part of the hall or the seats below the balcony, so additional ceiling loudspeakers are required [8.12]. To allow multi-channel input of amplified music, a further four systems are provided in addition to the monitors in the platform area. They consist of the Altec Lansing components, cabinet 817A, bass loudspeaker 515-16G, Mantary horn MRII594A, and driver 288-8K.

a)

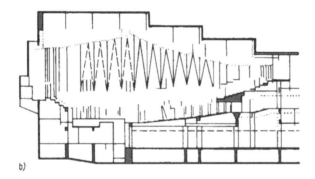

b)

Figure 8.35 Suntory Hall: (a) view of the platform area; (b) sectional view.

The sound control and recording room, which offers visual contact with the platform, is equipped with a Neve 5462/16 mixing console and Sony digital recording devices.

In addition to the large hall, there is also a small hall with 368 seats, which is also used for rehearsals.

8.4.3 *Casals Hall, Tokyo*

The Casals Hall, named after the cellist Pablo Casals, is a chamber music hall that was inaugurated in 1987. The hall has a width of 14 m at the

Figure 8.36 Multicluster in the Suntory Hall.

bottom and of 18 m on the balcony, and a depth of 34 m. There are 427 seats in the stalls and 84 on the balcony (Figure 8.37) [8.13].

ROOM-ACOUSTICAL CONDITIONS

The room acoustics were specifically designed for performances of music by Bach, Handel and Haydn, but so that eighteenth-century music could in

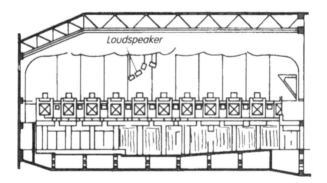

Figure 8.37 Sectional view of the Casals Hall.

principle also bē performed. The basic form selected is the so-called 'shoe-box'. The interior wall material is plaster; the decoration elements designed for sound dispersion (simulated curtains, etc.) were similarly formed of plaster.

The average reverberation time is about 1.6 s.

At the platform rear wall, space was provided in 1989 for a pipe organ, which for financial reasons was to be installed later.

SOUND REINFORCEMENT SYSTEM

The permanently installed sound reinforcement system is intended to:

- ensure intelligibility of the spoken word;
- ensure a high recording quality;
- reproduce amplified music and sounds.

To meet these requirements a monocluster was permanently installed (Figure 8.38). It consists of two Altec MRII-5124, three Altec MRII-594, five Altec 289-15 (drivers), five Altec 8126 (bass horns), and five Altec 405. To this there was added a loudspeaker system for monitoring purposes on the platform (one Altec MRII-594, one Altec 299-16 (driver), one Altec 8126 (bass horn in cabinet), and one Altec 405). In addition to portable platform loudspeakers there are 11 wall-mounted loudspeaker systems (Yamaha 5-20X) for realizing motion effects or localizable sound effects/sources.

In a control booth behind the rear wall of the hall there is the mixing console (Yamaha PM-3000) with 24 inputs, plus two delay lines, a reverberation unit, and a harmonizer. A portable recording system (Studer and Sony) and four associated tape machines (Otari and TEAC) are also available.

8.4.4 Philharmonie Berlin

The old Philharmonie Berlin, which – with its shoebox shape – boasted outstanding acoustical properties, was destroyed during World War II. With his late work inaugurated in 1963 the architect Scharoun won himself the merit of achieving a further enhancement of these overall qualities, not only in the acoustical performance, but especially also with regard to architecture and design. The amphitheatre-style accommodation of the audience (seated all round the orchestra platform), brought about acoustical advantages and disadvantages.

ROOM-ACOUSTICAL CONDITIONS

When specifying the acoustical concept of the house, L. Cremer took care that no seat was more than 30 m away from the platform. Since the large

Figure 8.38 Multicluster in the Casals Hall.

wall surfaces that exist, for instance, in shoebox-shaped halls were missing, Cremer introduced sound-reflecting terracing partitions, so that later on the phrase 'vineyard terraces' began to be used in connection with halls of this design. Above the orchestra ceiling elements are arranged to help the musicians hear each other. The volume of the hall is 21 000 m^3

Figure 8.39 Philharmonie Berlin: view into the hall with newly designed central cluster.

(characteristic volume/seat value approximately 9 m³/seat), and the reverberation time is about 2.1 s in the empty hall and about 1.95 s in the full hall (both values are for medium frequencies) [8.14]. Thus the sound reinforcement system is required both for speech transmission and for the reproduction of amplified music.

SOUND REINFORCEMENT SYSTEM

A sound reinforcement system that was in place when the hall was opened was replaced in 1982 by a large, central monocluster designed by the Philips company especially for speech transmission. Because of the more stringent requirements, triggered by the increased quality of domestic sound transmission (the CD technique was gaining increasing ground at the beginning of the 1990s), specifications for a new monocluster were drawn up in 1992. It was to be capable of transmitting not only speech but also music signals of high quality. Extensive computer simulations were carried out by ADA (Acoustic Design Ahnert) with the help of the EASE CAD program, in order to ensure selection of the right components.

In this way a monocluster solution was found that was additionally verified in test runs carried out by means of an array rigged up of five 602LS and two 902LS manufactured by d&b Audio AG. Nevertheless it still

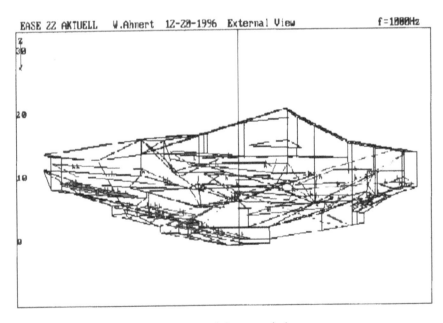

Figure 8.40 EASE drawing: coverage of the central cluster.

took more than two years before, in cooperation with the Monuments Preservation Authority, a visually acceptable enclosure was found for this monocluster (see Figures 8.39 and 8.40).

Additional loudspeakers were permanently installed in the platform area, and mobile systems made operable by means of new outlets. By installing four loudspeakers of type E3 it thus became possible to achieve a platform edge coverage enabling the perceived acoustic image to be lowered for the first rows (Figure 8.41).

Directionally-correct sound reinforcement was made possible by the installation of a modern DSS mixing matrix. The installation of additional cabling in sound control booth 1 and to the new sound control console rounded off the upgrading of the sound reinforcement system.

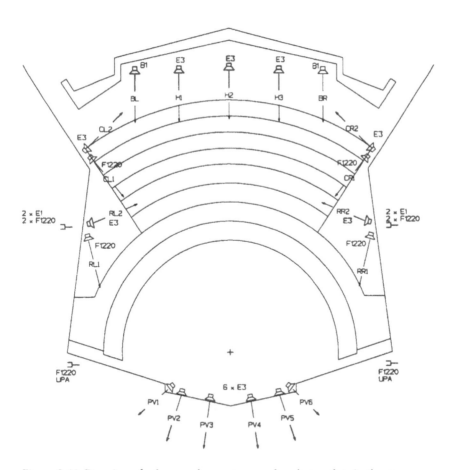

Figure 8.41 Drawing of other speaker systems and socket outlets in the stage area.

8.5 Theatres and opera houses

8.5.1 *Staatsoper, Dresden*

After its reconstruction, the Semper Opera was opened on 13 February 1985. The auditorium was rebuilt with little change in its original form, as can be seen in Figure 8.42. The width and depth were extended by about 4.5 m (by omitting the access to the boxes), while the stage proscenium was widened by 3 m. In spite of the volume increase to about 12 500 m³, the number of seats was reduced by about 300 to 1290, since many seats with limited visibility conditions as well as the seats on the fifth balcony were omitted to make room for a lighting gallery, and the spacing of the rows in the stalls was increased.

ROOM-ACOUSTICAL CONDITIONS

Because of the reduction of the number of spectators together with the increase in volume, the specific volume/seat value is 9.6 m³/seat, which is relatively large for an opera house, and thus also the reverberation time, depending on the stage setting, of 2–2.4 s in the empty hall, is rather long.

Figure 8.42 View into the auditorium of the Semper Opera, Dresden.

The height of the hall produces a spaciousness that is relatively high for an opera house, and almost equals that of a good concert hall.

As can be seen in Figure 6.44, the sound reinforcement system consists of a main reinforcement group divided into the following subgroups:

- a ceiling group (A_1), which can be lowered from the architrave above the stage proscenium, and which can be subdivided into one left and one right subgroup ($A_{1\ 1}$ and $A_{1\ 2}$ respectively), each of 175 W power-handling capacity (see also Figure 2.3);
- subgroup A_2, consisting of four sound columns installed in the left proscenium boxes and behind an acoustically transparent covering in the side of the proscenium arch, with an overall power of 400 W;
- subgroup A_3, arranged symmetrically in the same fashion at the right side of the stage proscenium;
- the music loudspeakers (type ETS 100) of partial group A_4, arranged left and right in the boxes, the low- and medium-frequency components of which are each supplied by one 100 W amplifier.

For the generation of effects in the auditorium, which helps to involve the audience more closely in the performance, single loudspeakers of the reinforcement system B (cone diameter 215 mm, maximum power-handling capacity 10 W) are used. Because of the historically authentic reconstruction of the auditorium it was necessary to disguise the loudspeakers and their casings. Figure 2.2 shows some of the solutions that were adopted; their acoustical functionality was proven by measurements in the free sound field. The switching pattern of these loudspeakers not only allows the formation of groups assigned to specific areas of the auditorium, but also makes it possible to pan a sound signal around the hall. To achieve this, up to 12 loudspeakers arranged side by side or one on top of the other are switched on in pairs. Each time a loudspeaker pair is switched on, the previous pair is switched off again. It is also possible to superimpose the panorama signal over the overall sound effects or a localized sound effect. A panel of LEDs provides the sound engineer with a visual read-out of the actual switching condition of the switched loudspeakers.

The effects loudspeakers of system C are located in the ceiling of the auditorium. This system consists of a group with a power capacity of 200 W concentrated in the central ceiling area (above the chandelier) combined with two groups of single loudspeakers uniformly distributed over the front and rear ceiling areas.

A system of relatively powerful loudspeaker groups is used for input from the stage area. These had to be positioned so that they were capable not only of providing localization to a specific stage area, but also of

(a)

(b)

Figure 8.43 View into the control studio of the Semper Opera, Dresden: (a) mixing
console and tape machines; (b) rack front with sound-processing
equipment and distribution panel.

simultaneously supplying as large an audience area as possible with direct sound. This is why the relatively powerful loudspeaker groups H_1 and H_2 were arranged left and right respectively in the area of the rear lateral technical galleries. Groups F_1 and F_2 were installed as 4 m long horizontal loudspeaker arrays at the left and right sides of the proscenium towards the rear stage. Above the proscenium loudspeaker group G is arranged below the technical gallery and loudspeaker group I is mounted on the backprojection booth in the rear stage.

In addition to the permanently installed loudspeaker groups there are numerous connection outlets for mobile loudspeakers. At the inner sides of the main, lateral and rear stages there are also loudspeakers to provide monitoring for the performers: for example, for transmission of the director's commands, or for improvement of orchestra audibility.

The sound reinforcement system is provided with a (at the time) well-equipped control studio (Figure 8.43) arranged behind the stalls for visual contact with the auditorium and the stage. The hall also contains connection facilities for a small mixing console for testing adjustments during staging rehearsals, and for a *hall sound mixing console*, which has proven its worth for musical stagings. In addition to the mixing console, sound recording equipment (studio tape machines) and power amplifiers, the control studio also contains the racks containing the sound-processing equipment (including filters, reverberation and delay devices, sound processing/effects equipment, and audio exciters), the distribution equipment (6 dB signal lines[1] and 100 V loudspeaker lines), and the acoustical and visual control devices.

8.5.2 Landestheater, Altenburg

The Landestheater, built in 1869–1871 in the Thuringian town of Altenburg, was refurbished and modernized in 1994–1995. In the course of these works the whole electroacoustical system was renewed during the 1995 season interval. With a volume of approximately 3400 m³, the house offers about 550 seats in the stalls and on three balconies. The uppermost balcony is used only in exceptional cases by the audience, so the new sound and lighting control facilities were installed in the central section of this balcony (Figure 8.44).

ROOM-ACOUSTICAL CONDITIONS

With its reverberation time of 1.1–1.2 s, the room boasts a high definition, which is particularly favourable for spoken drama. However, this

---

[1] Audio line level (0 dB) is generally taken in the USA and UK as being 0.775 V, often also referred to as 0 dBm (although this is only true when the load impedance is 600 Ω). In some studios and broadcasting organizations +4 dB is used as a reference level. In Germany 6 dB, i.e. 0.775 V, is used as the nominal line level.

Figure 8.44 Landestheater, Altenburg: view backwards into the hall.

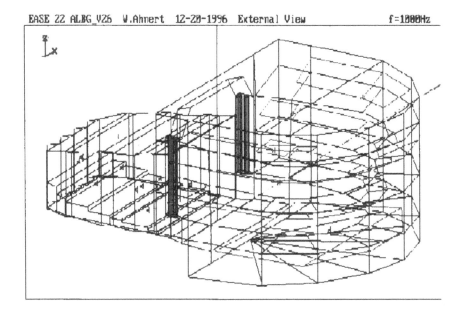

EASE ZZ ALBG_VZ6   W.Ahnert   12-Z0-1996   External View   f=1000Hz

Figure 8.45 EASE drawing: new orchestra shell on the stage of the Landestheater, Altenburg.

reverberation time is too short for concerts and other music performances, since it produces an insufficient spatial impression. With concerts it was additionally aggravating that the proscenium height was too low, and that the structure of the old concert shell was inappropriate for transporting sufficient energy into the auditorium. Thus the reconstruction work included, as a room-acoustical measure, a heightening of the proscenium. A new concert shell is actually being manufactured (Figure 8.45).

SOUND REINFORCEMENT SYSTEM

The main group of the new sound reinforcement system for covering the audience area consists of four passive two-way loudspeakers model ITS 304 from MEG (Musikelektronik Geithain GmbH), arranged left and right in the proscenium boxes and below at stage height. In the masked or undersupplied areas below the first balcony and in the third balcony there were additionally installed four smaller loudspeakers, model ITS 305, from the same manufacturer. For bass enhancement two TT 300-1 subwoofers are arranged on the technical floor above the decorative grid over the chandelier.

The effects system consists of seven loudspeakers at the side and rear walls of the stage, plus two further ones above the chandelier on the

Figure 8.46 Sound control booth in the Landestheater, Altenburg.

technical grid (as the 'voice of God'). These are complemented by 32 panning loudspeakers on the walls of the auditorium. These are ready for the connection of a panning control unit and an electronic reverberation-enhancing system.

All sound reinforcement systems are driven via delay devices in such a way that the resulting localization is largely towards the stage: that is, towards the lowest proscenium loudspeakers.

The new sound control booth is equipped with a digitally controlled mixing console with data storage (Figure 8.46) manufactured by the firm Tactile Technology. If necessary a second console in the stalls can be attached to serve as a hall mixing console. All adjustments made there can be stored and reproduced later on in the sound control booth.

The power amplifiers and associated delays are decentrally arranged in the proscenium side areas, and are controlled by an amplifier remote monitoring system.

8.5.3 Deutsche Oper Berlin

The Deutsche Oper Berlin, which was re-opened in 1961, is a multi-balcony opera house offering good visual and auditory conditions. As the sound equipment had become totally obsolete, measurements concerning the quality of the loudspeaker system were carried out in 1988, and recommendations were made for the installation of new equipment.

ROOM-ACOUSTICAL CONDITIONS

The house is exemplary for good acoustics, so that the preconditions were excellent for the functioning of a sound reinforcement system.

The measurements revealed that modifications were necessary in the proscenium wall and ceiling regions. Apart from a more differentiated arrangement of the reflecting surfaces above the orchestra, in the main the lateral walls in this area required remodelling to obtain improved sound guidance from the depth of the stage and from the orchestra pit. In particular, the stalls area was to be provided with more direct sound energy, to which end it was necessary to remove the iron lattices existing in the proscenium area and to replace them by inclined reflecting surfaces directed towards the hall.

All this was largely realized in 1989. The side walls in the proscenium area, however, were replaced not by reflecting surfaces, but by acoustically fully transparent coverings, behind which room-acoustical reflectors and new loudspeakers were arranged.

SOUND REINFORCEMENT SOLUTIONS

The first part of the new sound system was installed in 1989. In the proscenium side walls, four pairs of high-performance compact loudspeaker systems (F1220 from d&b Audio) were arranged, appropriately inclined and angled one on top of the other (main group 2). In the upper proscenium area, two pairs of such loudspeaker systems were arranged side by side (main group 1 in Figure 8.47). Additionally a third system (loudspeaker system main group 3) is planned in the region of the second lighting bridge in the hall to cover the upper balcony.

To cover the front rows in the auditorium, connection sockets for additional loudspeakers were arranged below in the proscenium area. These may also be used for connecting mobile loudspeaker systems (also speaker's desk systems and the like).

The main groups, 1 and 2, permit radiation of an acoustic power of 10–12 W, which corresponds to the requirement for the generation of a maximum sound pressure level of 110 dB in the diffuse field of the auditorium.

In 1991 new systems for reproduction and effects purposes were planned and installed in the stage area (main groups 4, 6, 8 and 9). Groups 5 and 7 are to be installed during the next development phase.

The panorama loudspeakers in the wall and ceiling areas of the auditorium (extensions are planned here) serve for the direct broadcasting of effects into the hall. In parallel, the sound transmission network in the whole house was planned anew and installed in 1994–1995. Based on digital sound transmission using optical-fibre cables (SPR System from Stage Tec, Berlin), it allows any required cross-connection to be established through a central audio/video control unit (Figure 8.48). Preferences provided in the innovative operating software ensure that routings effected at the sound control unit cannot be cancelled by other activities in other areas.

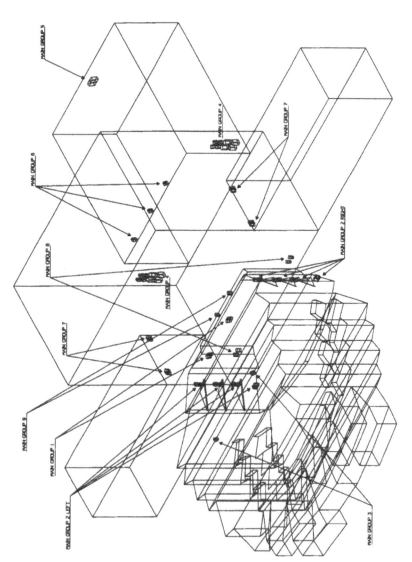

Figure 8.47 Recommended loudspeaker arrangement in the Deutsche Oper, Berlin: F_1–F_4, location of main loudspeaker systems; F_5, near-range sound reinforcement; F_6, planned location of ceiling loudspeaker system.

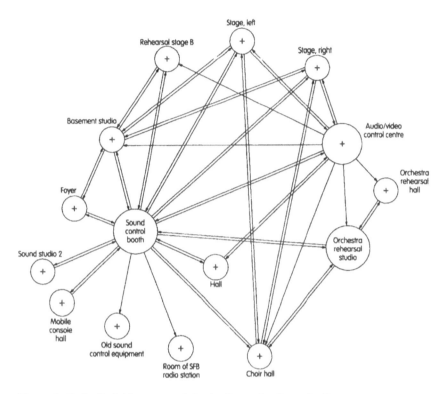

Figure 8.48 Audio/video network in the Deutsche Oper, Berlin.

Further remodelling phases include the installation of a new central stage manager system and the realization of a computer network enabling universal data exchange both within the house and, later on, with other houses of the City of Berlin. As a final stage of the extensive renewal of the audio and video systems in the house, replacement of the present analogue mixing consoles by digital ones is planned. Preliminary work to this end, such as the installation of a new sound control unit in a former spectators' box on the first balcony, had already been concluded in 1995. The mixing console in Figure 8.49 is still the old one. The digital signal distribution according to Figure 8.48 is already computer-controlled from the new sound control unit. The window in front of the console can be lowered to allow direct listening into the auditorium.

8.5.4 State Theatre, Tallin, Estonia

The double-housed building (opera house and concert hall) was erected at the beginning of this century in a typically traditional architectural style.

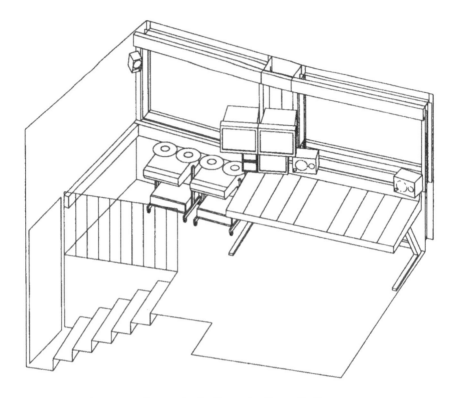

Figure 8.49 New control room in the Deutsche Oper, Berlin.

The horseshoe-shaped auditorium (two balconies) of the opera house has a volume of approximately 3800 m³. The concert hall has the shape of a shoebox.

ROOM-ACOUSTICAL CONDITIONS

The auditorium of the theatre has a capacity of 695 seats, the characteristic volume/seat value thus being about 5.5 m³/seat. This value is almost optimal for a drama theatre, but too low for music performances, for which the optimum value is 7–8 m³/seat. The low characteristic volume/seat value implies short reverberation times, so that with the mentioned volume a value of about 1.4 s ought to be considered here as an optimum. For frequencies below 250 Hz the reverberation time should increase additionally (to up to 2 s at 63 Hz), so as to produce a 'warm' sound impression with opera performances. Note also that the theatre hall is of very damped design (seats and balustrades), so that the reverberation time is bound to decrease still more.

These expectations, which result from studying the objective conditions, are corroborated by subjective experience, which has given rise to complaints about the acoustic atmosphere in the house being 'too dry'. Moreover, the sound level produced by the singers is insufficient to overcome the level of the orchestra, since the inadequate height of the stage proscenium tends to attenuate the level of the singers significantly, particularly for those at the back of the stage.

COMPUTER SIMULATION

In this auditorium a new sound system has to be installed. However, since we are mainly concerned here with room-acoustical problems, we shall discuss the sound system elsewhere. To illustrate the advantageous use of the EASE simulation software for this case, we are going to deal here only with the necessary reconstruction work in the proscenium area. It is certainly true that the lighting in the transition zone from stage to auditorium was significantly improved by the installation of a lighting bridge in front of the proscenium, but much to the disadvantage of sound propagation from the stage to the auditorium. Here the lighting bridge acts fully as a sound barrier (Figure 8.50). By modifying the position of the proscenium lighting bridge, and by adding a second one behind the proscenium (not shown in Figure 8.51) from which the rear of the stage can

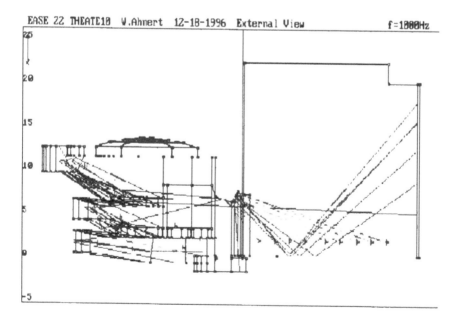

Figure 8.50 State Theatre, Tallinn: EASE drawing of sound radiation with existing proscenium bridge.

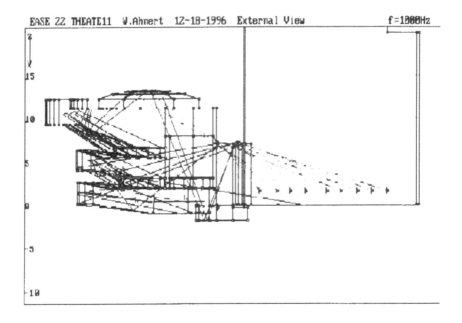

Figure 8.51 State Theatre, Tallinn: EASE drawing of sound radiation with changed shape of the proscenium bridge.

be lit, sound propagation from the rear of the stage is no longer hampered. The inclination of the surface above the stage proscenium could be optimized by means of computer simulation. This surface has to be acoustically hard-finished so as to enable short reflections.

References

8.1 Orth, D., 'Tonsäulen für Innen- und Außenbeschallung' (Sound columns for indoor and outdoor sound reinforcement), *Technische Mitteilungen RFZ*, 20 (1976) 4, 73–77.

8.2 Ahnert, W., 'Simulation of complex sound fields in enclosures and open-air-theatres', 77th AES Convention, Hamburg (1985), preprint. no. 2186.

8.3 Steffen, F., 'Elektroakustik im Plenarsaal der Volkskammer der DDR' (Electroacoustical equipment in the Plenary Hall of the Volkskammer of the GDR), *Technische Mitteilungen RFZ*, 27 (1977) 3, 70–71.

8.4 Fasold, W., Sonntag, E. and Winkler, H., *Bau- und Raumakustik* (Architectural and room acoustics) (Berlin: Verlag für Bauwesen, 1987).

8.5 Steffen, F., 'Akustische Probleme im Großen Saal des PdR' (Acoustical problems in the Great Hall of the Palast der Republik), *Technische Mitteilungen RFZ*, 27 (1977) 3, 51–55.

8.6 Hodas, B., 'Palast der Republik', *Sound & Video Contractor* (1987), 131–137.

8.7 Steffen, F., 'Beschallungsanlage der Sport- und Kongreßhalle Rostock' (Sound reinforcement system of the Sports and Congress Hall Rostock), *Technische Mitteilungen RFZ*, 25 (1981) 2, 25–29.

8.8 Furduyev, V.V., *Acoustic Fundamentals in Sound Reinforcement* (in Russian) (Moscow, Edition Swjas, 1960).

8.9 Steinke, G., Fels, P. and Hoeg, W., 'The Delta Stereophony System (DSS) in the City Hall of Stade and in the Open-Air Theater Trachselwald', 88th AES Convention, Montreux (1990), preprint no. 2921.

8.10 Peutz, V.M.A., 'Articulation loss of consonants as a criterion for speech transmission in a room', *Journal of the Audio Engineering Society*, 19 (1971) 11, 915–919.

8.11 Steffen, F., 'Die elektroakustischen Einrichtungen im Neuen Gewandhaus Leipzig' (The electroacoustical facilities in the Neues Gewandhaus Leipzig) *Technische Mitteilungen RFZ*, 27 (1983) 2, 25–30.

8.12 Naniwa, K., 'Suntory Hall', *Sound & Video Contractor* (1988) 26–32.

8.13 Nagata, M., 'Design problems of concert hall acoustics', *Journal of the Acoustical Society of Japan*, 10 (1989) 2, 59–72.

8.14 Beranek, L.L., *Concert and Opera Halls, How They Sound*. Published for the Acoustical Society of America through the American Institute of Physics, Woodbury, 1996.

8.15 Steffen, F., Fels, P. and Schwarzinger, W., 'Die tontechnischen Einrichtungen der Semperoper in Dresden' (The sound-technical facilities of the Semperoper in Dresden), *Technische Mitteilungen RFZ*, 30 (1986) 3, 49–56.

Index